THE ONION FIELD

Joseph Wambaugh

Quercus

First published in Great Britain in 1973
This edition published in 2007 by

Quercus
21 Bloomsbury Square
London
WC1A 2NS

A CIP catalogue reference for this book is available
from the British Library

ISBN 1 84724 174 3
ISBN-13 978 1 84724 174 0

10 9 8 7 6 5 4 3 2 1

Typeset by Deltatype Ltd, Birkenhead, Merseyside
Printed and bound in Great Britain by Clays Ltd, St Ives plc.

NOTE TO THE READER

There is no way to thank properly the sixty persons whose intimate revelations to me permitted the telling of this true story, nor would the book have been possible without the help of scores of others who provided the thousands of pages of evidence and the mountain of clues needed to understand fully the most maddening case of any detective's life.

The re-creation of events was at all times done as accurately as possible. The strange relationship of Gregory Powell and Jimmy Smith could be re-created in great detail, thanks in part to a frank, unpublished autobiography written by Jimmy Smith during the trials.

The courtroom dialogue was not re-created. It was taken verbatim from official court transcript.

The names of two minor characters have been changed by request: the priest, Father Charles, and the juror, Mrs Bobbick.

After having lived so long with this story, the investigator must implore the reader to respect the privacy of those men and women who let the truths be revealed, who let their story be told.

JOSEPH WAMBAUGH
Los Angeles, California

The wild insistent pipes and the marching feet defiantly answer that there is no more death.

A piper's incantation

The gardener was a thief. That's the thing that bothered him most. The trials didn't bother him so much anymore. It was strange how much he used to fear the trials, but not now. He just went to court and testified when he was told, then went back to his gardening.

For a while they feared the trials would continue clear into the new decade. But now they assured him it was almost over, and 1970 was still eight weeks away.

Sometimes the gardener wore a hat, the battered wide-brimmed straw he was wearing today. Mostly he wore it to keep the sweat out of his eyes, not to protect himself from the sun. The gardener loved the sun. He hated dark places, hated the night in fact, and sometimes sat up until dawn. No matter how tired he was and how much work he had to do the next day he was always glad to see the daylight come.

'Would you like some iced tea?' The old woman had come out on the porch without the gardener seeing her.

'No ma'am, I just have to trim back this juniper,' he answered, squatting in the grass, his shears poised.

'That's a blue pfitzer.'

'Yes, I know. It's one of the tallest junipers. It's very pretty,' said the gardener. He was trying to remember what he'd learned about the blue-green juniper while landscaping with his friend. His friend knew all the botanical names. For the last year, though, he'd been working alone. He preferred to be alone these days.

'You know about junipers? How nice,' said the old woman.

'I used to know the names better,' said the gardener, removing the straw hat and raising his sweat-stained dusty face to the November sun. Yes, I used to remember, he thought. But so many things were slipping away. It was getting so hard to remember.

Then the gardener glanced at a 'no parking' sign on the street. He stared at the red arrow on the sign and something flashed in his mind, an indefinable glimmer. He began getting afraid for no *reason at all, and a throbbing pain started at the base of his skull. He crossed the small yard to the mower because he didn't want to talk to the lonely old woman today. The fear was weakening him and the pain was ferocious. He wanted to work it off.*

The gardener yanked the rope and the motor caught and he was behind the mower, the engine roaring in his ears. The pain was spreading like fingers of blood, spreading from the tiny spot at the base of his skull. He knew it would flood over his head until his entire skull felt crushed by a merciless bloody hand. Sometimes then it would go away.

Even the pain would not stop the gardener from thinking about his crimes. As he walked behind the mower, the pain pulsing clear through his eyes, the gardener thought of how the first time had been.

He had been walking through the store as part of a job, when he saw something he needed: masonry drills. He needed them for some cement work he was doing on his duplex. He had picked up the drills and put them in his pocket. And that was that. He couldn't believe it until it was done. But his heart began pounding then, more from excitement than fear. Or was it excitement? It was so hard these days to recall all the emotions clearly. The crimes he could remember as if they just happened. The feelings were what eluded him.

All thieves start small, the gardener thought gravely, as the shower of grass in his face helped him forget the pitiless driving pain. He used to think about the night in the onion field all the time – before he became a thief. But now the crimes he had committed overshadowed all else, even the four men and the death in the onion field.

ONE

The night in the onion field was a Saturday night. Saturday meant impossible traffic in Hollywood so felony car officers did a good deal of their best work on side streets off Hollywood and Sunset boulevards. On those side streets, revelers' cars were clouted or stolen. F-cars also cruised the more remote commercial areas, away from intersections where traffic snarled, and the streets undulated with out-of-towners, roaming groups of juveniles, fruit hustlers, desperate homosexuals, con men, sailors, marines.

Nothing the Hollywood Chamber of Commerce said could camouflage the very obvious dangers to tourists on those teeming streets. Most of the famous clubs had closed, the others were closing, and Hollywood was being left to the street people. The 'swells' of the forties and early fifties had all but abandoned downtown Hollywood and were gradually surrendering the entire Sunset Strip, at least at night.

In spite of it all, Hollywood Division was a good place for police work. It was busy and exciting in the way that is unique to police experience – the unpredictable lurked. Ian Campbell believed that what most policemen shared was an abhorrence of the predictable, a distaste for the foreseeable experiences of working life. It wasn't what the misinformed often wrote, that they were danger lovers. Race drivers were danger lovers. That's why, after Ian and his old friend Wayne Ferber had crashed a sports car several years before, he had given up racing, though he would never give up police work.

He felt that the job was not particularly hazardous physically but was incredibly hazardous emotionally and too often led to divorce, alcoholism, and suicide. No, policemen were not danger lovers, they were seekers of the awesome, the incredible, even the unspeakable in human experience. Never mind whether they could interpret, never mind if it was potentially hazardous to the soul. *To be there* was the thing.

*

Karl Hettinger was newly assigned to felony cars and Ian was breaking him in. The partnership had jelled almost at once.

'You were in the marine corps too?' Ian asked, during the monotonous first night of plainclothes felony car patrol.

'Communications.' Karl nodded.

'Really? So was I,' Ian said, flickering his headlights at a truck coming onto Santa Monica from the freeway.

'The voice with a smile,' Karl said, and they both grinned and made the first step toward a compatible partnership.

Each man learned after two nights together that the other was unobtrusive and quiet, Ian the more quiet, Karl the more unobtrusive, but a dry wit. It would take two men like these longer to learn the habits and tastes of the other, but once learned, the partnership could result in satisfying working rapport. There is nothing more important to a patrol officer than the partner with whom he will share more waking hours than with a wife, upon whom he is to depend more than a man should, with whom he will share the ugliness and tedium, the humor and the wonder.

'You dropped out of college in your final semester?' asked Ian during their third night. 'So did I. What were you majoring in?'

'Agriculture, beer, and poker, not in that order,' said Karl, who was driving tonight, a slow and cautious driver who now wore glasses at night, finding he had some trouble reading license plates.

'I was in zoology and pre-med. Looks like we're both out of our elements.'

'I'm taking police science courses now,' said Karl.

'So am I,' said Ian. 'You must know something about trees, don't you?'

'Probably not as much as I should,' said Karl. 'An ag major has to know a little bit about tree and plant identification.'

'I'm really involved in trees now,' Ian said, becoming unusually garrulous as he always did when something interested him. 'I'm landscaping my house, or trying to. You know anything about fruitless mulberry?'

'Not much.'

'Well, it grows big and wide and fast. Instant shade. I like that. I get impatient waiting for things.'

4

'You have to be patient to make things grow.'

'Sometimes I think that's why I'm a policeman,' said Ian. 'Not patient enough. Antsy, my wife calls me. I guess I just have to be free and moving around.'

'I don't know why *I'm* a policeman,' said Karl. 'It just happened. But I like it. I couldn't have a job where I was closed up inside four walls and a roof. That's the latent farmer in me.'

'The best thing is that no matter how boring things get, like tonight for instance, something might be right around the corner. A little action I mean,' said Ian.

Karl touched his cotton shirt, open at the throat, and the threadbare sport coat. 'I'm glad not to go back to uniform.'

'One thing to remember is that all those working hours you spent in patrol refereeing family beefs and writing tickets and taking reports – we'll use all that time in felony cars for one thing: to find serious crime on the street. You're bound to run up against a hot one once in a while. You just have to be a little more careful working this detail.'

'Don't worry, I will.' Karl nodded. 'By the way, you ever cruise around behind the bar up here on McCadden? In the parking lot?'

'Parking lot? Don't think I know it.'

'You just go north on McCadden from Sunset till you smell it, then go east till you step in it. It's like a zombies' convention back there. When I worked vice I used to see a lot of activity at night. Probably hypes more than anything.'

'Let's check it out tonight,' said Ian, pleased to see that his new partner was energetic. Good police work made time race.

'Hey look at that,' said Ian on their fifth night, slowing as they passed a wooded acre in front of a white Spanish colonial home on Laurel Canyon. It was a balmy evening because the warm Santa Ana winds were blowing, and the canyon was a respite from the Hollywood traffic.

'Whadda you see?' Karl asked, twisting abruptly in his seat, tensing for a moment, as he peered through the smoky darkness in the woodsy residential valley.

'Liquid amber,' Ian said, admiring the foliage almost hidden by tall shaggy eucalyptus. 'You should see them in the fall. They change colors like flames. Beautiful. Just beautiful.'

Karl shook his head and grinned.

Ian Campbell never noticed the grin. He watched the trees. The eucalyptus reminded him of a park in the heart of the city where the smell of tar filled the air and had once ignited a boy's imagination.

Ian had been a bookish romantic youngster – a dreamer his mother called him – and even as a high school senior, loved to dawdle for hours by the pits and stare into the tar until he vividly imagined great Pleistocene creatures there.

The boy could guess how it was when Imperial Mammoth went to the tarpits to die. Or rather *went* to drink. The pool at night looked inviting to Mammoth and the ominous bubbles rising were of no consequence. Nor was the black slime that slithered between his toes and climbed sucking his ankles. Panic struck when, loin deep in water and having drunk his fill, he tried to take his first step out and found himself trapped in the tar.

Mammoth was bewildered after the first surge of terror. He stood fifteen feet tall and his curved tusks even measured a greater length. Yet with all his might he could not drag his hairy bulk more than inches through the tar. His fearful bellow paralyzed the other creatures of the forest.

The great bellowing pipe suddenly blew a plaintive blast, and upon hearing it some of the creatures were filled with grief and dread because they instinctively knew death was upon him. Many of the predators, despite their fear, were then drawn to him and themselves would die that night locked to his flesh, sucked down by the tar as they fed.

Ian Campbell heard Mammoth clearly as he lay there on the grass and stared into the dank pond, like ice varnished black except for the gaseous bubbles plopping on the surface in the moonlight. It was very dark despite the moon, and quiet, and the tar smell was everywhere. Ian heard how Mammoth sounded at the last: plaintive, yes, but defiant.

Somehow Ian *knew* that Mammoth would be defiant at the end. And Ian suddenly had the urge to jump to his feet and sound a call which he was sure somehow would drift across the ages to Mammoth who would sense what every piper knew – that there is no death.

Then to prove it he stood, adjusted the braces on his teeth to better taste the reed, and breathed deeply of the tarry chewy night air which could be blown into a tartan bag.

His silhouette there on the grassy knoll startled a little girl who was strolling with her father through Hancock Park along the path just north of Wilshire Boulevard. The child stopped and gasped as the silhouette took shape in the darkness. It had three horns which protruded from the side of it. It was tall, slender, erect, its head thrust back from a length of horn distended from its mouth. Then the sound came out of it – eerie, baffling – and she started to cry from fear. Her father picked her up and laughed reassuringly.

'It's a bagpipe, honey. It's just a boy playing a bagpipe.'

Ian Campbell never heard her cry. He was preoccupied, struggling to get the reeds vibrating the right way. Sometimes they just wouldn't snap in there. In their own way the pipes were much harder than the paino. With no chords you just couldn't put harmony into them, and the timing and grace notes which embellished the melody notes meant everything. He took a deep breath, moistened the valve, and was careful to keep an imperceptible pressure on the bag with his elbow, hoping to keep the constant flow through the reeds. He blew and hoped, and on top of everything else the reeds began to chirp!

Ian tossed the three drones off his shoulder and began pacing disgustedly. For this he had pleaded with his mother to sell his piano. For this crazy instrument! Three hundred years ago Pepys heard one and said, 'At the best it is mighty barbarous music.' He was dead right, thought Ian.

The boy glanced at the tartan bag. It was a Campbell tartan, of course, for his clan. As always it stirred memories of the race, of fighting men with huge claymores, and the Campbells who sided with the English king against Bonnie Prince Charlie, and who slew the Macdonalds.

Then Ian discovered that he was unconsciously marching the twelve-foot square, caressing the ivory and ebony shaft, pressing ever so lightly on the tartan bag with his elbow. So he boldly threw the drones over his shoulder and without a moment's hesitation played 'Mallorca.'

It was good. The best he'd ever played it. And he tried 'Major Norman Orr Ewing,' the song which would earn him a medal in the novice class of the coming Winter Games. He played and played and marched the twelve-foot square, lost in the music.

His mother did not allow him to play his pipes in the

apartment. But what did it matter? Living across the street from Hancock Park and the tarpits was perfect for a piper. What better place to march than here on the turf out in the open, under the stars and lights off Wilshire Bourlevard, with no sound but distant tire hum, smelling grass and ferns, and the tarry air so thick you could taste it. The seventeen-year-old solitary piper sucked the tar-laden air, and blew it through the blow-pipe, his fingers striking alertly, and imagined the bag would somehow be better if magically cured by tarry fossilized air from another age.

Chrissie Campbell sat outside on the porch of the apartment waiting for Ian and enjoying the evening. In the distance someone was playing the radio loudly and from time to time she would catch bits of music, and later the laughter when the debris inevitably crashed from the swollen closet of Fibber McGee and Molly. Then the station was changed and the dialer stopped for a moment on a program of classical music and she tried to identify the piece being played by the violinist. She was reminded of her husband, Bill Campbell, the tall, curly-haired doctor who had also played violin and was now dead five years. She sighed and wished for him. It was easy to wish and remember on nights like this, bright and balmy, when something like Indian summer comes to Southern California.

They had met at Manitoba Hospital where she worked as bookkeeper, she born in Saskatchewan, daughter of a railroader, her family even more Scottish than his Highlander people because hers were originally from the Hebrides and spoke Gaelic. It was natural that these two Scottish Canadians should meet there in the hospital and fall in love, and that in the hard times they should emigrate to America where things were said to be better.

They had good years in Valley City, North Dakota, the small college town where they lived almost on the bank of the Cheyenne River, on flat land near wheat fields and homestead trees.

The Depression was almost as hard on a doctor as it was on farmers and other town workers, but it was a very good life until after the war began, when the physician began to die from cancer.

He was in fact dead for the year he continued to draw breath. Many of their talks, their secret talks, were of death because he diagnosed his own illness and they had to prepare for it. The

Depression and the illness drained them financially and there were long serious conversations riddled with merciful lies from her.

'You're not afraid are you, Chrissie?'

'No, Bill, I'm not.'

'You're a strong capable woman, you needn't be afraid about making your own way.'

'I'm not afraid, Bill. Really I'm not.'

'The more we talk about California the better I like the idea.'

'Yes, so do I, Bill.'

'The war has made things boom out there. There's a great need for people. You're certainly not too old to find a good job.'

'I'll raise a strong son, Bill. I swear it.'

'You're not afraid, Chrissie?'

'I'm not, Bill. I'm not.'

And when she was alone with her thoughts during that year and for some time afterward, the fear would come. She never told him of the smothering fear which came always in the night and had to be defeated.

Chrissie believed she had some salvation in the inherited blood of dour and steely men. Her people were from the Isle of Lewis, the northernmost island of the Hebrides, tempered by the icy Atlantic brine which blasted their faces for centuries. She had their strength and she knew it. More than that, she had their capacity to endure.

It was Chrissie Campbell's theory that she could give Ian culture and discipline, and that these were two great gifts, perhaps all she could ever really give. After Bill's death the discipline was essential for them both.

'You're too strict with me, Mother. You're just too strict, and *don't* say someday I'll thank you.'

'I won't say that, Ian.'

'You're just *too* strict, Mother.'

Chrissie found work, first clerical, and then as an auditor and bookkeeper, and now with the war three years past, they were living in the Park La Brea Apartments, in what was certainly an upper middle class neighborhood. It was very hard to pay the rent, but what a marvelous neighborhood for Ian, who could spend every day across the street in Hancock Park. And he had to be near a park or he'd be arrested for the noise he made with those infernal pipes.

It had been disappointing to sell the piano for bagpipes, especially since she had started Ian on the piano at the age of four, and Chrissie had had a few of the passing dreams that mothers of young musicians have. The piano was the perfect instrument to help instill culture and discipline. The pipes were another matter, a wild, almost willful instrument which undeniably stirred your blood, but with a key (if you could call it that) which could never be duplicated on piano, so that she wondered what all the years of piano had really done for him except to win him a contest he cared nothing about, where he played Rachmaninoff so well.

But if he wouldn't be a piano-playing doctor, at least he'd be a doctor, that was certain. And what a physician he would be! Chrissie imagined him ten years hence, a tall, curly-haired intern like the one she married, but taller, even *more* intelligent, and sensitive, with a resolute mouth and classic jaw, one of the finest she had ever seen, and of which he was totally unaware. And that was another source of her great but secret pride – he was not aware of his considerable attractiveness. As though all teenage boys adored history and philosophy, had a wall full of serious books and read them over and over. As though all teenage boys had season tickets to the Philharmonic Concert Series.

And there was the other thing, she thought even more important. His natural facility with people. Quiet and nonaggressive, an impractical boy, but with strength other boys seemed to sense.

For five years she had guarded against the possibility of a fatherless only child becoming a mother's boy. So she had always tried to conceal most of her pride in him.

Then she thought she heard the skirl of pipes in the distance and she listened and imagined the smell of tar from the pits. Chrissie placed her palms on her cheeks and slid her fingers under her glasses and rubbed her eyes for a moment. She smoothed back the dark hair just starting to gray, and waited for the solitary piper to march through the darkness.

Ian Campbell was disturbed when the Communists swept south of the 38th parallel, but he was not as disturbed as Chrissie. He wasn't quite nineteen then, a dangerous age for decision making. As always he and Chrissie talked about what was bothering him. He was in the Naval Reserve at this time and was making uncertain plans for his college education. The more he talked about what was happening in exotic places like Pyongyang, Taejon, and Taegu, the more secretly frightened she became. There was the Massacre at Hill 303, then Inchon, and Chrissie knew further argument would be pointless. Everyone who ever knew him well knew that when he finally decided a personal issue, further discussion was futile.

Some time later, on a freezing winter night in the Punchbowl, north of the 38th parallel, a bleary-eyed marine corporal burst into a warming tent where a tall young marine radio operator sat reading an old stateside newspaper, drinking beer, and worrying the stem of a large curved pipe for which he could not seem to acquire a taste.

'Scotty, did you hear about the Scotchmen?' asked the corporal shaking the snow off his shoulders.

'What?'

'You know, those Argyll guys.'

'The Argyll and Sutherland Highlanders?'

'Yeah. Well I heard they took a direct hit on their equipment trucks. I heard it was a mortar attack. Guess what? You probably got the only set of bagpipes left in Korea now. Whatcha think, Scotty?'

Ian raised the bottle of beer and drained it in three gulps. He had drunk several bottles already.

'Well, Scotty, you gonna let the friggin gooks get away with it?'

'Is it really true?'

'Hell, I don't know, but let's show them mothers they ain't friggin well gonna get away with it. Get your friggin pipes. For once I wanna hear em.'

'Okay,' said Ian feeling light-headed when he stood. Twenty minutes later the burly corporal, carrying half a bottle of bitter rice wine and an M-1 rifle, and Ian Campbell, carrying only his pipes, were less than two hundred meters from no man's land on a moon-flooded rise of ground, marching a twelve-foot square in the snow.

'It's a bloody shame for a Scottish regiment to go into battle without their piper out front,' said Ian, surprised at the thickness of his words.

'That's tellin em, Scotty,' said the corporal, and tossed Ian the wine which was brackish but warm.

Ian deeply breathed the five below zero air and raised the bottle, the glass clicking against his teeth. Then he blew the first blast of a quick march, 'Scotland Is My Ain Hame.' It deafened the corporal for a moment. Then they both marched the square while Ian played 'Campbells Are Coming,' and soon Ian was sitting in the snow, shivering so hard he couldn't play. Then they heard Red bugles, and following the bugles, sporadic small-arms fire.

'Let's get the hell outta here, Scotty!' yelled the corporal.

'Right!' said Ian, scrambling to his feet and they ran back to camp, the corporal falling drunkenly every ten meters or so. The next day the story of the pipe and bugle war spread through the battalion and Ian's company commander told him to knock off agitating the gooks with his pipes.

Chrissie's favorite marine pictures of Ian were those he sent from Hawaii, one where he was standing in the center of an open, grass-roofed PX. Young marines – shirtless, dungaree-clad – were sitting around playing cards, gesturing wildly in exuberant youthful conversation. In the center of the picture and the pandemonium was Ian – tall, broad-shouldered and bony in his dungarees – eyes closed, oblivious to it all, playing the pipes serenely, while no one listened.

Ian decided not to take his pipes the second time he volunteered to go overseas, but finally after many weeks, at the conclusion of a letter, was the line: 'And Mother, could you please send me my pipes?'

A bug-eyed ketch, the *Jolly Roger*, twin-masted, shining in the sunlight. The Pacific, calm and blue. The horizon fiery. Knifing through the glossy isthmus under full sail, the crew of the tiny thirty-five footer was excited and happy.

The captain, Earl Schultz, a new friend, was a salty ex-merchant seaman, who looked like Popeye, a man partially crippled from a fall down a cargo hatch. The two crewmen were younger men, childhood friends: Wayne Ferber, the smaller one, eyes deep and close set, sharp chin, ears and mind. The other, a tall, twenty-five-year-old student of zoology and pre-med, a Korean vet, his short curly hair flashing auburn in the white sunlight. He was atop the bowsprit piping in celebration whenever the ketch tacked or changed course, each successful maneuver calling for a drink of wine from the goatskin bag they dragged behind the boat in the cold water.

And when they approached the harbor that fourth of July weekend, knowing Avalon would be jammed with tourists, the piper shouted: 'Let's go in first class!'

And no schooner crew ever passed the isthmus more proudly than this scaled down bug-eyed ketch with the young man on the bowsprit piping 'Campbells Are Coming.'

'Ian's having a whale of a time up there,' said Wayne to the captain, who was letting him steer the little boat. 'When he first came back from Korea he just wanted to sit around the apartment and play the bagpipe chanter. His mom would needle him a little and say, "Drowning in your own sorrow again, Ian?" Finally he snapped out of it.

'It's been a great outing for us,' Wayne continued. 'It's like our kid adventures, when we rode bikes clear to Lake Elsinore, and when we used to ride the Red Car to Santa Monica pier, or bicycle to Griffith Park and hike up to the observatory. I guess you could call us dreamers with these crazy schemes we've got to build our own boat and sail to Australia.'

Ian piped them into Catalina grandly and they had to anchor loose. The ketch had no radio, no engine. Schultz called her 'a real seafarin boat.' The three of them swam off the dinghy that day and hiked over the island, then joined a yacht club party at the harbor, almost sinking the dinghy on the way back. Ian piped in the hangovers the next morning. They considered it a real South Sea sailor's holiday.

It was at the yacht club party that Ian heard the story of the shark. A portly suntanned yachtsman in a navy, gold-buttoned jacket and immaculate white trousers was telling about it over a wet martini. 'They're *stupid* creatures you know,' he said, sucking a stuffed olive off a green plastic toothpick and shifting it from one side of his mouth to the other. 'They're *incredibly* stupid. You gut one and throw him overboard on the starboard side of the boat. And then throw the guts on the port side, and he'll swim around following the spoor and bite at his own guts. Savage and *disgusting*!'

'They say they have no nervous system,' said Ian quietly. He had been standing away from a group of yachtsmen listening. It was unusual for him to volunteer an opinion in a group of strangers, and he reddened when they turned.

'No nervous system, you say?' said the yachtsman.

'That's right, sir,' said Ian. 'Old shark hunters believe they're missing something *we* have. They don't die of shock. They're not necessarily stupid or savage, they just lack something we have.'

'Well, you can empathize with sharks if you want to,' said the yachtsman. 'I say they're disgusting beasts who eat themselves for sheer savagery.'

The squeal of tires. A skidding black and white police car. Two blue-uniformed policemen held a struggling, handcuffed man. He was moaning grotesquely, not like the physically wounded. Ian watched as he stood outside Unit Three, at Los Angeles County General Hospital with his friend Ray Sinatra and waited for Art Petoyan, who was an intern in Psychiatric Admitting. When the police left they entered.

'Doctor Petoyan, I presume,' said Ian to the dark young doctor with the Armenian nose and the sensual jovial mouth.

'Welcome to the snakepit, gentlemen.' Art smiled. 'Glad you could come tonight.' Then they put on the white jackets Art borrowed for them and headed for the open ward to await incoming patients and administer perfunctory physical exams. Between exams there would be time for coffee, cigarettes, and lots of talk, mostly of medicine and music, Ian's and Art's common interests.

'Why an Armenian bagpiper, Art?' Ray asked, knowing about it from stories Ian had told of the old days when Art and Ian were novice pipers.

'I got interested several years ago when I was an undergraduate and Ian was still in high school. I thought I was going to the University of Edinburgh Medical School but I ended up at USC.'

'So all that piping was wasted.'

'Wasted! Are you kidding, Ray? Listen, you just stick around Ian and me and we'll make a believer out of you. If an Armenian can be a bagpipe aficionado, so can an Italian. You have hairy knees? *I've* got hairy knees. That's the sign of a good piper. You should see me in kilts and a glengarry cap. And you should've seen Ian and me competing in the Highland Games. Here was this tall, straight, shiny kid, the epitome of Jack Armstrong, and here I was, a greasy slouchy little camel driver. Now who would you have picked?'

'Let's talk about medicine,' said Ian. 'That's what I'm supposed to be going to college for.'

'Did Ian tell you about *my* pipes, Ray?' Art persisted. 'They're from India. Good ivory and ebony, but Pipe Major Aitken, our old teacher, wouldn't go for them. They had to be Peter Henderson pipes, like Ian's. You see, in Scotland, after they drill the holes they point the pipes toward the Orkney Islands and let the cold wind blow through them for three years.'

'Ye cannot have the hot winds of Pakistan blowin through yer pipes, can ye, laddies?' Ian said, and he and Art smiled at the old memory.

'Why didn't you get the Scottish pipes?' Ray asked.

'Two hundred good reasons,' Art said.

'Scotsmen aren't the only ones who knew the value of a buck,' said Ian. 'Armenians shop around a bit. Tell Ray about all the money you're making now that you're a big shot doctor.'

'Oh yeah,' said Art, dropping easily into another subject. 'Know what they pay interns? Seventy bucks a month. Know what the rent is? Seventy-five a month. Know how I support a wife? Selling blood and gastric juices.'

'What?' said Ray, and Ian chuckled.

'First I sell my blood like a common skid row derelict, but that's nothing. Every Saturday morning I sell my gastric juices with the rest of the starving interns. I just go in there for about three hours, insert a hose through my nose right down into my stomach and shoot them the juice. I used to almost faint when I did it, now I just drop that old hose right down there. In fact, I got more gastric juice than anyone.'

'He's rather proud of it,' said Ian.

'They give it to some old geezer who's about ready to croak,' Art explained. 'For a few days he feels real good. The medicines he's taking orally get absorbed and he gets more nourishment from food. He feels pretty much in the pink before he checks out. I get fifteen bucks for about three hours' work.'

'Oh my . . . Oh, God,' said Ray, dropping his head to his folded arms.

'Even thought about selling my seminal fluid,' said Art. 'But I chickened out. I figured sure as hell, soon as I go into practice, some broad would come in for a check-up with a little Armenian kid and I'd start feeling responsible without knowing for sure.'

Then Art's animated conversation was interrupted by another pair of uniformed policemen entering with a handcuffed man. He was a young man, blondish, very pale, wet beads over the lip and around his temples, the sideburns shaggy, plastered to his scalp by the moisture. He seemed passive enough.

The young man was checked in by a nurse and given a glance or two by the uniformed deputy sheriff on duty. It was Art's job to treat such patients for any obvious organic ailments before they entered the psychiatric wards.

'It's a ruthless type of medicine,' Art apologized to Ian as they were getting the patient ready for the exam. 'It's trench warfare. Do you have any idea how many patients they bring in here each night? Most or all of them idigent? The hospital can't possibly treat or commit anyone unless he's stark staring nuts, and only then if he's an imminent threat to himself or somebody else. If they've got alcoholic breath or a history of d.t.'s, pow! We kick them out the door tomorrow. Just don't be disillusioned. This isn't the way I'll practice medicine when I finally get out in the world.'

'I won't be disillusioned,' Ian said as they entered the examination room.

The patient looked up from a chair and started to stand.

'That's all right. Sit down,' said Art. 'We're just going to have a look and see if you're all right.'

'I'm okay,' said the young man, looking from Art to Ian. To Ray. Back to Art.

'I'm sure you are,' said Art, sitting down beside the patient, attaching the sphygmomanometer to a very rigid and muscular arm.

'This is just to test your blood pressure,' said Art reassuringly, noticing that the sweating was becoming excessive.

As soon as the first slight pressure was applied to the bulb, the man moved. Without warning he screamed in terror and Art was on the floor and the sphygmomanometer was smashed against the wall, and the uniformed deputy was vaulting into the room on top of the pile of shouting thrashing bodies on the floor.

The young man was quite strong, and almost impossible to control. He threw them off, all the time shrieking for release from his private demons. Finally sufficient help came and he was controlled and led weeping through the double doors.

Art accompanied the patient while Ian and Ray drank coffee and rested, regaining their composure.

'I was talking to the resident shrink back there,' Art said when he returned moments later. 'He knows that kid. Said what happened to him here was homosexual panic. That's when the awareness of his condition suddenly bursts forth in a repressed homosexual.'

'Incredible,' said Ray. 'I never saw anything like it.'

'You can write a textbook after you're here a couple of nights,' said Art wearily.

'Poor guy,' said Ian. 'I really felt sorry for him when they led him away crying.'

'Keep studying. We really need some idealism in this racket,' said Art smiling, but without irony.

One of Ian's friends, a floundering Ph.D. candidate in psychology at UCLA, warned Ian he'd never finish college. 'Ian, it's an anthropology class!' said Grog Tollefson. 'You're supposed to provide a short definition of Homo erectus. The test says he lived half a million years ago in cave communities and was a cannibal. Why go off the subject and spend your time with Kant or Santayana and try to relate them to your course in anthropology? You'll *never* get through undergraduate school at that rate.'

'I appreciate the advice.' Ian shrugged. 'But grades bore me. And I *like* philosophy. What am I going to do?'

'Damn, I've already proved *my* academic instability by dropping out of two colleges, and being on probation here. But psych majors are supposed to be flaky and insecure. I always thought you were too much in control. As much in control as I'm out of it. Now I'll bet *you're* going to drop out first.'

'You're psychoanalyzing again, Grog.'

'So what? It's a psych major's prerogative. Now let me help you a little bit. What the hell was your social studies test all about?'

'Don't think I know for sure.' Ian grinned. 'There was an essay on bigotry.'

'Good, did you say something to the point for once?'

'I meant to, but then I got to wondering how a great composer like Wagner could have been such a narrow anti-Semite and I ended up writing about Wagner's music.'

'You're hopeless, Ian,' said Grog disgustedly. 'I think your big problem is your glands are calling and you want to marry Adah. I don't give you another semester.'

And Grog was right. Ian suddenly quit, and impulsively joined the Los Angeles Police Department, and married Adah.

Art Petoyan was to say to a fellow piper, 'I was shocked to hear that Ian dropped out of college. To become a cop! I just couldn't believe it. And it's changing him fast. He isn't the same guy. Maybe not embittered, but he's at least becoming cynical. Those idealistic dreams of bettering society by being a physician are gone forever, I'm afraid. Police work is changing him. He knows he isn't improving society now. It's just a holding action, he says. I'm sure he's become aware of evil – believes for the first time that it can touch him.'

Chrissie listened quietly when Ian first told her he wanted to join the police force. She nodded occasionally and did her best to smile. There had already been many talks about his getting married, and he knew she felt he should wait until he finished his undergraduate work, at least until he was partly through medical school. But now he was telling her there would never be a medical school. Never. And she knew his decision was irrevocable. And so she reassured him, and said she was happy for him if this is what he wanted. If he was sure. And that night, for the first time in many years, Chrissie wept.

'Adah isn't your typical showgirl,' was the phrase most often used by those wishing to describe her. The speaker would then invariably add: 'I don't mean she's an intellectual or something. She's not. A high school dropout in fact. She took a correspondence course to get a diploma. Very shy. Never opens her mouth. An innocent kid. I'm not kidding you, an *innocent* Las Vegas showgirl.'

But she looked like the others, busty, *too* slender, bony hips protruding from her costume. Taller than most, six feet one inch in spikes. Not a beauty but attractive, with blue-gray eyes like Ian's that are invariably referred to as hazel. With hair, dark like his, tinted red for the shows.

At first Adah was afraid of him, had never been around a man like him. And the manners he got from his mother intimidated her, but at the same time made him terribly attractive. She was even more terrified of Chrissie Campbell, for she found her too genteel, everything Adah was unaccustomed to in her own family. And she was afraid Chrissie would blame her for Ian's quitting college.

'Ian hated pre-med all along,' Adah confessed to a friend. 'Chrissie never knew it. I don't even think Ian knew it for sure. He's antsy. Wants freedom on his job. He likes working nightwatch because of the activity. Sometimes though, he just can't go to bed when he gets off work. He might go to see my old bosses, Jack and Marge. Being show business people they're up all night choreographing. Or he might go see Grog Tollefson, or Wayne Ferber. Or to Chrissie.'

After a few weeks of marriage Chrissie asked: 'What do you think of married life by now, Adah?'

'Oh, I think it's great except that Ian *reads* so much.'

At first the newlyweds lived in a Hollywood apartment to which Chrissie was often invited. Adah became less inhibited around her, more anxious to talk, although with strangers or in a group of people she was still self-conscious and would usually curl up in a chair and retire from conversation. But she *was* changing. Becoming less withdrawn and physically more attractive as she toned down her showgirl appearance. She'd been a showgirl only in fact, not at heart. Now she became what she was born to be, a wife and soon a mother.

Ian Campbell would still drive by the Park La Brea Apartments to see Chrissie, sometimes at midnight when he finished nightwatch. And if he was disturbed about anything she would know it, and through gentle turns, the conversation would turn to the disturbing thing. It had been that way all his life. After transferring his problem, Ian's face would light up, the brooding look would be gone, the eyes would go more from gray to blue. Then Ian could sleep but his mother would stay up half the night with the problem.

'Ian's always been such a castle builder,' Chrissie told Adah. 'He was always unrealistic about helping people and always expects more from life than he gets. I'm afraid police work'll somehow awaken a person like him too harshly.'

It was hard for Ian to find any time for piping now. The job and court time were making great demands. Sometimes, early in his marriage, while living in the Hollywood apartment, he would in desperation go inside a walk-in closet, close the door and play for an hour, or as long as he could bear the confinement. Adah would find a cloud of steam and puddle of sweat on the floor when he came out. About all he could do in those days was play the practice chanter, which went from G to high A and sounded like an oboe. It was subdued and didn't frighten small children and didn't disturb the neighbors or his wife. But it wasn't the same. Not to a piper.

It was at this time, when he couldn't often play the pipes, that Ian became obsessed with the idea of playing the pibroch, the classical bagpipe music. The sonata of bagpipe, the incredibly difficult musical exercise. Darting, striking movements, riddled with grace notes, played for the clan chief when he was brought back from battle on his shield. But it meant more: the ultimate music to demonstrate the qualities and technique of the piper. A solo piper. A virtuoso.

He had been working on 'Cha Till McCruimein,' a famous pibroch, anglicized as 'MacCrimmon Will Never Return.' It was melancholy with a pastoral air. It was a baffling experience, frustrating, but exciting beyond anything he'd ever done.

But soon he was the father of two little girls and like most small children, Valerie and Lori were terrified of the sound of pipes. They would scream whenever they saw their father blow up the bag and he would try to reassure them. And then there were the neighbors, so close when they moved to the new housing tract, and there was Adah. So the piper would put sugar in the bag and rub it in with a little neatsfoot oil and only dream of playing the pibroch.

For if he could not play it in fact, he could play it exquisitely, majestically in his imagination. On monotonous evenings, riding in the black-and-white, during dark and quiet early morning hours, he could play the pibroch. When sitting at a desk in college, now a part-time student in law and police science, tensely awaiting the exam, he could be soothed by playing the pibroch. While lying still in the night on those sleepless nights which followed the cruel or distressing thing he had seen during a tour of duty, in those moments he could certainly play the pibroch. Always it was the melancholy, pastoral, plaintive pibroch: 'MacCrimmon Will Never Return.'

The gardener was unhappy with the job he'd done on the old woman's lawn. The grass croppings lay in heaps in several places where the catcher had missed. He went to his truck to get the leaf rake. The old woman deserved an absolutely perfect job today to make up for the fact that he hadn't listened to her gossip.

While he was raking he thought about the time he stole the hardware. As always, he could relive the act, but not the feelings he had when he stole. He wiped his hands on his blue workshirt and got a better grip on the leaf rake. For a moment he thought of the testimony he would give tomorrow in court. He never had to stop working when he thought of the trials, not even when he thought of his crimes.

The November California sun was wonderfully warm today. He had a fine jacket of sweat under the workshirt. Soon it would be winter and the sun would not be as good to the gardens he tended. That much was certain, the gardener thought. As certain as death in an onion field.

TWO

The felony car partners ate Irish stew in McGoos on Hollywood Boulevard on their eighth night together. Ian was even more quiet than usual and then he said, 'You don't have kids, Karl?'

'I've only been married a short time,' Karl said, 'but we got one in the hopper.'

'That's great,' Ian said soberly. 'A marriage is something that takes caring for. Like plants and trees. A marriage is a good thing for a man.'

Karl glanced up from his stew, his glasses in his pocket now, and saw Ian with fork poised, staring at his own plate, preoccupied. Karl had heard that a few months before, Ian's wife had gone to work in Bimbo's nightclub in San Francisco supposedly to earn enough to buy furniture for their new house. And their children had gone to stay with her mother. When some of the other policemen, inquisitive ones, asked Ian if he were separated from his wife, he denied the estrangement. But Karl didn't believe him, especially now, with the way he talked so wistfully about his marriage.

'You have two girls don't you, Ian?'

'Two girls,' Ian answered. 'Valerie's three and a half. Lori's just two. Getting old enough to really enjoy now. They were staying with their grandmother for a while but now I've got my girls home. All three of them.'

Karl did not question his partner's last remark. It was not his affair.

'Valerie's already learning to read. I'm teaching her.'

Karl nodded, imagining that his partner was probably a bookworm.

'I like to read,' said Karl. 'Someday I'd like to have time to do a lot of it. But I don't care for novels.'

'I never read novels either,' Ian said, not knowing that Karl's tastes were far different and ran to fishing and outdoor literature, and 'How To' books. 'My wife used to say reading was *all* I cared about.'

'You like growing things,' said Karl.

'And bagpipes.'

'I heard about that,' said Karl. Then he added, 'You having dessert?'

'No, I've finally got my weight below two hundred,' said Ian patting his stomach.

'At least you're tall enough to carry it,' Karl said, feeling his own bulging middle. 'Married life is turning me into an avocado with feet.'

'Don't worry about it,' said Ian.

'Oh yeah?' said Karl. 'I gotta hold my arms out these days to see if I'm walking or rolling.'

Pictures of Karl Hettinger would verify that he was much heavier now than he had ever been in his life. The pictures would reveal the fullness in his lightly freckled face, the crew cut would make his face even fuller. The picture would be on the front pages.

When it came time to pay for meals or leave tips, neither noticed another thing they most certainly had in common – frugalness. Neither man had the slightest difficulty remembering where every dollar was, and where a wholesale price could be obtained, or a police discount.

So the two young men in felony unit Six-Z-Four were becoming more and more acclimated to one another, heading toward what might be a significant friendship. It was truly a bonus when a partner could become a friend. Karl unconsciously relinquished leadership of the team to Ian by virtue of the other's seniority on the car and on the department, and also because Ian was almost three years older at thirty-one. But most of all because Ian had a certain bearing and quiet self-confidence. It was natural that he lead when a leader was required.

Yet Karl Hettinger was not without his traits of leadership, even had his preeminence predicted by his junior college yearbook. He often remembered that prediction. Who could've dreamed he'd end up a cop? he thought, as they resumed patrol.

As long as he could remember, Karl Hettinger had dreamed of being a farmer, a unique ambition for a boy raised in the megalopolis of Los Angeles. But his mother had filled him with wonderful farm stories as a boy. She was a vivid story teller and used her talent with growing things to full advantage on their one-acre piece of ground. And the visions of the clean and tidy German farm where she was raised came to the small boy and lingered.

Karl had come late to his parents, well after his sisters, Miriam and Eunice. His father, Francis Hettinger, was in his mid-forties by the time the boy was ready to learn. The father was unaccustomed to a child's demands, never having had to surrender his working time to the girls. And he had no time to spare. He was a quintessential working man, one who worked not just for wages, but for fulfillment, and for profound personal pleasure. He was a carpenter and a fair plumber, and he knew wiring and cabinetmaking, and was an adequate backyard mechanic, a man who knew the worth of a pair of hands. His were big and rough, knuckle-heavy, hanging from big wrists, strong, fully capable hands. He was a quiet, sought-after workman, a perfectionist, deeply proud of what he produced, a trait rare in those war years when good workers were scarce. He knew how to boss a job when necessary, but he did not know how to teach a boy.

'Dad, do you think I could help you sometime at your job? School's out this week.'

'I can't, Karl. The boss wouldn't like a boy hanging around. Maybe when you get older.'

'Well, do you think you might be able to show me how to use the miter box? I mean at home? On Saturday? When you're not at work or anything?'

'I . . . well, it's just that I have to get that job right, son. You're just too young, right now. I just have to get it perfect. It's important to get the job right. When you get older, Karl, I'll show you. But for now, son, I wish you wouldn't use my tools.'

And the tow-headed boy would look with sky blue eyes at the big hands of Francis Hettinger and wonder how the hands worked so wonderfully. He longed to see his own become large and adept and even bruised and broken like those of the carpenter. Somehow he never did develop those hands.

And as Karl got older, his father did try to teach him, but they were ineffectual attempts. He was not a good teacher. He would overexplain, repeat himself. In the end he could not hide his displeasure with the boy's efforts. They rarely talked together, the carpenter and his son. And the carpenter and his wife rarely talked together. He expressed his family devotion the only way he knew, through work.

Sometimes the boy wished they could sit and that words would flow. But when they sat, eyes would turn shyly away. Father and son usually fell silent. It was ancient inherited shame of fathers and sons.

But farming, that was something he could learn. And he would beg his mother to retell the stories of the clean prosperous farms. When he was still a very young boy he was sure of his vocation and became interested in the things his mother grew in their garden.

In fact, all of Karl's hobbies had an outdoor flavor. There were team sports, like baseball. And he learned to fish at an early age.

'Would you like to go fishing this Saturday, Dad?'

'I'd sure like to, but you know I gotta work six days.'

'Sunday. Then how about Sunday?'

And his father would say, 'A man's got to rest sometime, son. I'm just gettin too old to be goin every day.'

'But you'll be working Sunday I bet. On the car or something.'

'Oh, now putterin at home, that's not work. That's restful. But fishin? That's work, son.' And he would laugh and timidly paw at the boy's shoulder with a hand full of blackened split fingernails. And that was as demonstrative as he would ever get. Karl would usually go fishing alone.

But his childhood was by no means lonely. With two sisters so much older, it was as though he had three mothers, one of them always ready to give him attention, and there were other boys to play with. His mother would always brag that Karl was so good she never had to whip the boy. And once when Miriam tried, Karl looked up at her sadly with his pale blue eyes and made a silly face and they both burst out laughing.

Only once was he disciplined significantly. It was when he was six years old. Karl and another child named Grant would leave first grade and stop by the five-and-dime for a five-cent pickle to share, or, if they were flush, for one apiece. It was on a pickle-buying trip that they saw some penny erasers they thought they needed, and not having enough for the pickle and the erasers, each stole a few pennies' worth. When Elsie Hettinger happened to see the erasers at home and mentioned them to her son, Karl blurted out the bad thing they had done. His mother threaded a piece of string through each eraser and hung them on the wall over his bed for many weeks. The erasers dangling over the boy's bed like avenging daggers were enough. The youngster could not even think a dishonest thought for many years.

Though his parents rarely socialized, Karl was gregarious enough and always popular. He grew to be self-sufficient. If he had a problem he learned to keep it within himself.

'There was no need to burden others with your personal problems,' his sister Miriam was to say many years later. 'It was not the family way.'

His sister Eunice laughed when she returned home a bride, explaining to the Hettinger household what a shock it was to be hugged and kissed by her groom's family. Not just at the wedding, but almost every time they saw her, and her husband's people would talk to each other about the most personal things. It was strange to her. It occurred to Karl that he had not been kissed by his mother since he was a tot. Nor by his sisters. Then he suddenly realized that his father had *never* kissed him. It was not the family way.

Karl Hettinger was certainly not going to confess to his mother when, during his twelfth year, he and another boy stole some fishing plugs from the Sears store in Glendale. The two boys were caught by a store employee who saw their terror and pitied them.

'I ought to call the police,' the man said sternly, and now both boys were having difficulty holding back the tears. 'Would you sacrifice your freedom for a quarter's worth of fishing plugs? Your freedom is very precious.'

And he released them, and Karl Hettinger knew for a certainty he was fated to be a more than honest man. He would not abide a dishonest thought or word from that day forth, believing he had miraculously escaped the terrible wrath of the law. Nothing could be more fearful than losing one's freedom. To be confined. Never to see a golden cloudburst or rivers of sunlight on dark flowers. Never to walk your own cultivated furrows. And the memory dangled over his heart like the sword of Damocles.

His total honesty would be one of his hallmarks. He would not so much as tell a lie from that day on. His conscience would not permit it.

There was only one school in the area for Karl upon graduation from high school, Pierce Junior College, in the San Fernando Valley. It rested against a row of hills with the wide valley expanding below, and soaring mountain ranges to the north. White stucco buildings had multicolored Spanish tile roofs, and student landscapers of years past had generously planted the campus in palms, evergreens, and willows. It was where a Glendale boy would go if he wanted to learn agriculture. It was where Karl Hettinger, only sixteen but a college freshman, found everything he'd hoped.

He flourished in college. Any shyness he retained from his adolescent years all but vanished in a burst of campus activity. At first it was working in the college fields, picking corn, shucking it there where it grew, eating it under the hot summer sky, never tasting better because he had helped its growth. And there were other truck crops: radishes, onions, carrots, tomatoes. Perhaps the tomatoes were best – fat, heavy, red and dew-wet on cool mornings. Karl would kneel among them in earth moist and spongy, richer by virtue of care and cultivation than any land in the valley, and with little imagination he would see vast sections of glossy tomatoes, himself walking lovingly among them on his own land, relying on no man for a wage, only wanting a little rain and luck, and lots of sweat, which frightened him not at all.

It may have been his affinity to growing things, or it may have been living in the dorm, truly self-sufficient for the first time, or it may have simply been his ripening past adolescence, but whatever it was, Karl began to burgeon as surely as the crops he tended. He was a varsity baseball player, a member of the student council, and finally the student body president. Karl, though only of medium size, had strength, agility, and most of all the determination to pursue any task to the end. Often he would be found irrigating in the fields in his rubber boots long after classes, earning extra money. The dorms were thirty dollars a month and his father worked night and day for the properties and goods he possessed, and believed his children should do the same.

With his diligence and love of basic field toil he was a natural to be chosen as a college leader at this agricultural school which still clung to the Protestant ethic of hard honest work.

Life in dorm three was a pleasant interlude he would look back on at a later incredible time of life. He would remember the marvelous poker games. Karl would sit there in a nickel-dime game, the light shining off his crew cut, now darkened to red blond with the passing of years, puffing nickel cigars though otherwise he didn't smoke, his blue eyes also darkened to a deeper color, expressionless as long as the game was in progress. And Iranian students, sons of sheiks, would play as fiercely as he, as though every dollar they owned was in the game. In Karl's case it often was.

But there were disadvantages to a school like Pierce in these years: for one, a dearth of women students. The few there were were attracted to Karl, not only by virtue of his student body leadership but by reason of his diffidence around them.

The year and a half he spent at Fresno State College was less pleasant. Though most of the old Pierce boys stayed together, they played too much, and the grades began falling. Like Ian Campbell, Karl Hettinger dropped out of college in his senior year.

Karl joined the marines and was stationed at Twenty-nine Palms, California, for most of his enlistment.

His childhood friend Terry McManus talked of Karl: 'We went to a beer bar one night right after he got out of boot camp. The broad behind the bar was a tough old bat and some guy was coming on pretty rank with her. And she put up with it as long as he was spending his money, but finally he was getting awful aggressive. She was the owner's wife and had probably handled a thousand guys like him, but all of a sudden Karl gets up and goes over and tells the guy to leave the lady alone. Imagine, in this day and age, and in a joint like that, defending a lady's honor! Well, Karl was always a wiry guy, and hard-muscled, but this other guy knew all about handling his fists, and after Karl took his lumps and I helped him out of there, he just grinned and said, 'Guess I thought I was a marine tiger.' But it wasn't the marine thing at all. He would've always done something dumb like that. He was like so many of those farmboys I met at Pierce: stubborn, quiet, and determined about right and principle, and just about thirty years out of touch with the city life of Los Angeles all around them. In Los Angeles, only a farmer would shrug shoulders at adversity and bear up without complaint and hope for fair weather. And for sure only a hick would defend a barmaid's honor.'

After the marine corps, Karl Hettinger was suffering the same adjustment pangs that all vets undergo – restlessness, uncertainty – when one day a friend talked him into accompanying him to City Hall where the friend was going to take the test for policeman. On impulse Karl took the test himself. Ultimately it was Karl and not the friend who entered the police academy. It was a surprise to his family, to everyone who knew him, most of all to himself. He wasn't at all certain this was to be his life's career or even how he got there. But, typically, he put forth maximum effort and did well and was named class valedictorian.

The first two years on the police department were for Karl like they are to most young policemen: incredibly boring, intensely exciting, mostly bewildering. Having a quiet but sharp sense of humor, Karl tried to slavage the funny moments for old friends who wanted to hear cops-and-robbers stories. Karl shared an apartment with Bob Burke, a policeman three years his senior, who had helped break him in at Central Division. They were compatible roommates, both meticulous housekeepers, who stayed out of each other's way and worked different shifts. Bob was more of a ladies' man than Karl, who was still an avid outdoorsman. Anything to get out in the mountains or deserts, or especially the lakes.

Then came the transfer to Hollywood Division, and with only two years' experience, an assignment to the vice detail. There was seldom time to think of the old dreams of tomatoes glistening in the sun.

Vice work was both enthralling and appalling. He found the bookmaking detail best, and was surprisingly good at working bookies. The prostitutes were impossible for him, at least as an operator. He could never master the lecherous role a vice operator must play. He was not an actor, could never pretend to be someone else.

The omnipresent Hollywood homosexuals were perhaps most shocking, and the sordidness of the assignment was best mitigated by dark humor.

'Look at the whanger on that one,' whispered his partner, John Calderwood, at midday, as they sat on a hillside in Griffith Park and watched a well dressed, middle aged businessman leave his company car two hundred feet down from the restroom and disappear into the bushes just below their place of concealment. There he conducted a bizarre ritual with a shoe, a jar, and a nylon stocking, ending in five minutes of violent masturbation.

'He reminds me of a tripod I used to own,' said Karl dryly.

'What a maniac,' said Calderwood, scratching his mane of blond hair. 'What the hell can you do about people like that?'

'Circumcise them,' said Karl.

'What the hell's that gonna accomplish?'

'With a chain saw,' said Karl.

On and on went the defensive vice squad comedy until partners began learning their lines so well they began playing straight man to each other in front of other members of the squad.

One vice arrest of a homosexual was to have significance at a later time in his life.

'We was up off the path in Ferndale Number Nine one afternoon,' said John Calderwood. 'I always liked to work with Karl because he had more jokes than any man I ever met in my life. Clean jokes and dirty jokes, he had a million even though he was a little quiet. Anyways, Ferndale Nine is the famous restroom up in Ferndale Park. All the fruits congregate there and do everything you can think of in the restroom.'

Karl and Calderwood were up on the hill that day directly east of Ferndale Nine. There was a trap where vice officers watched for homosexual acts and made arrests for lewd conduct in a public place. Though most of the homosexuals who frequented Ferndale Nine knew that vice cops lurked there, they still came because the possibility of being caught excited them as much as the sexual contact. In fact one effeminate young man with a puckered nervous face was seen writing 'Peek here' on the wall of the restroom just under the observation window of the trap. So the homosexuals and the vice officers engaged in an endless dreary war there in Ferndale just below the Griffith Park Observatory.

The homosexual who was to emerge later in Karl's life was cruising that day and spotted Karl standing on the hillside by Ferndale Nine. He stared at Karl with a feverish panicked gaze. This kind of look often signals danger to vice officers. In terms of physical injury, vice work was more hazardous than any other police function with the exception of the motor squad. Panic made it so, and in terms of injury to the spirit, the vice squad produced more casualties than all other police details combined.

Karl was anticipating danger and watching slant-eyed and poker-faced when the young man shook back a lock of brown hair, smiled, and without hesitation reached for Karl's crotch.

'Police officer! You're under arrest!' said Karl, grabbing the man's wrist. The burly young man drew back in disbelief. His bright round eyes blinked and filled with tears. Karl heard the crackling of branches as Calderwood ran from the bushes. Then the young man erupted in a blur of fists and feet, punching, kicking, screaming in outrage and betrayal, and the three of them were on the ground fighting. Not a movie fight where one stands back and resoundingly cracks homeruns cleanly off the other's chin. A *real* fight – a choking, gouging, biting, kicking fight. Hate-filled, blood-smeared faces. Desperate vice officers. A burly neurotic filled with panic. Muffled curses. Blows thrown by the policemen which often as not hit a partner. Each of them with one thought in mind: a headlock, a choke hold, the crook of the forearm closing on the carotid artery. A hold which has saved more policemen than all the sophisticated self-defense holds combined, because anyone with average strength can get a choke and squeeze until the antagonist falls limp. But they couldn't get a choke. He was thick and strong and hysterical. He was making animal sounds and Calderwood was around his neck bending him backward away from Karl whom he was stomping down on his back.

Then the hold was firmly achieved and he was choked. They handcuffed him when he was flopping on the ground, face like a spotted plum, mouth gaping idiotically, gasping, swallowing his breath. Then, incredibly, he was on his feet, hands cuffed behind his back, running blindly across the hill. The two policemen did not have enough strength left to catch him, but tried, until he ran straight off a fifteen-foot embankment. They heard him shriek and fall. They ran to the brink and saw him at the bottom on his stomach, head twisted, eyes closed. They thought he was dead, but he was not. He only suffered a broken collarbone.

Their hands were still unsteady twenty minutes after the booking when they sat in a Hollywood drive-in restaurant drinking coffee.

'I don't care if I never bust another fag,' Calderwood groaned, holding the coffee mug in both hands, gingerly shifting his weight in the chair. 'I'd be tempted next time to use a magnum on an asshole like that.'

'Ordinary magnums wouldn't help much with that one,' said Karl, massaging his ribs.

'Why not?'

'You'd need silver bullets.'

'Hope to hell he pleads guilty,' Calderwood moaned. 'I don't even wanna see that beast again.'

But he did not plead guilty. They did see him again. In court. And they heard from him again. The man wrote letters complaining of brutality to the police department, and to the municipal court before and during his trial. Finally after his conviction he threatened in writing to kill Karl, Calderwood, the city attorney, and the judge. Then he was all but forgotten.

By the time his vice tour was ending, Karl was ready for a less sordid assignment. He was sick of the intimate contact with wretched people, had had enough of the sweet sick smell of wine and stale sweat and vomit that permeated patrol cars until after a while you thought you smelled it everywhere. He hoped to work juvenile, was sure he would be good at it. He was patient and enjoyed children and was troubled by the tragic wandering teenagers he saw in Hollywood at night. When he mentioned to a partner that the young wanderers bothered him the partner grunted and said, 'Wanna save the world, huh?' And Karl smiled self-consciously.

By then he had met Helen Davis at one of the many parties Bob Burke arranged in their apartment. And though she was only twenty years old, Helen was a girl who knew what she wanted. She jokingly said that it was time a man almost twenty-eight years old got married. Karl laughed and treated the remark lightly but saw in her hazel eyes, or thought he did, a certain strength to match his own.

In December of that year, Karl's police friend Jim Cannell routinely reminded Karl of the fishing trip they had planned to Lake Isabella and asked him if he would be bringing Helen.

'Can't make it this week, Jim,' said Karl.

'Why not?'

'We're going to Las Vegas to get married,' he said casually, and they did. Characteristically, he had told none of his friends.

At the Hollywood Division Christmas party in 1962 they were more happy than they had ever dared dream. They glided across the dance floor, Helen's light brown hair tumbling across his face when they whirled, and both had enough to drink to be sure they were the best dancers on the floor. They vowed to go dancing often, at least once a month, and to attend many of the police department's dancing parties. But the ninth of March was less than three months away. This would be their last police department dancing party.

Now the gardener looked up at the sun and without his watch knew he was an hour and a half from lunch. That was certain. The certain things of life pleased him. He was sure for instance that if he cared properly for the old woman's Aleppo pine it would grow to be a soothing pale green with lacy needles.

After lunch when he would leave here maybe the headache would stop. He would get to do a beautiful yard this afternoon, the best on his route. The yard had some fine specimens of Monterey pine lining the driveway. They were so stately and deep green and symmetrical they hardly ever needed trimming.

The gardener looked at the old woman's petunias. They had good color: reds and whites and off-shade blues, but the old woman didn't understand petunias. They bloomed profusely but they were delicate. You must never forget how delicate they are, he thought. You must be kind to them. Life was a tenuous thing to the gardener.

And then, there in the petunias he caught a picture of himself in a store by an untended counter. Saw himself glancing to his right at a clerk whose back was turned to him. He hadn't stolen anything big as yet, only pocket loot. Now, though, the bulkily packaged electric knife looked irresistible. Where could he secrete it? Under his coat? Inside his belt? Perhaps in his hand? Yes, brazenly in his hand. Store security men looked for telltale bulges, furtive behavior, but what if you just held the large thing in your hand and simply walked out with it? People don't see the obvious. That was the way, yes.

Now, suddenly, the crimes were almost too painful to recall. He'd continue to think of them later. He had all afternoon. When he started thinking about his crimes, he had to start at the very beginning and think of all of them, the same as when he used to think about the night in the onion field.

He always thought of that night from the very beginning to the end. If it weren't for the trials he wouldn't think of that night at all. Why should he? It was seven years ago. Besides, he had all his crimes to think about. That was the real horror.

THREE

It was Ian Campbell's turn to drive on the ninth day of the partnership: Saturday, March 9, 1963. He and Karl were dressed, as always, in old comfortable sport coats and slacks.

Ian's coat was wearing through where it rubbed against the butt of his revolver. They talked about clothes during the early part of the night, both of them ever thrifty, wishing that the department gave a clothing allowance to plainclothes officers.

There were two other young men driving toward Hollywood that night in a maroon Ford coupe, who had begun a partnership on exactly the same day as Ian Campbell and Karl Hettinger.

The two young men in the coupe also talked about clothes. The blond driver, Gregory Powell, admired the black leather jacket and cap he wore. The darker man, Jimmy Smith, hated his, but wore it because Greg insisted.

'Fuckin Saturday night traffic,' Jimmy Smith muttered.

'Stop griping, Jim,' said Gregory Powell, his gaunt face turning toward his partner on the swivel of an incredibly long neck. 'Wait'll we take off that Hollywood market. Wait'll you get your hands on some money.'

Jimmy Smith grunted and adjusted the Spanish automatic he carried in his belt.

'I just wanna git it over with, is all,' said Jimmy Smith in his soft voice, and Gregory Powell smiled, his front teeth protruding slightly over his lower lip, and dreamed past this robbery to much bigger ones. And dreamed of having enough money to go into legitimate business.

Imagine what they would say then, thought Greg. All the bitterness would be forgotten. They would come to him for money or favors. Gregory Powell chuckled silently as he thought of them – his family.

*

'I didn't really mature until I was thirty-three years old,' Rusty Powell often admitted, but by then his eldest son, Gregory, was already twelve years old, and had attended a dozen different schools before moving to Cadillac, Michigan. Rusty Powell had been pretty much of a drifter during the Depression and early war years, playing in small dance bands whenever possible, while his young wife had a baby every three years until after Douglas was born in 1942.

At age thirty-three, Rusty Powell enrolled in the Cincinnati Conservatory of Music. But now, with new time-consuming zeal, and guilt-relieving work, he knew his children less than he did in the drifting dance band days.

For his son Greg though, there had always been the Governor, his maternal grandfather, whose picture he would carry even as an adult, a sturdy old man with deep creases in his cheeks, with white hair and a white moustache, a pipe smoker, fond of comfortable baggy pants and suspenders, and a shirt open at the collar. A grandfather who looked as grandfathers should: strong, wise, patriarchal. Greg would spend hours talking with the old man and many of the talks would be about Rusty Powell. The Govenor would try to make the boy understand his father.

'He doesn't have an easy time, your dad. Why, your mom's had every disease and operation known to medicine almost, and what with four kids and all, your dad just hasn't had an easy life. And sure, he's an easygoing sort of man and maybe you think he lets your mom run over him too much, but a mountain is quiet too and I suspect there's great strength in your dad. Maybe just going along with her and not riling her is a way to have peace in this world.'

And the boy would nod and say, 'Tell me about my dad when he was courting my mom, when he had the fight with the boy that insulted her.'

And the old man would tell him once again of how Rusy Powell crushed the other boy in his arms, breaking a rib or two, and Greg would say to the Governor, 'Do you think I'll ever get as big as my dad?'

When Greg's mother, Ethel Powell, would unreasonably revoke an agreement that he could go ice skating, or to the movies, he would tearfully drag his father into her bedroom and confront the invalid woman with her broken promise. But she would deny that she had ever made it, and would rail at both of them and remind them of her thyroid and of the Bright's disease and of her nervous condition. Invariably, Rusty Powell would shrug his large sloping shoulders, shortening his long muscular neck with the gesture, and he would yield – to her. And for an instant the boy would hate him even more than he hated her during those years of illness and 'nerves' and he would try to think of his father as a boy whipping the tar out of Greg's uncles, showing that he was a man. Greg would make allowances and resume his duties: the shopping, the minding of the three younger children, the ordering of the home.

He didn't always hate his mother. When she was even partly well, she could manage to look incredibly young and alive. They would go for walks or even skate together and she could *be* incredibly young, as young as he, and totally enter his child's world. He was intensely proud of her during these times and wanted everyone to see her. He would forget the unreasonable nagging and the hateful lying she seemed unable to control. But then she would get sick and nervous again.

'Listen, Greg,' the Governor would say. 'I know it's hard with your mom and dad, but a man can abide. It don't matter what others do to you. Why, it just don't matter at all because you can always come to me. The Governor's always here to help you.'

And Greg would nod and be comforted because it was true. The Governor managed to abide all these years, and his wife, Greg's grandmother, was a Christian Scientist, to Greg a fanatic. He thought the old man was right, he could live in peace with someone he didn't respect as long as he just didn't show them how he felt. He only had to pretend to go along and do things his own way when he could. The Governor was always there to help, except that the Governor died in 1947.

They were in the big two-story wood frame house in Cadillac then. It had a large yard and there was a fish pond in back, room for dogs and cats and birds and fish, and even without the Governor things were bearable. Except that his mother started to get well.

At first it was a subtle change. Douglas, the youngest, was now well past the toddler stage and with Rusty teaching music, life was indeed much easier. Ethel Powell began fixing her face every day, paying attention to her hair, and her clothes and person. Then there were touches to the house here and there. Bright things, a preponderance of reds, lots of bright cloth-wrapped wires which only remotely resembled plants, and would later give way to a taste for plastic grimcracks. And then she, not Greg, started disciplining the younger ones: Sharon, Lei Lani, Douglas. Ethel Powell began to do the shopping, and assumed responsibility for paying the bills, and suddenly it was all too much for Greg, who was now fourteen. The fights started and were only bitter at first. Finally, outright warfare ensued.

'Why *should* you listen to her?' Greg would say to the younger ones. 'I always told you right, didn't I? I took whippings for you when you were bad and never opened my mouth, didn't I? I took care of you all your lives, didn't I? How come now you got to do what *she* says? You always did what *I* said, didn't you? It's not gonna change around here, you hear me?'

'But Mom says, Greg. Mom *says*.'

And tearful battles were waged over the dinner table for many months to come.

'You're just gonna have to learn who's the boss around here, young man,' Ethel Powell would warn. 'You don't just run things around here no more.'

'I don't, huh?' the boy would answer, his blue eyes sparkling, his head turning on the swivel of a neck longer than his father's.

All of the children were blonds and rather fair, and the other three were better-looking than Greg. They would remain silent during the flare-ups, not certain whose side to take, not certain whose authority was supreme.

'Now all of a sudden you start bossing the house, huh?' he said, tears of wrath spilling. 'Well how come you wasn't bossing when Doug needed his drawers changed, or when the girls needed help with arithmetic, or when somebody had to get up an hour early every morning so's to get them all off to school? I wanted to be a crossing guard and couldn't because I had to get the kids off to school. You wasn't the one taking care of them. It was *me*. *Me*.'

'You sound like you don't care that I'm well now. Like you wish I was still sick.'

'I don't give a damn either way.'

'That's enough,' his father would thunder. 'No one's had an easy time around here.'

'You can't side with her now, Dad. You can't. Who'd you always come to with the money for the shopping? To me, that's who. Who'd Sharon and Lei Lani and Doug always run to if they was hurt? Huh? Did they go to her? To you? No, they damn sure didn't. To *me*, that's who. To *me*!'

Greg would sit in class that year, in junior high school, and his thoughts would be in the big two-story frame house: She's a liar, that's what she is. She's always been a liar and he's a coward and takes anything she hands him and now they're trying to turn the kids against me. Against *me*.

It was about this time that his performance began to suffer both academically and in extracurricular endeavors. He no longer asserted himself in sports. He'd always considered himself a good athlete, especially in winter sports, and now he didn't seem to care. He even lost interest in his saxophone and in music in general. He began losing weight and dropped off the football squad.

Then, when he was fifteen, he ran away. There were many times on the road that he regretted his decision, especially when he stood on the mountain highway in Kentucky in the rain, and the rain turned to sleet and made heaps of gray-brown slush, and the sky blackened before his eyes so that the boy had a feeling that the sun would never return. All the cars passed without slowing and he counted his money for the tenth time, but it still totaled three dollars and some pennies. He was wheezing, rattling, ripping phlegm from deep within. Then a big sedan stopped, skidding a little on the wet asphalt.

'Want a lift?' asked the man, holding the door open and Greg splashed through a puddle and fairly leaped into the car.

'Thanks,' said the boy when his teeth stopped chattering.

'Going far?' the man asked, and for the first time the boy looked at him. He wore a black topcoat and black pants. He had dark hair and eyes, wore glasses, was both tall and big.

'I'm going to Florida.'

'Well.' The man laughed softly. 'You have a ways to go. Do you have money?'

'Enough,' the boy said suspiciously.

'Where do you come from?'

'Cadillac. That's in Michigan.'

'Thumbing all the way?'

'No, I came by train most of the way.'

'Where're your parents? Michigan?'

'I don't know where my parents are and I don't care. Now maybe you better just let me out if you care so much.'

'Hold on.' He laughed. 'Don't get angry. I didn't mean to pry. What's your name?'

'Greg.'

'I'm Father Charles, Greg,' the man said, and it startled the boy. For the first time he noticed the Roman collar barely showing beneath the black topcoat.

'I never met a priest,' said the boy. 'Do they call you "Father" or what?'

'Most people do.' The priest laughed. 'Some less charitable Protestant neighbors call me other things. I have a parish in Georgia. You can ride a piece of your journey with me.'

And then the priest began suggesting, gently at first, that Greg should at least call his parents, and he was saying something else, something about traveling like this, and the conversation seemed to have religious overtones but Greg couldn't tell. His eyelids were closing and his head was nodding forward onto his chest. He woke up in the state of Georgia.

'This is where I stay, son,' said Father Charles as Greg rode with him to the rectory, planning to leave after a promised hot meal. It was a small poor parish in a region of Baptists and Methodists, but the parish house was clean and warm, and there was a part-time housekeeper to help the priest keep things tidy. Despite his long sleep in the car, the boy was glad to accept the invitation to stay another night. He slept thirteen hours in a warm clean bed.

The next morning Father Charles said, 'I can't persuade you to wire your folks?'

'No sir.'

'You're determined to go on?'

'Yes sir.'

'Well then, would you consider staying here for a while? I know a man who has a job putting up galvanized sidings on buildings. I spoke to him about you. He has a job for you.'

'He does? Well I . . . well . . . yes. I guess so. Yes, I can stay. For a little while.'

And the boy went to work that very day, and in the evening after dinner, when the priest returned from visiting a sick parishioner, Greg surprised him by joining in when the priest sang a popular song as they washed the dishes.

'Greg, you have a nice voice.'

'My dad's a music teacher. He went to the Cincinnati Conservatory of Music. I been playing music and singing since I was a little kid. I can just pick up an instrument and you tell me the scale and I can play. I got perfect pitch.'

'Oh you do, do you?' The priest grinned, and took off his glasses, looking at the boy more closely.

They spent the rest of the evening singing together and Greg thought that his voice blended nicely with the clear booming baritone of the priest.

The next morning at breakfast the priest said, 'There's going to be a dance in the parish hall next week. Would you like to go?'

'Sure.'

'I can arrange that you escort a young lady. We've got lots of pretty belles around this part of the country, you know.'

'Thanks, but I'd just as soon go stag.'

'Can you dance?'

'Sure.'

'Have you ever dated a girl, Greg?'

'No, I guess not.'

'Really? And why not?'

'I been too busy raising my sisters and my brother. And going to school and working.'

The priest seemed to notice the catch in the boy's voice and didn't pursue it.

'It's time you became interested in girls,' he said, picking up the dishes, turning his back as he walked to the sink.

'I don't care about them,' Greg said.

'You should. You're not a bad looking boy. A bit skinny but we can fix that up.' He laughed. Then he came over to the table and put his hand on Greg's head. 'You have very handsome hair. Most girls are partial to blond wavy hair, you know.'

A few nights later the priest took Greg to bed with him.

'Was that the very first time you've done that with a man?' the priest asked afterward, lying beside him.

'Yes. The first time with anyone, Father.'

'Greg, God permits men . . . people . . . all people . . . to express love in many ways. What I've . . . we've . . . done is a gesture of love. Shame and delight ... well ... these are man's responses, not God's. With us it was just our way of loving, a moment of love. The only emotion man can ever know for sure he shares with Our Lord. Do you understand, son?'

'Yes, Father. I understand. I'm not sorry. I feel the love. I really do.'

And the priest looked sadly at the boy, then turned his back. Greg was puzzled, provoked, impassioned. He had difficulty sleeping.

One evening after a dinner discussion about the intimate sensual beauty of Christ and his world, Greg suddenly craved the darkness and the priest's bed, and there in the small living room of the rectory, he threw his arms around the tall man and touched the priest's fine maple brown hair.

'Father. Father. Oh, Father,' he whispered, and was startled when the priest roughly pushed him away.

'Greg, I've got to talk to you.'

'What is it, Father?'

'You've got to leave here. You've got to go home to your family.'

'Why? What did I do wrong?'

'Nothing. You've just got to go. Your parents have undoubtedly notified authorities about you and it's not right for you to be here.'

'I done something wrong, didn't I?' asked Greg, eyes already wet.

'No. Yes. Greg, it's becoming obvious ... I mean, about us ... what ... the way we express love. It's that ... you're so clingy. You're getting like a girl. People are bound to notice. Already have, I fear. You've got to leave here, son.'

Greg left the next day when the priest was saying mass. He had been awake all night thinking of the virulent letter he would leave. Now he hated the priest, and couldn't understand how he could have felt anything for this Judas. He had a delectable vision of himself walking into the church during mass, mounting the altar, and addressing the congregation, telling them about their priest who seduced fifteen-year-old boys. But two things stopped him: his vanity for one. He couldn't bear to think there *had* been others. He had to have been the only one. And if so, he couldn't have Father punished for it. Secondly, he knew he had not been seduced. It had taken no coaxing. He had been ready.

Gregory Powell took to the road early that day, bitter, betrayed, bewildered. He strode along the shoulder of the highway kicking stones and gravel until his toe was battered, enduring the pain and confusion, hating the dismal Georgia countryside and the cold rain starting to fall, taking solace in the blackened stormy sky which aped his mood. Then he saw a dot in the sky, suspended, shimmering against a lighted streak of cloud like the eye of God. He watched it become bigger, yet still it hung. Then it dropped suddenly, without warning, taking shape as it fell – down, down – and then the wings spread majestically. Not twenty yards away in the field it silently struck something vulnerable. Seconds later Greg saw it swoop up: coppery triumphant wings, hooked bloody beak, soft furry thing dead in the talons.

By the time he hitchhiked into Lake Wales, Florida, he had learned things, and it was a more cynical manipulative boy who affected what he hoped was a passable southern accent. He prepared a story for meddlesome adults that he lived in the next town and was just hitchhiking to his grandfather's house.

Greg was not even remotely effeminate in appearance, but there was something, some hint in the promising gaze of the boy that was a sign for even a mouse-faced bell captain who hired him as the hotel elevator boy at first glance, and who propositioned him the first moment they were alone. Greg accepted with lip-jutting brazen defiance, but he learned that there was great danger in homosexual encounters. He barely escaped the bell captain with his wolf breath and his studded leather belt.

His sexual appetite was not nearly as sharp then when finally he trudged into Orlando, Florida, neglecting his southern accent when stopped by the local police.

Gregory Powell didn't want to go home, but it was either that or a Florida reformatory for runaway boys and he reluctantly agreed to tell them to contact his mother, who, with her newfound health, set out to be a solicitous mother. She came personally to Orlando and took her eldest son home to Cadillac, but things were no better at home. In fact they were infinitely worse.

In Greg's absence his mother had totally dominated the household and the boy could see that any resistance his father may have offered previously was finished now. Greg tested his authority with the children and it was the same. They turned to her for approval or usurpation of his commands. He still found her to be a consummate liar, yet he made allowances, always at night, to himself. Filled with confusion over his feelings toward his mother and father, he made allowances. His mother had been ill for years, and weren't some of her lies just harmless fantasies? And wasn't she just now beginning a normal life? And his father, well, he had to get by, didn't he?

And then he would hate himself for excusing them, and would actually despise them for a long blistering moment. That would lead to thoughts about himself, the way he was, the undeniable hatred of school and the attraction for a curly-haired boy named Archie. So he and Archie stole a car in Big Rapids and drove exultant and free to Indiana, where he engaged Archie sexually in the back seat of the car, afterwards falling asleep, only to be awakened by Indiana police. Once again he was returned home by out-of-state authorities. Archie, being older, was given one year's summary probation for the car theft.

This time Greg could live at home for only a week. The Powells had taken in a roving orphan boy named Harold – a boy slightly older than Greg – big, blond, silent, a masculine boy close to being movie-star handsome. Perhaps to the Powells he was a surrogate for the son they had all but lost, another chance with another troubled boy. But it was too late for the surrogate as well as the actual son.

By now Gregory Powell's weight had fallen from one hundred and fifty to one hundred and eighteen pounds. He was succumbing to one virus after another and when Harold told him he was going to join the navy it was more than Greg could bear. He idolized the older boy, wanted only to be with him, was sickened and worried by his secret desires, telling himself a thousand times he was no queer.

A few days later Harold stole some money from a sister and Greg stole a car and they were eventually arrested in Yellowstone National Park. This time Gregory Powell was not released to his mother. Barely sixteen years old, he was convicted under the Dyer Act and sentenced to a juvenile facility at Englewood, Colorado. Harold was by law an adult and was sent to an adult facility.

Five months later, Gregory Powell, found to be bright but emotionally unstable, was transferred to the National Training

School for boys in Washington, D.C. Ten months after that he escaped, was recaptured in two weeks, and sentenced to the federal reformatory at Chillicothe, Ohio. He was released the next year, having served twenty-two months in all, and returned home just before his eighteenth birthday, a swishing, emaciated, self-proclaimed faggot.

His first weeks at home were a nightmare. He walked with iron in his back to correct the exaggerated queen's step he had consciously acquired in the reformatory. And he thought of the reformatory and wondered how he had found it so hateful, why he had wanted to escape. There were worse things. This, for instance. His nerves were ragged. He was concentrating on being a man, afraid every moment he would betray what he was. He began experimenting with marijuana and would use it sporadically, mixing it with liquor all of his adult life when he wasn't in prison. And this time he would make what he would later confess to the others in group therapy was a conscious effort to get back inside. He admitted and surrendered at once and irrevocably to a hateful notion: he was in fact an institutional man.

So Gregory Powell stole a car, was caught, and sentenced to the state prison at Jackson, Michigan. This time there was no doubt in his mind. He wanted to stay, but they put him out when he was twenty years old. He discovered it was easier to get in than to stay in. You only had to steal a car and cross the state line to assure entry.

This time out he was to have his first sexual experience with a woman. She was six years older and taught him about heterosexual love, and though he was to maintain in later years that he was a confirmed homosexual early on, it is clear he was not. Greg would, during his years of incarceration and freedom, rebound back and forth between men and women, never knowing what it was he pursued. When he was in prison he was always certain that repressed homosexuality was at the root of his troubles and he would be as overt a homosexual as prison authorities permitted. By now, jailhouse tattoos adorned both arms: 'Greg' on one arm, 'Mother' on the other, the two people about whom he felt most ambivalent.

In Leavenworth Greg was to meet a willowy Indian boy called Little Sheba. They formed an alliance. Prison was their home and they would remain together. By now Greg had reverted to a masculine type. But they were caught by guards in a sex act and separated, and then Little Sheba was suddenly 'given a date.' He was going home.

After his separation from Little Sheba, Greg was temporarily quite mad. He began chewing holes in his wrists, sitting on the floor of his cell bleeding surreptitiously into the toilet, flushing the blood away until he became too ill to continue, inflicting wounds which would heal into ugly ropelike wrist scars. Finally he was hospitalized, given transfusions, and tied into bed so he would not strip the tubes from his arms.

As always, Ethel Powell was soon by his side. Nothing stopped her, not prison walls, guards, nor wardens. She would invariably enlist the aid of a prison chaplain, travel to whatever prison her son was in, and practically camp out at the prison gates until she got whatever she felt her son needed. This time she felt it was a transfer, and he got it. Her boy was sent to the prison at Milan, Michigan, where he could receive that which she believed would undoubtedly restore him to health – visits from his mother. But upon his arrival there, and five minutes after the removal of his handcuffs, he chewed the stitches out of his wrists. He was resutured and placed in a padded cell.

The next several years were repetitious of the others: in and out of prisons, gonorrhea, psychotherapy, liaisons with women, including a promiscuous black juvenile who accused him of fathering her child. Cruising for gay men in between and during heterosexual affairs, always masculine in appearance now, the queen years far behind. A passionate two years in and out of prison with a black drag queen called Pinky who could be used to rationalize later problems: 'If only they had paroled me to Los Angeles where I could be with Pinky . . . '

And yet the heterosexual relationships were the most lasting during his short months of freedom, belying his rationalization. As always, his mother was arduous in her visits and letters to the various penal institutions, while the other children were faring only slightly better than the eldest. Douglas was to become for a time a heroin addict. Lei Lani, the most vivacious of them all, was to run away to an unhappy life of bad luck and tragedies. Sharon was to have a stormy marital life.

The Powell family was finally to move to Oceanside, California, which permitted an occasional visit to Greg, who was in prison in

Vacaville Medical Facility where he had undergone a craniotomy to determine if calcification of the brain found in an X-ray could have been caused by a tumor – a tumor which could explain his behavior. The exploratory surgery revealed no tumor but the neurosurgeon reported that he found 'mild atrophy.' That finding would be used as a defense at a later trial to mitigate Greg's volcanic behavior, but other experts would challenge it as impossible to recognize in exploratory surgery, and actually 'mildly' present in many prudent, reflective, prominent persons.

Regardless, the craniotomy provided a necessary rationale for Rusty and Ethel Powell. It comforted them. Greg's troubles were organic, they were sure of it. The old guilt became at last tolerable. It was a blessing.

Greg was almost twenty-nine years old when, in May of 1962, he was finally paroled from Vacaville to his parents' home in Oceanside, California. He had spent ten of the past thirteen years in penal institutions, only nominally able to shape the family's affairs through scolding extravagant letters. His prospects were not good when, shortly after release, one of his sisters introduced him to Maxine.

But then neither were Maxine's, and had not been for years. She was twenty-six years old, plain, with worn-out eyes and three children. But her smile was pleasant enough and Greg began seeing her often. Soon he had all but moved in with her.

Maxine, like Greg, had been through years of family wars and was currently involved in a separation from her soldier husband from whom she drew an allotment, and was in a court battle with her parents, who wanted custody of the children. In September, by court order, the children were given to Maxine's mother and partially blind father, who proved their daughter's failure to care for them. Some months later the army discontinued Maxine's monthly allotment check, which she and Greg had been spending, and diverted it to Maxine's parents for the children. Maxine was upset and wrote to the army, to no avail.

'You and me'll soon get married, honey,' Greg promised her.

'Yes,' she sobbed, 'and we'll get the kids back from my folks.'

'You bet we will,' said Greg, hugging her close.

'And then the allotment check'll stay with us.'

'Well, not if we're married, honey,' Greg reminded her.

'Oh, that's right. But I'll get something, won't I?'

'Those things have to be worked out. But don't worry, my luck's gonna change.'

And Greg tried a car wash and gas station and finally a scheme to open his own automotive garage, and for one reason or another all ended in failure. Then the final battle of Gregory Powell's family war drove him out of the house for good. It evolved over an anonymous letter one of his sisters received accusing her of various marital infidelities and of having had incestuous relationships with Greg. The girl took the letter to the family council, and Greg's father found some notes in Greg's pocket written by Maxine and, after a family conference and homemade handwriting analysis, it was decided that Maxine was indeed the culprit. And further, that the vile and libelous letter should be brought to Greg's attention. It was, and Greg drove Maxine straight to San Diego to a detective agency where he paid sixty dollars for a polygraph examination and the examiner said he was satisfied Maxine was telling the truth. Next, Greg returned to the family with his polygraph results and demanded they turn the letter over to a handwriting analyst for further clues as to who wrote the despicable letter. But they had already made up their minds. 'You've got no feeling for my private life,' Greg told them, his jaw muscles throbbing. 'You went in my pockets and took notes from my girl to me. You had no right. None at all.'

'Look what she wrote about your sister. And about you.'

'I don't believe she wrote it. She said she didn't. And now the point is that you'd treat me like this! Just go into my pockets!'

'It's for the good of all of us. It's for the family.'

'I oughtta tell you what I think of this family. That's what I oughtta do. But I won't. I'm leaving this house as of now. Oh, I'll still see you, don't worry about that. I won't abandon you. Somebody's gotta direct you or you'd all walk off a cliff or something. But I'll never live in your house again.'

And he was right. He never would.

Toward the end of 1962, Greg and Maxine decided to go to Boulder City, Nevada, to care for his sister Lei Lani, who had been in a traffic accident, leaving her neck broken and her leg terribly burned by battery acid. Gregory Powell found he still couldn't get away from them, and it was not only his sister but his mother who became disabled by illness and again announced that she was going to die. He found himself making trips between Boulder City and Oceanside with Maxine, his father, brother, and mother when she was able.

In January they knew Maxine was pregnant, and Douglas, by now a heroin user, had come to Boulder City. Greg told Maxine that he could never escape the family. Never. And once again he was right. But he was always to wonder if he really wanted to escape.

Greg was growing restless at his sister's home surrounded by his family. One day he concocted a bizarre scheme in which he and Doug were to drive to Oceanside and kidnap Maxine's children from their grandparents and bring them to Boulder City and their mother's loving care. After several hours of planning the brothers drove to Oceanside but they returned the same day. Greg once more had changed his mind.

Then on January 29, Douglas Powell, using the identification of a cousin, Thomas Powell, residing in Michigan, bought a Beretta 7.65 automatic from a Las Vegas pawnshop.

On January 31 a service station in Las Vegas was robbed by a lone gunman. On February 6 a Las Vegas drugstore suffered the same fate. On February 9 the original service station was robbed again. The Powell brothers announced to the other family members that they had acquired a night job driving a truck and unloading box-cars. The job would last about four hours on each night they worked. The pay was surprisingly good.

Then Douglas and Greg both left Boulder City separately for Los Angeles, Greg explaining he wanted to collect some money owed him. On February 15 Greg made two 'collections' at a West Covina liquor store and another at a Santa Monica liquor store, but by then Douglas had returned to Oceanside. In late February Greg's 'collections' were interrupted when his sister broke into his apartment and took his automatic. Every dollar he had stolen also disappeared, more than six hundred. It was the final ignominy.

'I know why she did it,' he raged to Maxine during the telephone call to Boulder City. 'She's trying to stop me from robbing, Max. She's trying to straighten me out by stealing my gun. But goddamnit, why the hell did she take the money too? She could at least have left me the money. It'd take the whole goddamn staff at Vacaville to figure out just one goddamn move by one goddamn member of my family! They'll drive me to a little rubber room, I tell you!'

Greg drove to Las Vegas, bought a Colt .38 revolver with a four-inch barrel, and returned to Los Angeles with Maxine, where they moved into the apartment of a black drag queen he had dated in the past. One day a knock at the door startled Greg, who ran into the kitchen, drew the revolver from his waistband, and accidentally fired a shot through the floor. The drag queen suggested that they find other lodgings.

Then Greg met a little black man named Billy Small who helped them find an apartment in a black neighborhood on 65th Street. Billy and Greg were to keep very busy for the next few nights.

After the gardener had finished the old woman's yard, he went to his next stop, an old California home, with a line of yuccas out front. They were sound healthy yuccas with long stout spiny leaves bending from massive trunks. Out near the street was a leaning Japanese black pine. In just the right place, so that the majestic plants didn't overwhelm it, was a dwarf nectarine with droopy shiny leaves. The gardener wished he had the training and license to landscape places like this, not just maintain them. A landscape architect, that would be the thing to say when people asked what you did for a living.

When the gardener used to do his job with an old friend it was always he, the gardener, who had the touch for slipping plants. The friend often admired the way the gardener could make things grow from cuttings.

'He has a way of making things live,' the friend would say. 'He cares about living things and that's something a license can't give you. I'd rather have him landscape and care for my yard than someone with more imagination. He cares about living things.'

And now the gardener knelt beside a potacarpus and wondered if perhaps this weren't the only error the owner had made in this otherwise magnificently landscaped property. A touch of greater delicacy was needed on this side of the yard, not this evergreen with its illusion of fullness. The Italian stone pines around it were enough. You mustn't be afraid to have a little spot of bare earth with nothing growing there. You must know when you're finished and then stop. That's the way he committed his crimes. He had a regular route. He stole from each place on his route and he didn't stop for the day unless he got something of value from each of them.

Then he unloaded his mower from the bed of the truck and with the sun straight up and hot, rolled up the sleeves of the workshirt and took off the hat to wipe the sweat from his forehead and neck. The headache wasn't so bad now. As he stood there on the lawn he saw a mailman walk down the street carrying a leather mailbag. The mailman looked familiar, but the gardener couldn't remember. He started getting stomach cramps and hoped it wasn't another attack of diarrhea. It was so hard to remember faces anymore. Maybe the mailman was from the other life, back in those days. But how could he be? Then it struck him. The mailman looked like the man in the yellow shirt, the one was was

50

surely a security officer, the one who watched him that day when he had his pockets loaded with loot and a set of wrenches under his coat inside his belt, and even his hands full of packages of screws and bolts and other items of hardware.

The man in the yellow shirt had followed him, never taking his eyes from him. He had walked slowly toward the door, waiting, tensing, waiting for the sound of running footsteps or a shout: 'Store security officer. You're under arrest.'

He had walked through the door with his loot and across the busy parking lot to his car. There had been no footsteps, no voice. He had made it.

The gardener watched the mailman pass by and cross over to the other side of the street. The gardener thought: No, that's not the man in the yellow shirt. Strange, up close he didn't even look like him. Not at all.

FOUR

The partners in the four-door Plymouth and the partners in the little Ford coupe were both battling traffic at that moment.

Ian Campbell, the driver of the Plymouth, was turning east on Hollywood Boulevard but decided momentarily it was a mistake and quickly got out of the traffic. Karl Hettinger was cleaning his glasses, feeling his belt binding him and wondering if a hamburger would add to the bulge.

'One good thing about Saturday night traffic is it gives you a better chance to get lost after a job,' Gregory Powell said as they began to ride from Wilshire Boulevard to Hollywood in the little maroon Ford.

His partner did not reply but continued fiddling with the gun in his belt, twisting in his seat.

'Goddamnit, Jim, relax,' said Greg. 'You know I'll do all the real work once we get inside.'

Jimmy Smith grunted and watched the cars that passed them. The March night air was cold, but he was sweating, and his mouth and throat were dry and hot. He lit a fresh cigarette with the butt of the last one.

'After tonight we'll have a stake, Jim,' Greg said and Jimmy Smith wished his partner would shut up for five minutes. He had to think.

'I've got a feeling you and me're gonna score big tonight,' Greg said. 'Our little family's gonna get well tonight.'

Family, thought Jimmy. If I hear another fuckin word about *family* I'll . . . Then he looked toward his partner's belt. He couldn't see the Colt .38 in the darkened car but he knew it was there.

He hated it when his partner talked about his family and how Jimmy was now part of it. Jimmy Smith had never been part of anybody's family, never wanted to be. And if he did, there was

52

always his Nana. I'll go see my Nana one of these days, thought Jimmy. Soon as I cut this crazy bastard loose, that's what I'll do. I'll go find my Nana.

'Jimmy is not really my son,' his Nana would one day tell a jury, 'but he's the onlyest son I'll ever have. I'm really his auntie, his great-auntie. I raised his mother from when she was a little girl and then she went and had her baby in Crowell, Texas, when she weren't but thirteen years old and she couldn't take care of him so she give him to me in Fort Worth. I think his daddy was a fifteen-year-old white boy but I ain't even sure of that.

'When he were just a little tiny boy about three years old, I had this accident with a Colt .45 revolver, and he were there with me when it happened. What happened is my husband's gun was under my mattress. I slept on the floor. I just stuck it under the mattress on the floor, and when I got up to put the mattress on the bed, I rolled it up, little skinny mattress as it was. I looked back and Jimmy were there and I were afraid he would get the gun so I got the gun and the mattress all in one hand. I guess I got it by the trigger when I started to the bed with it. The mattress started to slip and I gripped the mattress. I guess I pulled the trigger.

'I were shot in my left leg and I had to wear a cast for a year without turnin over. It went up to my waist, and over my left leg all the way down, foot and all. And I laid there a year with that cast on.

'Well, you might say Jimmy took care of me. We had nobody. I had been makin six dollars a week before this happened, but after this I was crippled for life. And Jimmy would give me water and do things for me. Jimmy turned the gas on at three years old. We had one of those open stoves and if it would get too cold I'd tell him, "Son, you have to try." And I'd tell him to light a match and lay it on the stove and then turn the knob because I weren't able to get up at all.

'And the same way with the lights. We had lights that swung down from the ceiling and we had to turn them off from there, and Jimmy would pull the breakfast table out of the kitchen, get on the table, and turn the lights off. Nobody was there with me daily, just Jimmy and I. My first husband finally left me, you see, after I got shot.

'After I was shot I were never able to get Jimmy to take a nap anymore unless he could wrap his head up and just smother hisself practically, so I just had to discontinue his naps because after that he was afraid to go to sleep in the day. It was all right at night, but takin a nap in the day, Jimmy refused to sleep unless he could wrap his head up.

'When Jimmy were about seven, a playmate got hit by an automobile. Jimmy thought he was dead. It didn't kill him, but Jimmy ran to the woods. It happened at school, and Jimmy told them he was goin to the woods and never come back.

'Well, the whole town, practically, went lookin for him. They found him with his head and shoulders under some bushes, and only his feet was stickin out, and he jumped up cryin, and he come runnin and said he was sorry this boy was dead. He just could not stand to see anyone hurt.

'He were afraid of a gun as far as I can remember because the lady I worked for, she bought him a tricycle and a little cap pistol for Christmas. It scared him to death, and I didn't want her to know he couldn't play with it, so I put it under the mattress and hid it. But he found it one day and I finally had to give it away. She bought it for him, but Jimmy was extremely nervous ever since I got shot. And he wouldn't fight. The children would hit him, fight him, but he wouldn't fight back.

'It were hard with a little baby, and me already past forty and never had a child of my own. And my new husband, Sylvester, never gave Jimmy a dime in his life. All he did was nag and fuss at him.'

The only man in Jimmy Lee Smith's early life was his Nana's second and last husband. Sylvester was a gambler who would leave for weeks at a time, then return to abuse him and his Nana, and to take what little money she had. Once he beat her severely when she said the money was for Jimmy, for school shoes. And during the beating Jimmy called Mr Ed, who was a neighbor. The giant black man stopped Sylvester. Then when his Nana limped off to work, Sylvester tied Jimmy to a bed with a pair of stockings and beat him with a board from a crate. His screams and pleas were heard by all the neighbors and when his Nana came limping into the apartment, she had already heard about it. Without a word, she got Sylvester's gun, and shoved it in his face crying. 'You ain't never gonna lay hands on my boy agin. Never!'

She didn't know how to cock and fire the old revolver and

Sylvester got it away from her. But he retreated, never again to strike Jimmy in anger, and after they went to Los Angeles he left their lives for good.

Whenever in later life Jimmy Smith tried to remember triumphs, it was somehow the defeats which came to him. Like the time the Goodwill Industries in Fort Worth had given him a pair of candy-striped overalls for Christmas and he was so proud of them he wore them to the movie that very Saturday, and Flatnosed Riley surreptitiously unbuttoned the flap. Jimmy wondered all the way home why the other boys were snickering and pointing at him. When he discovered the truth, he burst into tears, crying, 'You all played the dozens on me. You all played the dozens on me!'

Jimmy was never very popular with the other boys. There were distinct disadvantages to having light skin and mulatto features. 'You ol yella-faced ponk. You ol half-breed nigger. Thinks you is white, dontcha? You oughtta git a ass kickin.' And then they would do it.

But the advantages came later. With the girls, especially the very dark girls. 'My, Jimmy, you a fine bright-lookin boy. Take the shirt off, Jimmy. Mmmmm, ain't you a *fine*-lookin bright boy.'

But he couldn't be too bold about picking the prettiest girl even if he knew he could get her. Not when he was resented anyway because of his fair skin. So Jimmy decided to choose the ugliest chick when he was with the other black dudes, and to score on the pretty ones when he was alone. And it worked out fine. Picking the ugly chicks made the other boys approve of him.

Jimmy learned to hustle in Fort Worth during the Depression. There was treasure in tow sacks and even in rags, but nothing compared to what junk dealers would pay for zinc fruit jar tops and old copper. Even rich white folks hoarded such items. It took boldness to sneak into their garages and creep onto their screened back porches to get the tops off the fruit jars which they were using for canning. But if he were caught he would throw himself on the ground and sob as though his heart were breaking and tell them how hungry he was, which was true enough. Inevitably they would look down at the dirty, yellow-skinned boy with the tears streaking through the grime and give him only a mild scolding. Sometimes a lady would even open her purse and give him a dime and a motherly pat on the rump.

He learned that, with variations, the same thing worked in the

A&P market when he got caught. But he couldn't pull that too many times, so the best thing to do was steal only what he could eat and duck down behind the counter and eat it right there in the store and walk out with a small legitimate purchase. The thing to do was become popular with the old people of the neighborhood who needed a boy to run errands and buy fifteen or twenty cents' worth of foodstuffs for them. The penny he got for the errand was nothing compared to the quarter's worth of food he had the opportunity to consume in the store.

But as always, he got caught.

He could never forget the beefy face of the clerk who found him behind a display shelf while he crammed cookies and bologna into his mouth. He had been jerked to his feet by the neck and carried through the store, his feet not touching the ground, a huge hand strangling him. He had felt a devastating pain between his buttocks as a kick sent him sprawling on his face. But big Mr Ed Dixon at the other market across the street had seen it and run across and paid the outraged clerk from his own pocket.

'I'm sorry, sir,' Mr Dixon apologized to the white man. 'I been knowin this little boy all his life and I'm terrible sorry. His auntie ain't the kind of woman who would appreciate this stealin. This yella nigger boy is gonna get one hell of a whuppin.' And later to Jimmy he said, 'You don't have to steal, son. If you ever gets that hungry, come to me.' And he gave Jimmy a nickel and sent him home.

So Jimmy Smith learned that if he cried, or looked like he was going to, or just didn't act like he wanted to be Number One Dude, well then, somebody would always take care of him. And it was the same on the corner, the same hustling on the street. Just let Number One *think* he is Number One and he has all the headaches and ends up doing what Jimmy wants him to do.

It was only a short walk from Fifth and Stanford to Second and San Pedro. To a tourist passing through Los Angeles both streets would look very similar – downtown, shabby, in or on the fringes of skid row – forbidding.

To a crippled fifty-three-year-old black woman there was all the difference here at Second and San Pedro. With the Japanese returning from relocation camps this commercial neighborhood was thriving. She didn't believe there were dangers on this street for a teenager. Jimmy had his own hotel room now two doors down the hall from her room, and no longer had to sleep on the floor. But Fifth and Stanford was only a short walk away.

Fifth Street, from Main Street east, is the heart of skid row, the one street in Los Angeles which can compete with the worst of eastern slum neighborhoods. Bars line both sides of the street, and numerous liquor stores stock more Sneaky Pete and Sweet Lucy than the rest of metropolitan Los Angeles combined, most of it in short dog bottles which the derelicts can afford.

The derelicts were of all ages and races, a plentiful number of women among them. Fat and bloated squaws, emaciated toothless whites, with flaming swollen hair-covered legs, and faces like red balloons. Ancient bony black women past the age for ordinary hustling. all of them prowling, snuffling through blackened hotel hallways, conducting business there on the dirt-crusted slimy floors with a rag of a dress pulled over the face because who wanted to see what she had to feel and smell? Until at last she could no longer feel or smell.

Most were thankful that oral copulation was what the men wanted, because it was so much easier, and some of the derelict whores could line up three men in an alley in broad daylight and be through with them in less than ten minutes, making as much as a dollar for their work. Often, an impotent alcoholic would plead dissatisfaction, and a drunken business dispute would erupt between the two staggering wretches. But more often, these women were able to protect their earnings from the male predators.

It was a dangerous neighborhood to grow up in, and if you had lived there since arrivng from Fort Worth you would of necessity have grown very tough or very cunning. Jimmy Lee Smith was cunning.

The true derelicts, the down-and-outers, gave up Fifth and Stanford, surrendered it to the blacks, and seldom roamed that far east. There was no need. What could be begged from a black neighborhood? The winos roamed west toward downtown, toward the live ones, and left Fifth and Stanford to the blacks, the pimps and prostitutes, the bootleggers, dope dealers, the thieves and con men.

After Jimmy and his Nana moved to Second and San Pedro there were many reasons for Jimmy to run the few blocks back to

Fifth and Stanford, back to a life infinitely more exciting than a lonely hotel room when his Nana was at work at the laundry folding sheets. For one thing, at Fifth and Stanford there was the shine stand and the promise it held.

Before they'd moved up to Second Street Jimmy had persuaded his Nana to let him work in the shine stand across from the hotel, convincing her that she could look through the window and see him shining shoes. Finally she agreed. But of course much more could be made by running wine across the rooftops for the bootleggers than by shining shoes. Jimmy could go up a fire escape on one end of the block and come down through a skylight at the opposite end with a case of wine and never set foot on the street where the cops were. The bootleggers were small timers who brought their wine from Delano in barrels and diluted it for sale on the street. An enterprising boy like Jimmy Smith knew that every hotel roof in the neighborhood had caches of wine for the taking if you were sly and quick enough. So the men came to admire the clever boy and began calling him 'Blood' and 'Youngblood.'

The jukeboxes of most of the east Fifth Street establishments hunkered out on the sidewalks in those days, and raucous music mingled with car horns and tire sounds and laughter and occasional screams, the ordinary sounds of skid row.

Then there was Carole Lombard.

She was big, brown, and smooth, with loose full breasts and a blue satin dress which was tearing at the seams because there was so much woman to cover. Jimmy didn't care that she was thick through the middle and past her prime. To him she was beautiful.

He was thirteen when the big car stopped and a white man beckoned him to the curb. He saw at once it was not a plainclothes police car. It was a Buick, with snowy whitewalls, and a fatcat white man in an up-to-date, wide, hand-painted necktie.

'Hey, kid, know where I can get me a colored woman?'

Jimmy just shrugged, not walking too close to the car.

'I'll give you a quarter, boy. A quarter just to point me in the right direction.'

'Over there,' said Jimmy quickly. 'In back of that shine stand. There's a room. Jist go on in. They gonna find you.'

'Thanks,' said the white man and tossed him a quarter.

It was so easy he could hardly believe it! So easy! Was that how pimps got started? And he had always thought there was something magical and infinitely complex to learn before you could stand tall in front of the hotel, hands in your pockets, leaning forward at the waist, with marcelled hair, in orange gabardine – billowing, sharp-creased slacks, pegged tightly to the ankle – pointed black and white wingtips and a long sport coat with big shoulders, and a broad-brimmed felt hat with a feather. Not too much feather, just enough to be lookin sharp, lookin *good*. And cool was what you had to be, pretending you didn't know everyone was admiring you, just busting out with a hee haw once in a while to show you knew you were the boss of this corner. You had the girls, six in your stable, you had the big Caddy, you had the *power*, man.

But he never really believed he could be one of them, not a real pimp. He was already resigned to a secondary role in life, a number *two* man, making small stings, someday hoping to drive a decent transportation car, taking a back seat to the big timer. It had its advantages. The big man was always ready to do a favor for the smaller man who was no threat. If something bad went down, like from a police bust or from an infringing rival, well the big timer was knocked off, but the number two man did business as usual. It was much safer not to aspire to high position here on east Fifth Street.

But when he got that quarter from the white man he saw how it *could* be done, and he ran across the street to the shine stand and waited. Soon the white man came out of Carole Lombard's room, and Jimmy worked up his courage and ran around back and knocked at her door.

'Who's knockin?'

'Jist me,' Jimmy said softly, and that was another thing he learned about being a number two man. To talk softly. It was more cool and safer. And everyone liked it. Talk *softly*.

'Who the fuck is jist me?'

'Jist the one that sent the trick.'

The door opened and there she stood, not completely tucked into the blue velvet dress, barefoot, her breasts all but in his face, just buttoning the front.

'Looky here.' She smiled cheerily and another black woman,

leaner and darker, also half-dressed, came out from the closet carrying a roll of toilet paper.

'Come in, honey,' said Carole Lombard, and Jimmy obeyed, his eyes furtively glancing from the toilet paper to the bared bosom.

'Ain't you a pretty child?' Carole Lombard said, putting a soft, long hand on his head. 'He got us the trick, honey.'

'Well bless you, little boy.' The other one smiled and went back to the tap to run the water.

The smell was everywhere in the shabby little room. Overpowering, sickening, yet more exciting than anything he'd ever imagined. And he thought of them in this little room, these two big steaming women and that sweating fat white man. He looked at the walls almost expecting to see the wallpaper soppy, curling off the walls. Muggy as a jungle, that's what lovemaking with a grown woman would be.

'Guess you deserve fifty cents for what you done,' said Carole Lombard, and sauntered to a closet where he heard coins clinking in her hand. He watched the big shivering buttocks as she walked. When she gave him the two quarters she put her hand on his head again.

'My, you got nice soft hair,' she said. 'You ain't woolly at all. Martha, come here and feel this boy's hair.'

'I ain't got time, less he got three dollars,' said the voice from the other room.

'You got three dollars, honey?' asked Carole Lombard, looking down at him.

'No,' he whispered, gaping at the breasts.

'You so pretty,' she said. 'You got features kinda fine, know that? And so bright. You know, you put some pomade on, that's all, don't need no process, and I bet you could almost pass. You so pretty.'

Jimmy was trying to think of something to say. Something about working a deal where he could send them maybe two, three tricks a day. Only he wanted seventy-five cents a trick, not fifty. He was trying to get up the nerve to say it, when she pulled his face into her breasts.

'You pretty boy, you run along now, hear me? You send me a good trick once in a while and when you get big you come see ol Carole. Carole wanna take you on your first ride, little jockey. But right now you ain't no bigger'n a doodle bug, and I gotta turn you loose.'

'I'm . . . I'm . . . big,' said Jimmy, and his throat and mouth were so dry, his lips popped. Carole Lombard held him back, and laughed uproariously.

'You devil,' she said. 'You pretty little devil. Git on outta here.'

And she opened the door and swatted him on the bottom as he stepped out. The swat humiliated him and he felt a rush of anger and turned to see her watching him.

'Don't you go abusin yourself too much while you growin up,' she said, still chuckling. 'You save it up for ol Carole Lombard and then you come back. Soon as you old enough to pay three dollars. Hear me, pretty child?'

'I *got* three dollars now,' Jimmy lied. 'I got it right now.'

'Then come on back, little jockey.' She laughed.

'No. I wouldn't give it to you. I wouldn't spend it on you. You ain't the best around here.'

'Now you know that ain't true, little jockey.' She laughed, and Jimmy ran out into the sunshine holding back tears of anger.

'You gonna see me agin, bitch,' he whispered. 'I'm comin back to you with *twenty* dollars someday. And you gonna do everything to me I want. Everything, nigger. Jist like a dog! Everything!'

There were other things he learned at the shine stand. Like the eye game. 'Jist stare em down, Jimmy,' said the boy they called Hip-enuff. 'Jist look a whitey in the eye and stare and watch em look away. See, they's *afraid* of you, man.'

'Afraid of *me*?'

'Uh huh,' said Hip-enuff. 'This ain't Texas. You livin in Los Angeles, man. They's afraid of *you*. They thinks you is some wild nigger. Stare at em and watch. Especially the pussy. Stare at some white bitch and watch how quick she go shaky in the lip and look the other way. Hot damn!'

'But why I wanna do that? What do I git from that?'

'Git from it? Sheeeit, you know you the *man*, that's what you git from it. Huh! Git from it. Boy, you ain't got no sense, you know?'

But Jimmy never played the eye game. Never stared anyone down, white or black, still unconvinced that it would show that he was the *man*. He proved to himself he was right one afternoon while walking by a Main Street shoe store. A young white girl with a pretty page-boy hairdo was trying on shoes. When the salesman went in back Jimmy walked in the store, hoping he could grab a good pair of men's shoes off the rack and run. But instead of paying attention to business he found himself staring at the girl, who was no more than nineteen. She didn't see him and she had pulled her pale blue pleated skirt up to mid-thigh and was touching up the orange-tan leg makeup just above the knee. It was duirng World War II and nylons were all but impossible to get.

Her legs were shapely and elegant and Jimmy completely forgot what he had come for. And then she saw him and they locked eyes and Jimmy suddenly remembered the admonition to stare down a white bitch like a wild nigger. But he didn't. He dropped his eyes. And deliberately he tried his own ideas by saying, 'I'm sorry for starin, miss. You're an awful pretty girl and all. I couldn't help lookin at you.'

'Why, that's a nice thing to say!' she said. And he looked up and she was beaming at him. 'That's not something to apologize for.'

Then he turned and ran out, but waited outside, and in a few minutes she came clicking out in white, toeless wedge heels with little taps on the soles.

'Can I help you carry your things, miss?'

'Why, yes,' she said, handing him the paper bags with women things inside, and Jimmy walked two blocks with her to the bus station. He was oddly at ease and not self-conscious, and told her eight or ten lies as they chatted during the walk. He stood with her while she waited for the bus and when it came, he was bold enough to say, 'If you ever comes down here agin, let's go to the movies.'

She laughed and looked him in the eye for a moment and then shook her head still smiling and said, 'Thanks for the help.' She handed him fifty cents which he put in his pocket.

And that night he lay in bed in his room at the hotel and thought of a slim white girl with shiny hair and painted legs and the way she looked at him and gave him money. And then he laughed aloud as he thought of Hip-enuff, who claimed he got his from staring them down.

We all got our ways, Hip-enuff, Jimmy thought. You git yours by starin em down. Some guys git theirs by jackin off in a black leather glove. I git mine with my smile and my soft voice, and my curly hair, and let's see who's the biggest fool after all. He lay there giggling until he fell asleep.

'And one day the police brought Jimmy to my door,' his Nana told a jury. 'They knocked on my door and told me that Jimmy and three other boys had broken in a warehouse and stolen some Boy Scout things. Jimmy had been a Boy Scout when he was younger but he never had any Boy Scout things. Then they left Jimmy there with me and told me to bring him to court.

'Jimmy was raised in church. Church and Sunday school. Church and Sunday school is something that any kid live with me gets because I believes in God. I told them that.

'Yes, I took Jimmy down there. There was six other boys and they asked him why. And well, he said, he never had any Boy Scout things and he wanted them. I don't know anything about the Boy Scouts, but he got up and he recited several verses in the Boy Scout code he still remembered, and the man asked if Jimmy had any people that I could send him to to get him away from this bunch. And I sent Jimmy to Wyoming to his real mother.

'Well, his mother didn't understand. She didn't understand children. They called me up one night and Jimmy was cryin that she had choked him and he wanted to come home. I wired him the money that night.'

Jimmy's Nana did not mention that he had conceived a child on this trip to Wyoming. He was to deny paternity, but two years later Jimmy was to see a photo of the handsome, yellow-skinned child named Ronnie, and there was no doubt. But he never saw the boy and wondered vaguely about him only once or twice over the years.

Now Jimmy was back at the shine stand. It was also a good place to score a few cans of pot which could be cut and resold. The stolen wine could be diluted to water and as long as there was a bit of color someone would buy it. There were lots of ways to make a sting at the shine stand and his Nana would never take any money from him, only cautioning him to save what he earned. She was proud of him when he got a job selling papers. It didn't last long, because he got too greedy and stole too much from the paperman on collection day. Yet it wasn't bad while it lasted because he had the opportunity to stuff cotton in some pretty good pay phones along the paper route, and later collect the coins that didn't fall through. And it was nice riding a bike around his route. Of course he had the finest, because why steal a cheap one when people were so careless with good ones?

The pot started getting to Jimmy. Eventually he wanted nothing but a can of pot in his jeans. He became a stone grasshopper, ditching school, wearing a Western Union cap so that if a cop stopped and asked why he wasn't in school he could point to the cap and say he was working part-time and therefore wasn't truant.

'The weed smoking made me so lazy and lackluster that I hardly done anything except lay around the streets and panhandle the pimps and whores,' he was to say.

'Well, I went to Bakersfield in '48 and I stayed there until '51, and Jimmy weren't with me all the time there,' said his Nana. 'He struck out on his own, doin farm work, but he would come and see me from time to time. And then Jimmy started gettin into trouble, in and out of prisons. But he never changed toward me. He never talked back or raised his voice to me. I don't believe he'd do it now. He doesn't, he just doesn't. He'd cry before he'd even talk back to me. He'd rather cry.

'Jimmy would never do a violent thing,' his Nana would plead to a crowded courtroom when she was seventy-eight years old. 'Why, once he sat up with his hurt dog and cried all night. He was always cryin cause he would get hurt so easy.'

When he was twenty years old, Jimmy Smith fathered his second illegitimate child, a girl. He was in jail when he heard of the birth in Bakersfield. The light-skinned boy and the girl named Helen Marie were the only two children he knew for sure that he fathered. He was to say of his daughter: 'Somewhere along the line I heard she was livin in L.A. but I never thought to look her up.'

For a time Jimmy worked close to the land, in the sunshine and the dirt of the San Joaquin Valley. He hated it. All of it. The dirt in his teeth, and the sweat like turpentine in his eyes, the sun burning his skull from the inside out. The land was a hateful place, intimidating, defeating, dehumanizing.

'No more nigger work for me. No more pickin cotton in Bakersfield. No sir. No more pitchin watermelons in Blythe. Uh-uh, not this cat. No more swampin potatoes and pickin grapes like a fucking wetback. Not Jimmy Smith. Not this dude. Catch you later, baby.' So he returned to the city to something he understood.

He was to say that heroin gave him the finest hours of his life, the loveliest of all. He was now an addict, a shoplifter, a petty thief.

Once he was convicted of stealing a suit out of a car and pawning it for two dollars. And there were state prisons. Hard time. Soledad. San Quentin. Then he was caught stealing a nine-dollar toaster to sell for heroin, a felony crime because of his prior convictions. He ended up at Vallecito Honor Camp fighting a forest fire, but ran away from the camp. Five months of freedom, the longest he would have as an adult, and then he was rearrested on a narcotics charge and convicted of the escape from Vallecito. Then the real jolt: the dreaded Folsom Prison. The theft of that toaster ultimately would cost him more than five years of his life.

FIVE

Ian and Karl had not gotten around to discussing where they would eat tonight. Few policemen brownbagged on a Saturday night. It was the night to eat as well as they could afford.

Felony car Six-Z-Four could easily have been in a restaurant during the time the little Ford passed through Hollywood.

Gregory Powell, the driver of the Ford coupe, was turning north toward the Hollywood Freeway. His partner, Jimmy Smith, was considering putting his automatic in the glove compartment until they arrived at the market.

'You know something, Jim?'

'Say what?'

'You know what I was thinking? I was thinking it seems like you and me been partners a long time.'

'Uh huh,' said Jimmy, thinking: Amen, motherfucker, and it ain't gonna last much longer.

'When you find a partner you can depend on, you gotta be loyal to him. That's one thing I admire.'

'Uh huh,' said Jimmy Smith.

First time he ever made sense, thought Jimmy. It *did* seem like they'd been partners for a long time. Nine days? Jesus! Would he ever be free of him? It was worse than being in the joint. Less than two weeks ago he *was* in the joint. Had never heard of Gregory Powell. Had never been shot at, or carried a gun. Less than two weeks ago. He was *freer* then. Jesus!

The young man lying on his bunk that night in Chinon Prison had thought vaguely of the past sixty-two months and wondered what his future would bring. He'd learned house painting in the joint this time and fancied himself a good workman. There would be painting jobs out there. And he'd heard about factories hiring ex-cons.

No more sissies for Jimmy Smith. No more visits to a 'little friend.' The sissies were always 'my little friend' even if they were six feet four. No more pot partners parading around in their tight tailored pants eating ice cream and making the jockers guess how they were. Who the fuck needs em now? Jimmy thought. Now it's *women*. Broadway. Out there.

No more bulls coming by at midnight with their count and shining their lights in your eyes for the fun of it. No more endless, boring convict discussions. There were only three things to talk about in the yard. First, legal issues – cases handed down or expected to be handed down – motions, appeals, writs. Second, the dangerous dudes: who to watch out for, who to flatter, who to follow. Who is *so* dangerous that it is best to avoid him completely. Who will hit you with his piece at the least provocation, the piece being a knife, not a gun. Third, the sex, the canteen punks: who is doing it to whom.

He no longer thought of the small successes he had in crime, the good things. He no longer thought very much about the armload of dope, the thing he once could not drive from his mind for even an hour.

It was Broadway for good. Broadway! The Great Out There! This time he wouldn't be scared of it. No need to ram a spike in his arm this time. He was a mature man now, thirty-two years old, grown up emotionally. He'd been in enough group therapy encounters to know that heroin was an escape. Lots of his Fifth Street partners grew up with the same crutch. Now he was cutting it loose, shining it on.

He did not think of the old street partners, Smooth-move Wilbur, Hell-fire Jack. He realized he didn't even know their true names. And they called him Youngblood and didn't know his.

All of these things had been important to Jimmy over the years, but not now. Not in February 1963, and his parole a few hours away. He lay on his bunk smiling and began thinking of one thing: women.

He was dragging a tail the next morning – five years – a long tail.

The nervousness and excitement mingled with dread as he saw the big bus pull up out front to unload the guys who were coming back on a 'straight violation.' Will I ever be takin that bus again? he wondered.

He thought of the papers he'd just signed with some fifty rules of parole, rules which would be harder to follow than the ten commandments. The thought of being violated or even busted for some new beef had proved less and less frightening over the years, and even now, the thought of being returned here to a place like Chino didn't frighten him at all. But Chino had been a stopping-off place. Ninety days before parole they had sent him here for special treatment as a paroled drug addict. No, he didn't dread the thought of being returned to Chino, or Soledad, or San Quentin, which he had never properly appreciated while he was there. The thing he feared, which made his stomach churn and his palms sweat, was the knowledge that he was a *Folsom* parolee. Folsom was where he'd spent the past five years. Folsom was where he would undoubtedly be returned if he was violated by his parole officer. Folsom, where he often saw himself in a dream – aged, insane, like the one they called the Flea – an ancient faggot, unshaved, unbathed, slouching in the yard, exuding revolting odors, making your skin crawl when he came close, mind mercifully destroyed, tolerated by the bulls, loathed by the cons.

Jimmy shook off the memory and walked toward the twelve-foot storm fence and handed his two little identification pictures to the bull inside the tiny stucco shack. The bull pressed a button on the electrically controlled pulleys and Jimmy Lee Smith was breathing free air.

Jimmy and two other parolees decided to take a cab to Los Angeles instead of the bus, and forty-five minutes later, after paying his share of the twelve-dollar fare, he was seated in the parole office with sixteen dollars and a new sport shirt, coat, pants, shoes, and underwear, which altogether were worth thirty-five dollars.

After a bull session, in which he was rebuked for spending money on a cab instead of a bus, Jimmy was taken to another room for his first encounter with naline.

He'd heard about it of course, and had thought that it was something phony to scare junkies with, that the naline wouldn't really prove conclusively whether you fixed or not. He was sure it would have no euphoric effect. He was wrong.

The needle itself excited him. The moment the doctor approached with the glinting silvery spike, his heart raced, and he felt a quick thrilling memory. It had been a long time. And when the needle hit, and dug, and slipped through the flesh and the naline surged into the blood, it was there. The delight. Then came a buzz, not the kind of high he would have spent money on but a high nonetheless, a floating feeling which was to last almost two hours.

The doctor shone a light into his eyes, made marks on a chart measuring pupil contraction, and released Jimmy. He learned that many addicts looked forward to their visits to the Naline Center, came to love the stuff, and would have taken a test every day if permitted.

Then there was more jive, like a reenactment of a job interview, with a parole officer playing the part of the employer. Jimmy smiled drowsily and played the game. Play the game, play any fuckin game they can dream up, he thought. Then they cut him loose.

He was free. Out on the streets. Free. Bursting. Floating with natural excitement and a naline high, fantasizing in the light of day, just as he had done so often in the joint. He thought about women, big-busted women who somehow all looked like Marilyn Monroe in the photograph he had seen in Folsom where she is standing over a subway and the wind is blowing up her dress. But he settled that night for a fat ugly five-dollar whore, and after eating a steak and paying for his hotel room a week in advance, woke up the next morning with forty-two cents in his pocket.

That morning, Tuesday, Jimmy went back to see his parole officer as instructed and was given an additional twenty dollars, the second half of his draw. He was also given the addresses of some car washes where he could surely get a job.

I'm in for a screwin, Jimmy thought at the first car wash he came to. He was made to sit for more than an hour while the proprietor supervised the line and moved the row of cars which had backed up during the lunch hour rush.

Jimmy Lee Smith had astigmatized vision but did not wear glasses. He had a cool soft junkie voice and a pleasant, even handsome face. When he raised his eyebrows his forehead wrinkled and his face looked very melancholy. People would tend to trust or at least pity a face like this. No one was wary of that face. Only one jailhouse tattoo was apparent, the L-O-V-E on his right hand.

But the car wash boss was a bald sweating white man and had

the look of a man who knew ex-cons, hired many, and could deal with them. He would not trust or pity a face like Jimmy's or any other.

'You're looking for a job, I hear,' said the man, and he threw a pair of rubber boots and an apron at Jimmy's feet. 'Okay, you got one. Hit the line.'

Just like a motherfuckin Pat O'Brien movie, Jimmy thought. The tough but kindhearted coach. Or priest. Or cop. Well I ain't one of the Bowery Boys, motherfucker, he thought.

'Let's hold it right there, sir,' said Jimmy. 'What's the pay?'

The man looked hard at Jimmy, smiled and shook his head, muttering something under his breath. Finally he said, 'A dollar an hour *if* you work until nine o'clock tonight.'

'Yeah, well, uh, I want the job, but I can't go to work today. I'll see you at ten tomorrow, okay?'

'Sure,' said the man, looking right past Jimmy, and Jimmy knew the man knew he would not be back tomorrow or ever.

Motherfuckin dollar an hour, Jimmy thought, as he walked toward downtown. Ain't that a bitch! I'd expect better pay for watchin flies fuck.

The next two days passed. When you figure it, Jimmy thought, I ain't doin nothin but time even out here.

First he discovered that his painting course which he took in the slammer wasn't much use when you had to be a member of the union, so he tried a body and fender shop owned by an ex-con he'd known in the joint. Jimmy got a mouthful of sympathy and a glad-to-see-you-back pat on the shoulder, but no job.

The walking continued in the sunshine and smog of downtown Los Angeles. He walked miles of concrete – wondering, thinking, dreaming, fearing, cursing the cheap prison shoes which were already wearing out.

Three days were enough. He was ready to make some bread and there was only one thing he knew for certain he was good at. So he was ready, more than ready, for what was waiting at a shine stand on Hill Street, between Sixth and Seventh. He would not cease thinking about it as long as he lived. That shine stand.

This was not the first trip by the shine stand since he got out. The first time he had strollled by he recognized a black man loafing there, looking seedy and alcoholic, a cheap cigar drooping from oily lips. His name was Small, and so he was, five feet seven, and slender. Jimmy knew him from the old days as one of the best creeper thieves in town, but now he was a drunken shine stand proprietor.

But this day, March 1st, Small might have been six feet tall. His hair was straightened and pushed up in glistening waves and he was sipping a bottle of soda which Jimmy figured contained whiskey, and he was wearing a suit and tie. He was lookin sharp, lookin cool, lookin *good* – lookin like *new money*, thought Jimmy.

Small was sitting up high getting a shine, and he grinned as Jimmy approached, and after greetings and handshakes he offered Jimmy the bottle. Jimmy swallowed the whiskey and soda pop and grimaced. Small said, 'Say, Blood, what's happenin to you these days?'

'Nothin to it,' Jimmy said, feeling self-conscious now, staring at Small's fine imitation alligators, aware of his own shoes, scuffed and dusty and little more than cardboard, and aware of his own odor, his clothes soiled and soured now. There was another shine man there, taller, darker, quieter.

'Want him to glaze your skates for you, Jimmy?' asked Small, pointing to the shine man in the blue apron.

'Like, I'll do the job my own self, it it's okay,' said Jimmy, not wanting the other man to see the holes in the shoes.

'Once a shine boy always a shine boy.' Small giggled as Jimmy buffed his own shoes, and Jimmy thought, that's right, nigger, but watch this shine boy scam a little bread off you soon as you have about three more drinks outta that bottle.

Small stepped down, examined his own shine and straightened his tie. 'Buy yourself a trip to Hawaii, my man,' he said, handing the shine man two dollars.

Jimmy looked knowingly at Small and said, 'Your ship musta came in, ol thing.'

'How you like my alligators, Blood?' Small grinned.

Jimmy was thinking about a five-dollar touch just as two men walked around the corner. Jimmy paid no notice to the shop-keeper who trailed behind the younger man. It was the young white man he was looking at.

'Whadda you think of the jacket?' the young man asked Small.

'Groovy, partner,' said Small, still giggling, obviously more than a little drunk now.

'Here, Billy, I'm buying this for you,' said the young white man, handing a leather belt to Small.

'Say, thanks, Greg, that's all right.' Small beamed, admiring the belt. 'See how my friend treats me, Jimmy?' He winked.

'The jacket's forty bucks,' the young white man said. 'Think it's worth it, Billy?'

'Worth it?' asked the shopkeeper. 'Worth it? It's an eighty-dollar jacket, so help me. This is the biggest sale of the year.'

Now Jimmy looked at the young white man square on. He wished the young man would hurry up and pay the Jew and get rid of him.

The young white dude was a lame, Jimmy was sure of it. It only took a few seconds to size up white men who associated with blacks. Sometimes they talked more hip than any black man, just to prove they could dig it, and could score. But hard as they tried there was always something missing, some little thing missing in white men. They were never *quite* hip to it, never *quite* made it, and could always be manipulated by a cunning black man. And, he thought, Jimmy Smith is every bit of that.

'It fits like a glove,' said the little shopkeeper. 'I got one myself.'

'It's a real nice jacket,' said Jimmy to the young man. And the sale was made.

'This is Jimmy Youngblood,' said Small. 'I been knowin him for a long time, ever since he was a kid.' And to Jimmy, 'This is my partner, Greg.'

'He was a funny-lookin guy,' Jimmy was to say later about Gregory Ulas Powell. 'I mean, you could tell he wasn't eatin right, kinda lean and hungry-lookin, but then so was I. And his dirty blond hair was cut short, real short, and this left his face lookin sharp. Like, his ears looked sharp, and his cheeks were sunk, and his nose was crooked, and overall he was sorta mean-lookin, in a Mickey Spillane sorta way. Actually he was really funny-lookin in the body. He stood real straight, and had kinda narrow shoulders, and though he was skinny he had pretty thick hips. But that neck, now that's somethin you would never forget. It was like a third longer than the neck on a normal man. That and his eyes. Like I say, he was kinda funny-lookin, but he had mean blue eyes.'

Jimmy saw Greg eyeing his shoes and started feeling self-conscious again. If they thought he was too far down on his luck it would make it hard to scam them.

But Greg wasn't thinking how shabby the shoes were. 'Say, Jim, those look like hot dogs,' said Greg, nodding toward Jimmy's plain-toed brown shoes. 'Just got outta the joint, huh?'

'Yeah, that's right,' said Jimmy, somewhat impressed that Greg had picked up on the hot dogs that quickly.

'How long you been out, Jim?' asked Greg, lighting a Marlboro.

'Couple days,' said Jimmy, trying to guess what their game was. One thing certain, he couldn't tell them he'd been in the joint for petty with a prior. A chickenshit petty theft would probably turn them off right away.

'What was your beef, Jim?'

'Robbery,' Jimmy lied, and Greg's smile proved to Jimmy he was right. Fuckin lames, he thought. He knew they couldn't be boosters or creepers, not flashing their bread like these two were doing. If they were surreptitious thieves, they'd be sly and sneaky, close-mouthed. He figured these two had to be stickup men.

'This guy was a complete fool if there ever was one,' Jimmy said later. 'I mean his mouth starts flappin and braggin and flashin green right in front of those two square shine boys. A complete fuckin fool. You couldn't tell him though and I didn't try, neither.'

'Let's go across the street and get somethin to eat,' said Small, and Jimmy tagged along hoping.

'Order whatever you want,' Greg said to Jimmy after they got inside. Jimmy wanted a complete meal, but thought better about showing them how much he needed it.

'Gimme a roast beef sandwich and coffee,' he said to the waitress. Jimmy tried not to attack the sandwich, because that too would betray him, and he nodded politely once in a while when Greg mentioned black guys he knew in the joint and asked Jimmy if he knew them. He knew few of Greg's acquaintances, and had never been in a federal joint or in Vacaville which he knew to be a hospital prison. Guy might be a fuckin squirrel if he did time in that place, Jimmy thought, as Greg talked uninterrupted. He had an uncanny memory for names, that was apparent.

After they ate, Small went to pay the rent on the shine stand and Greg took Jimmy next door to the drugstore, bought himself a pair of the best sunglasses they had and another pair for Jimmy who didn't really want them. With his astigmatism the sunglasses made things worse. But Greg said it was a gift and he wanted

Jimmy to have them. Then they rejoined Small and the three of them got in Greg's station wagon for a ride to Second and Grand, to Jimmy's hotel.

When Jimmy got out he figured he'd never see either of them again, so what the hell, he hit on them.

'Say, brother, how about lendin me ten till I get myself together?' Jimmy asked, expecting nothing, only hoping.

'Sure, Jim,' Greg said, pulling a five from his pocket. 'Give him the other five, Billy.' And then to Jimmy, 'This's not a loan, it's a gift. How you planning to make money, Jim?'

'The same way you guys make yours.' Jimmy smiled.

'See you, Jim,' said Greg cheerfully as they drove off.

'Later, man. Catch you later,' said Jimmy.

Amazed at his good fortune, Jimmy changed to his other shirt. He washed and shaved and felt relatively hopeful again. He went down to the lobby to sit and smoke and pass time. Then the two young white chicks came in. He'd seen them before and knew they lived somewhere on the third floor. He hadn't gotten around to any serious fantasies about them, but now here they were, coming into the lobby, and him with ten bucks in his jeans.

The brunette was a bit hefty but the little redhead had the nicest ass he'd seen since he got out. The big one as usual was talking and didn't notice Jimmy sitting there finishing the coffee in his paper cup.

'My, I think I'm gonna like livin here,' Jimmy said, looking at the little redhead as they passed. They both smiled, did not reply, and went up to their room. Jimmy watched the hips sway as they climbed the stairs.

He waited, and just as he was giving up, they came back and walked toward the front door as though they didn't see him, but with secret smiles which made his blood heat. Jimmy hurried out behind them and the big one turned and looked him dead in the eye.

'I didn't catch that remark you made. But it sounded like a pass.'

'Can you catch it?'

'Do I look like a football player?'

'Yeah, you sure as fuck do, you fat bitch, thought Jimmy, but said, 'Tell you the truth, you are two of the finest-lookin pretty young things I've talked to in a long time.' And at least that is the damn truth, he thought.

'I'm Pat,' said the big one. 'This's Linda.'

'I'm Jimmy. What say we go to a movie?'

'Just like that?' said Pat.

'Sure, I ain't been to a movie house in quite a while. Let's go see somethin sexy.'

'Well, why not?' said Pat, and Jimmy wondered whether the fat one was going for him or whether it was the chance for a free movie and maybe a hamburger later. He decided it was the hamburger.

The downtown theaters were only four blocks away and Jimmy steered them toward the marquee which showed a half-naked Roman girl. The stills outside suggested orgies on the inside: nude young women, men in loincloths. Maybe it would arouse them, Jimmy thought. Maybe he could even cop a few feels.

'Linda's got a steady guy,' Pat said as the little redhead wriggled up the aisle during the first reel to get a soft drink.

'So?'

'I seen you eyeing her, Jimmy, but it ain't gonna do you no good. She really digs this guy.'

That's a damn lie, thought Jimmy, who had seen each of them going to their room with different men during the three days he had been living at the hotel. Still he humored the fat girl.

'Well, I don't much care, baby.'

'Now I ain't got a guy,' said Pat, scooting a little closer.

I know what you like even better than me, thought Jimmy. 'Here, Pat, here's a dollar. Go get us some popcorn, huh? Extra butter. Oozy. And chocolates. Lots of it.' And Pat was gone in a flash, even before Linda returned.

There was no time to be cool, thought Jimmy, and dropped his hand on Linda's hot little thigh the moment she sat. She didn't object and in fact moved down a little and sighed. Jimmy was on fire at once, reaching under her dress just as Pat came back with two boxes of popcorn and a handful of candy bars, but Jimmy's throat was so dry and constricted from Linda's presence, he couldn't swallow his popcorn without choking. Pat was by then too suspicious to go back up the aisle for sodas when he suggested it.

*

One of the best times in Los Angeles is between eight and ten at night, Jimmy thought – when the smog is blowing away and it's cool, and the streets are almost quiet – when there's no pushing, no shoving. They strolled slowly back to the hotel, Linda holding his one hand shyly, while Pat tucked an arm through his and leaned so heavily she almost caved him in.

When they reached the hotel lobby Pat waved to two young white men who looked to Jimmy like ex-cons. He automatically looked at their shoes but they weren't wearing hot dogs. When Pat was gone Jimmy whispered to Linda, 'Looks like Pat's gonna be talkin awhile. How bout you and me goin on up?'

His lips touched her ear when he whispered and now he was really burning and almost trembled like a kid when she gave a demure nod and said, 'Okay, let's do.'

It was disgusting for a man to be in such a state, Jimmy thought as they climbed the stairs. He was so excited, so agitated by Linda's presence and the taut red wool of her sweater, he had to grab the handrail. Damn, he thought disgustedly. Weak in the fuckin legs.

'You got a radio in your room?'

'Uh huh. Wanna listen to music?'

'Yeah, yeah.' And Jimmy smiled weakly and swallowed twice, but the constriction in his throat would not budge. 'Yeah.'

The room was cluttered with useless and valueless female treasures and these too excited Jimmy. The room was clean and that was a surprise. Linda turned on the radio and Jimmy was glad she found some dreamy music because he was planning to ask her to dance, and all the fad dances like the twist had passed him by. He'd been keeping up somewhat with the new rock music, but he didn't really know how you danced to it. He realized the last time he had danced was when people dug on rhythm and blues.

'Wanna dance?' he croaked and tried to swallow again.

'Uh huh, I sure do.' She smiled, and in a moment he was holding her close in the little room where there was barely room to turn around.

She was so short his face was in her dyed red hair, not against her cheek where he wanted it. But it was all right. Yeah, it was *all right*. He wasn't scared now and let his hand slide down over the great little ass.

'What a pooper,' he was to say later to Gregory Powell. 'What a fuckin pooper on that little snatch. Hot damn!'

So the dance was not much of a dance. They just turned slowly, pressed together, kissing, nibbling, groaning, and within a few minutes they were in bed, disrobed and finished.

Be damned, thought Jimmy. That really wasn't as good as I thought it was gonna be. And then he chalked it up to her inexperience. She needed teaching. She didn't really know what the hell to do. Well, she was willing at least. And he had the time.

Then they talked and told their dreary little stories. He about being an ex-con and she nodding sympathetically, saying lots of guys in the hotel were ex-cons. And she telling him that she and Pat were both nineteen and had run off from their El Paso families to live in the big city. She talked of her two-year-old boy she had left in El Paso. When she said that about leaving her little boy he thought about another wild young girl leaving her little boy for her aunt to raise thirty-two years ago. Then he thought just for an instant of the children *he* fathered when he was a teenager and whom he never saw again. What the hell, he thought. Everybody fucks over everybody else. So he pushed it out of his mind. He gave her a couple of what-the-hell kisses which promised a next time under better circumstances.

Then Pat showed up and gave them both a naughty look and although they were by now dressed, Jimmy became embarrassed, for the first time aware of the overpowering aroma of their lovemaking. 'Some guys tell me they know broads where you can't smell nothin,' he said later. 'But that ain't the case in my experience.'

'You naughty naughty people,' said Pat, smiling coquettishly at Jimmy, and he thought she was a pretty good sport after all. A pretty good ol sport.

Jimmy returned to his room that night and slept well enough. The next morning he found Pat in the lobby on her way to work and she stopped him for a moment.

'I hear you just got out of prison, Jimmy.'

'Yeah.'

'That must be awful. What'd you do?'

Now he had her pegged and answered casually, 'Robbery. Stuck up a couple banks.'

'You gonna do it again?' she whispered.

'No more small time bullshit for me,' Jimmy said. 'Naw, I got connections. Gonna take it slow and easy til I get set up by my connections.'

'See you later, Jimmy. Gotta go to work.'

'Catch you later,' Jimmy said, thinking, I got connections all right. With a drunken nigger shoeshine boy and a big-mouthed long-necked paddy who has got to be the lamest squarest motherfucker I have ever met. I got connections – with two fools.

But they were all he had, so he walked down Seventh Street toward the shine stand.

They were not there, and it was despondency more than hunger which made him waste his money on the greasy fried chicken in the greasy Broadway restaurant. He had enough money for dinner and then he'd be broke. He went back to his room and slept away the day, waking up in the late afternoon in a sweat puddle in a hot and smelly room.

After taking a bath, Jimmy went down to the lobby and read the paper, sitting there until dark, eyeballing a fine little Mexican chick who was passing time with another girl and a bull-dagger. Jimmy smiled and said a few words which the bull-dagger didn't appreciate and he measured her before proceeding because he knew for a fact there were some bull-daggers who were willing to fight harder than any man to protect their territory. That was all he needed now, he thought: throwin blows with a fuckin dike.

He went to bed early and alone that evening, watching a fly bang against a hot naked light bulb, looking out the window at the city, smoking, hoping that Linda would come to his room.

Jimmy lay there relaxed at first, enjoying the breeze which snapped the dirty muslin curtains, making shadows on the wall. Then he smoked and watched the moon climbing. It looked to him like a gold coin. Like a gold coin *must* look, ought to look.

Then he lost the mood in the boredom, and again his thoughts returned to Folsom, and the same question, the old question, the tired question he had asked himself a thousand times these thirteen years. What if . . . What if he really *couldn't* live out here? What if he was what they talked about during group sessions at Chino Prison? What if he really was, like they say, an *institutional man*? Since he was nineteen years old he'd been in one institution after another, and even before that there was detention camp. In fact, if you added it all up he was on the streets maybe a year or a little more since he was nineteen. One year since he was a man full grown. This August he'd be thirty-three years old. What *if* . . . ?

But the thing that always dissuaded him was the memory of Folsom. Sure, he might go for that institutional man stuff if it was some other joint you were talking about. I could do nothin but time in a place like Chino, he thought. *Nothin but time.* And some of these other joints too where they have what they call 'counselors' and the food is good and they call them 'dorms' where they lock you up, and they have TV where everyone can see it, and all the ice cream you want, and it's warm in the winter and air-conditioned in the summer, and almost unlimited visitin. But then, fuck the visitin, because nobody visits me anyway. On the other hand, it's nice to see pretty girls millin around. And they have lots of groupin where you sit around and tell lies to each other and have a ball-bustin good time. And enough money for plenty of smokes, and where once in a while you can score a little taste, maybe a few reds, even a real fit if you got enough bread. But fuck the fix, no, that's way back, *way* back. It's too good a memory. Too good. Makes you feel too good to remember it. That's way back there. Better to remember the *first* fix. When I gave two and a half bucks toward a number-five cap, shot three drops, and puked all the way to San Bernardino.

Maybe I could get a job in the visitin center of a nice slammer, he thought. Yeah, he heard of those suckass jobs in joints like that, and shit, he was sly enough to con his way into a job like that, and then he could really make lots of stings in lots of ways. Shit, I could jive visitors by tellin them how nobody cares if I live or die and they'd lay some bread on me. Yeah, they'd forget who the fuck they came to visit. Or how about tourists passin through? Oh yeah, they'd give me smokes and coin. And who knows, probably in a job like that the bulls look the other way, or maybe I could slip them a couple bucks once in a while. Once every other Sunday maybe, I talk real sweet to some little bitch who's just come from visitin her old man, and they been talkin hot and she's just oozin. Yeah. And then the bull looks the other way and I slip her into a back room there. There must be a back room there, and I fuck her right down through the floor. Oh yeah!

But that ain't the way it is. Not for Jimmy Lee Smith. Not for me. Cause now I got a Folsom jacket. It's back to Folsom for me.

To the Big Gray Frog on the grassy green banks of the American River. Oh yeah.

He shivered when he pictured Folsom, squatting there, huge and gray, ghastly in the fog. Built from enormous, rounded, slimy wet and ugly moss-covered stones. He always dreamed of it that way. In the fog, when the stones were clammy and dark, ugly gray. It was like the prisons he saw in the old movies when he was a boy. He thought of his cell. The old spring cot that sagged like a hammock. The wooden table built from scavenged lumber, the chipped coffee cup which always cut his lip, the Bugle rolling tobacco and the sardine can ashtray, the cardboard taped over the bottom of the door to keep out the draft and the dust.

There were the bulls, old hands, immovable, old like the institution. Not 'counselors,' not 'correctional personnel,' but *bulls*, just fuckin bulls with cop uniforms and guns. Jesus, guns. Like they say, a water-cooled machine gun on one tower. Go ahead and try to make the wall and fence back toward the river. Oh yeah, go ahead. They want you to. And the other towers, rifles on those. And spotlights. And this is maximum security, baby. Like, forget it, you ain't leavin here. Not till the man says you're leavin.

Unlike at other joints, nobody was pressured into a job occupation, not at all. Like they say, in Folsom you don't gotta do nothing but *time*.

And as always his thoughts returned to the Flea – that filthy, prowling phantom Jimmy could not seem to escape. He'd see him in the yard, skulking around the housing unit, anywhere.

Jimmy knew it was probably imagination, but he seemed to be looking at him. Like he knew him. And smiling, or at least showing brown teeth, twisted and broken, and drooling. They let him alone because his mind was half-gone, even the bulls let him alone, and he prowled alone, this slinking bag of bones with tufts of hair hanging from each nostril. His only interest was oral copulation, and Jesus, he *found* guys! Jimmy felt the shivers surge through him. He *found* guys!

And Jimmy had nightmares about the Flea. In one dream, Jimmy couldn't seem to move through the yard, but it wasn't like he was walking with heavy feet, he just couldn't walk because he was old. He was an old man and had been inside most of his life like the Flea. Suddenly he was *so* old he could hardly walk through the yard. He was terrified because his life had gotten away. He couldn't remember how, but he could see by his hands, by his veiny brittle hands, that he was old! Then with a mighty effort he shuffled across the yard and through a door because a monstrous ugly con was eyeing him with lust and he didn't want them to do what he saw them do to the Flea. He ran to his cell and screamed for the bull to lock the door, and he fell across the cot, overturning the little wooden homemade table. The mirror crashed to the floor and he lay there, gaping in horror into the shattered pieces. He could see plainly in the slivers of glass. Clearly. He *was* the Flea!

Sunday afternoon, March 3, the house phone rang.

'Hi, Jimmy,' said the cheerful voice which Jimmy recognized immediately.

'Hello.'

'What're you doing?'

'Shavin. Come on up.'

'Come on down. Billy and me'd like to take you to my house. We'd like to show you something.'

'Be right down,' said Jimmy, and for the rest of his life Jimmy Smith would also mark *this* moment.

Why the fuck am I wastin my time with these two fools? he thought.

But what else do I got right now? I can at least go with them this afternoon. They ain't gonna stick up somebody this afternoon. And maybe I can get next to the paddy and scam a few dollars off him. And then just cut them loose. Maybe I can scam a whole mess of dollars.

For the rest of his life he would wonder if he could've made another play, played his own hand, or drawn different cards. Or was the hand played around him? Does somebody stack that fuckin deck, he often wondered, and ain't nothin you can do about it? It was something he never decided.

Jimmy saw that Billy Small was half in the bag even that early in the afternoon, and they drove aimlessly in Greg's six-year-old Ford station wagon making small talk about mutual acquaintances they'd known in prison. Small at last insisted on stopping at a liquor store for a half pint of Schenley's.

When they got around to going to Greg's three-room duplex

apartment, Jimmy was surprised to see it was near Sixty-fifth and Figueroa, in an all black neighborhood. Jimmy was later to recall, 'We went in the pad and he introduced me to this dumpy little pregnant broad named Maxine who he called his wife. She wasn't black like I expected, she was white. Fishy white, paler than him even. But she was pleasant enough in a plain sort of way and she had a quick smile which made her not quite so bad lookin.'

'Max, step in the bedroom a minute,' Grey said as Jimmy and Small sat down on the couch. Jimmy tapped his cigarette on the top of a cheap glass coffee table and tried to ignore Small's drunken babbling. He watched Greg gesturing angrily to Maxine and whispering, and then Greg bent over a dresser and turned sideways. Jimmy saw the revolver in his hand.

Greg flipped the cylinder open, checked it and flipped it shut with one hand as they do in movies. Then he came in the living room smiling cheerfully at Jimmy. Small sipped his third whiskey oblivious of them all.

'Why did you take that roll of dollar bills that was in the bedroom, Billy?' asked Greg.

Small looked at Greg, down to the gun in his belt, and back up. 'You gotta be kiddin,' he said finally. 'I don't know what you're talkin about.'

'Yes you do. You took a roll of one-dollar bills that belongs to me. Max said it was laying on the dresser yesterday morning when you and Max were here alone. She said she's looked everywhere for it and can't find it.'

Now Greg's tone was changing, but only a little. The gun was in his belt but Jimmy was watching the blue eyes carefully.

'Also, Billy, I'm reminded of the fact that yesterday downtown you were flashing an awful lot of one-dollar bills. And I know for a fact that when we split up the last take, I only gave you about twenty one-dollar bills and kept the bulk of them to make up that roll.'

'I swear I didn't, Greg,' said Billy, getting more sober by the minute.

'Furthermore, Max says that yesterday while I was gone you were fooling around the bedroom and even patted her on the fanny. Now, I don't care about the pat. We're sort of a little family, the three of us. I'm sure it was done in a friendly way and doesn't mean anything, but don't make it a habit.'

'No, Greg, I swear . . . '

'What I *do* care about is the dough, and it's missing. I want it.'

Small picked up the Schenley's, took a long gulp of courage, pulled his wallet from his back pocket and slammed it on the coffee table.

'All I got in that fuckin wallet is tens and twenties. She's a liar!'

So that's the way to talk to this weird guy, thought Jimmy, this oddball with his long neck and his funny talk – with his alsos, furthermores, moreovers, and all his other bullshit. He talked like some hick from Nebraska just come to town and begging to be flimflammed.

'Max never lies to me,' Greg said, and turned, taking a step toward the bedroom. Jimmy thought that even a drunken black shoeshine man could handle the likes of this paddy. Then Greg wheeled.

He crouched slightly, taking shuffling steps toward Small, the gun in his hand, pointed at Small's face, finger tight on the trigger, hand trembling.

Jimmy saw all of it, but mostly he saw the eyes: blue, cold, without life, so that the voice full of rage only exaggerated the promise of death in those eyes.

'I ain't never gonna forget those eyes,' Jimmy was to say. 'With a gun in his hands, something happened to them. Give him the gun and he could scare the shit outta *anybody* with his weird looks and those eyes.'

'After all I done for you,' Greg whispered, knuckles white, gun muzzle weaving a tiny circle. 'I split with you fifty-fifty. All I done for you and you dare to steal from me?'

Then Greg put the muzzle six inches from Small's forehead and cocked the hammer.

Small was frozen. He looked like the stunted rat Jimmy had seen in the hotel as a child. It was paralyzed when the gray cat cornered it, impotent before the great yellow eye.

Jimmy realized *he* was panicking and tried to think of something to say, something . . .

But Greg spoke: 'I oughtta blow your goddamn brains out.'

Now Jimmy could smell the sweat, his and Small's. His breath was coming hard but he sensed that Small was not breathing at all.

Greg stared for another ten seconds and slowly let the hammer down, the muzzle still pointed in the face of Billy Small.

'I love you like a brother, Greg,' Small was finally able to croak. 'Greg, we *are* brothers. I wouldn't steal your money.'

'If I was one hundred percent positive you did, I'd kill you now, you son of a bitch,' said Greg, putting the gun back in his belt.

'I didn't, Greg. I didn't, brother.'

'We're going out tonight, Billy,' said Greg, 'only this time you're gonna pull your own weight. *You're* going in alone, not me. Jimmy's gonna drive. I'm just gonna wait outside. And if you come out without the money, without doing your work, I'm gonna kill you.'

'Sure, Greg, sure, I'll go in. Sure.'

'And maybe it's possible you didn't steal my roll, but we're going out and make it all up tonight anyhow, and you're gonna do all the making.'

'Sure, Greg,' said Small, sagging back on the couch.

'Can you drive, Jim?' Greg asked.

'Sure, drivin's my game, Greg,' said Jimmy, beginning to calm down, shakily lighting a Chesterfield. 'Like, I'm a pretty good wheel man, you know what I mean?'

'Come into the kitchen, Jim,' Greg said. 'Let's have a drink. Billy, I don't think you should have any more.'

'Anything you say, Greg,' said Small. 'Anything you say.'

Maxine mixed them both generous portions of Schenley's, and Jimmy was grateful to get it.

'I can understand how you feel,' said Jimmy after a few sips. 'But honest, Greg, I been knowin that guy for a long time, and maybe he *didn't* take it. And if he did, maybe he was drunk. I'm glad you gave ol Billy a break. Shoot, it's just the whiskey.'

'I thought of that, Jim,' said Greg, with a friendly smile again, his front teeth protruding slightly, a partially filled silver tooth just right of center, much like Jimmy's own. 'I thought the whiskey probably made him do it and I made allowances. Otherwise, I'd have busted a cap in his goddamn head.'

'That's right, Greg,' said Jimmy, spilling a little of the drink. 'Stealin from a partner is jiveass bullshit, is all it is.'

'Well, it's getting dark now,' said Greg. 'I gotta get ready for our night's work.'

Greg went to get ready and when he came back he was wearing dark clothes, a pair of half-boots and a stingy brim hat.

'How do I look, Jim?'

'Uh, fine, Greg, you look jist fine.'

'I mean the difference. Don't I look different?'

'Oh, sure, Greg,' Jimmy said, getting nervous again, because he didn't want to make Greg angry.

'Do you know *why* I look so different?'

'No, well, yes, no, not really, Greg. You see, you got me fooled. You look so different, but I can't exactly figure why you look so different. If you could maybe give me a little hint I could sure as hell tell you why you look so different.'

'It's this,' said Greg triumphantly, pointing to his ear. 'I put a mole on my earlobe, and I stuck three little hairs on it. You see, Jim, people facing a gun will remember outstanding characteristics about the guy, and they'll all remember my mole. But after the job it'll be gone.'

'That's slick, Greg,' said Jimmy whistling, and holding up his glass in a toast. 'I gotta hand it to you. That's slicker'n snot.'

'And that's not all. You notice what else?'

'Else?'

'Well, goddamn, Jim, my appearance is completely different and you can't even see how I've done it!'

'Well, it's just the job you done, Greg. Honest. And my eyes are real poor. I should wear glasses but . . . '

'The eyebrows.'

'Eyebrows?'

'They're darkened and made larger. The witness'll say the guy had bushy eyebrows and a mole on his earlobe. And I got a thin pencil-line moustache here. See it?'

'Oh yeah, yeah, now I see it.'

'Well those're the things they'll notice.'

'Oh yeah. I get it. Right. Righteous. I get it!'

'Just one more thing, Jim,' said Greg, going back in the bedroom. He returned wearing a trenchcoat, a Marlboro dangling from his lips.

'I ain't lyin,' Jimmy was to say later. 'A trenchcoat! So now he looks like Bogart, he figures, and he's ready to go to work. You couldn't tell him though, and I didn't try neither. And I still coulda cut them loose, coulda made some excuse, even coulda weasled out, pleadin sheer lack of guts. But I didn't. I thought, shoot, one good score could give me the stake I need. Just one job with this crazy sucker.'

'Get behind the wheel, Jimmy,' Greg said, and Jimmy watched him kissing Max goodbye. 'We'll cruise around for a while until you get the hang of it. Billy, you sit next to Jimmy.'

Then Greg sat on the far side and Jimmy started up the tired old station wagon. He hadn't driven a car in years and the clutch slipped so badly it was impossible to do more than crawl from the curb.

'I had the clutch fixed that way on purpose, Jim,' Greg said, turning up the collar of his trenchcoat.

'Oh yeah?'

'Yeah, it keeps someone from speeding away from the scene of a job and drawing the heat.'

'Oh yeah,' Jimmy said, 'that makes sense.' But he later said, 'It really didn't make no sense at all. To have a slippery clutch, you know what I mean? Like, he was the only guy to drive the Ford, you know? Small didn't drive, period. And since he was the only guy to drive his car he could control his own speed, you know? I figured he just didn't want me to think he had a car with a bum clutch.'

'Runs pretty good when you get her in high,' Small said as Jimmy pulled out onto the Harbor Freeway and kicked it up to seventy with ease.

'She's a good car,' said Greg. 'I picked her out with care.'

'Oh yeah,' said Jimmy.

After what seemed like aimless driving on the freeways and surface streets, Greg directed Jimmy to the first job of the evening.

'This is the store, Jim. I've cased this one and it's ripe.' But when they pulled up in front of the large market they found it closed.

'Well, I cased it, Jim, but you know, it's after nine. All that time we wasted at the house with Billy, that's why we're late. Market probably closes at nine. In fact, I'm *sure* the market closes at nine. Well, that's Billy's fault. Let's just hit a liquor store. There's plenty around here. Make a U-turn, Jim.'

It was 9:30 P.M. when Jimmy pulled around the corner from a small liquor store near Washington and Leo Streets, turned out his lights, and parked in the shadows.

'Now listen, Billy,' said Greg. 'The fifty bucks is our main concern here. That's what you probably stole from me, you know. Now we're gonna hit this place together and maybe that'll prove to me that you didn't take the money from the bedroom.'

Jimmy Smith said later, 'I couldn't understand his logic in forcin Small to pull a job to prove he didn't take the money from the bedroom. You couldn't tell him though, and I didn't try neither. The only thing that made sense to me about the whole deal was that durin their past robberies Greg had did all the gun handlin and that's the way it should be. But Small nodded at Greg, so I thought, what the fuck, it made sense to *him*. But when we got there Greg changed his mind and went in alone.'

'Okay, Jimmy,' said Greg, 'don't panic when I come back. Just start nice and smooth and don't try to make a quick getaway.'

'Okay,' Jimmy said as Greg walked toward the store. Jimmy felt the sweat rolling down his body and he and Small did not talk as they waited. Now you're in, Jimmy baby, he said to himself. Now you're in.

'Give me a bottle of Schenley's,' Greg said to the proprietor.

'What kind do you want? Schenley is a distributor.'

'Then give me a bottle of Seagram's Seven. Half a pint.'

The proprietor got the bottle and returned facing Greg's gun. Greg cocked it.

'Open up and give me your money,' he said, motioning toward the register. And to a man who was shopping in the store and didn't notice the holdup, 'You, get around to the back.'

The customer threw up his hands instinctively and Greg glanced toward the windows and said, 'Get those hands down or I'll blow your brains out.'

Minutes later Greg was in the wagon, and Jimmy was in gear, panicked, ready to snap the clutch.

Suddenly he heard terrifying explosions! 'The guy's shooting at us!' Jimmy howled, and the wagon crawled out into the traffic lane and Jimmy forgot the clutch, forgot the money, forgot everything except getting the car going, trying to drive the gas pedal through the floor.

The engine was racing and the clutch was whining and the car was going so slow the liquor store proprietor could have outrun it. But he didn't.

'Bum couldn't hit the side of a barn.' Greg laughed excitedly. 'Firing blind at us with his hand around the corner.'

'Are we hit, Greg?' asked Jimmy.

'Are we hit? What the hell, Jimmy. You'd feel it if you were hit.'

My first robbery, Jimmy thought when they'd gotten two blocks away. And there's shootin. That's a sign, that's a fuckin sign to cut this guy loose. But then he thought of the money. Greg felt Small owed him fifty and he'd probably take that first. But how much was left? He wondered if he'd split the remainder three ways. But the take turned out to be only forty dollars.

When Greg saw the amount he got angry again. 'Bastard,' he said. 'Taking pot shots at us for a lousy forty bucks. Probably wasn't even his store either. Some brave boy protecting the owner's money. I should go back there and blow the bastard's brains out to teach him a lesson.'

Jimmy and Small did not reply, but both were relieved when Greg jammed the paper sack full of money into the glove compartment.

'Billy,' Greg said thoughtfully, 'what do you think about that place over on Western Avenue where you get your hair fixed?'

'The process parlor?'

'Yeah, think they got plenty of money in there? I believe they're open at night.'

'Maybe,' said Small, who looked like he'd rather be anywhere but here at the moment.

They left the freeway at Santa Barbara Avenue and got caught in the Sports Arena traffic. Horns, headlights, streets jammed from a basketball game, traffic backed up a quarter of a mile in all directions. Jimmy made two turns and got out of it but now they were going in the opposite direction from the process parlor.

'Ug, Greg, we're goin the wrong way,' said Small.

'Don't worry about it, we'll hit a liquor store down the street here. I think there's one.'

Jumpin fuckin Jesus! thought Jimmy Smith. We're goin the wrong way so what the fuck we just hit a place this way, and if we come up on a 'no left turn' sign, we just turn right and look for another store. This is how they case a job? These're professionals? Jumpin fuckin Jesus.

They parked in the lot next to the first liquor store they came upon.

'Okay, Billy,' said Greg. 'Here's your chance to prove that you didn't do what you did.'

Jimmy later said, 'I was gettin scared again and mad too. I didn't understand what he just said to Billy. I didn't know if I was stupid or just scared, but it burned my ass that Billy understood the things Greg said and I couldn't.'

'Just simulate a gun,' Greg said. 'Pretend you got one in your jacket.'

'I can handle it,' Small said in a drunken slur and staggered across the street toward the liquor store.

'Greg, you think you oughtta let him go in?' asked Jimmy. 'He's pretty drunk.'

'He knows what he's doing,' said Greg.

Then Greg got impatient waiting and stepped out of the car and started walking across the street. Then he turned and came back.

'I'm not giving Small any more breaks, Jimmy,' Greg said as they waited. 'If he can't knock off the little liquor store he ain't much of a robber anyway.'

Minutes passed. Greg was getting impatient, Jimmy frantic. Then Small appeared.

Jimmy knew he hadn't done it by the way he walked slowly across the street with a bag in his hand. There was only whiskey in the bag. And soda pop.

'No dice, Greg. There was three clerks in there and one guy who knew me from a long time ago.'

'You sure?' asked Greg coldly.

'Sure. Sure as rain. Even Jimmy's been knowin this cat. From down on Fifth Street. He used to have a record store down there.' And Small looked toward Jimmy, who saw the fear on him.

'Don't get tight jawed, Greg,' Jimmy said before Greg had a chance to speak. 'I can go check that story.'

'Do it,' said Greg with the same voice he used on Small at the house, and Jimmy leaped from the car and jogged across the street. He went in the store, saw only two clerks, total strangers, bought a pack of Chesterfields, and ran back to the car saying, 'He's right, Greg. It's that guy all right. Same guy. Owned a little meat market on Fifth and Towne.'

'I thought you said a record store, Billy,' said Greg.

'That's what I say, Greg,' Jimmy babbled. 'A record store.'

'I thought you said a meat market,' said Greg.

'He said a record store,' said Small.

'Oh fuck it,' said Greg. 'Let's get the hell outta here.'

So they went to their next job, this one on Jefferson. 'This time you get the piece,' said Greg, handing Small the four-inch Colt revolver. 'Put it in your belt. This is a ripe one. Don't muff it, Billy.'

Small got out of the car and reeled drunkenly for a few steps, righted himself, and with the instincts of an alcoholic, proceeded on a reasonably straight course to the door. Greg followed and stood on the sidewalk outside. This time Billy Small came out fast, almost running.

Jimmy slid behind the wheel, jammed it in low gear and once again crawled from the curb, engine racing, clutch slipping. As the gears finally caught up with the racing engine, Small reached into his coat pocket and pulled out a handful of crumpled bills. Then he reached into his other pocket and did the same. Then he threw the whole pile into Greg's lap.

'When I tell you I'm gonna do somethin I'm gonna do somethin,' Small announced. But he didn't give the gun back to Greg. Instead he reached for the whiskey and took a drink.

'Looks like a few hundred, partner.' Greg grinned. 'You handled that like a real pro. Was there a safe?'

'He was in the fuckin safe when I got the drop on him.' Small giggled.

They rode silently, Small drinking, Greg counting the money, and Jimmy unable to take his eyes from the green in Greg's lap.

'This is the last job I pull with you along, Jimmy,' Small suddenly blurted.

'What?' said Jimmy.

'You tryin to break up the partnership I got with Greg. Greg and me is partners and you can't come between us with your jiveass ways.'

'Listen, you bastard,' Jimmy said, careful to see that Small's hands were not near his waistband, 'I ain't gonna take that from you. Jist cause you pulled your first job and you got a gun in your belt and a gut full of whiskey. I don't need to get your okay to be friends with Greg, damn you.'

'Take it easy. Easy.' Greg laughed, looking strangely elated. 'Don't fight over me, boys. Let's not go off half cocked. Look, we've done enough for the night. I think we got maybe six hundred here, and that barber shop's closed by now.'

That was what Jimmy was waiting for. He made a right turn, hit the freeway, and headed for the hotel and his cut of the money.

'We could knock off four more places,' said Small, full of bravado now.

'No, it's been hectic tonight,' said Greg. 'And Jim's a little nervous. After all, we only took him out tonight to get him broke in.'

It was midnight when they reached the hotel and Greg handed Jimmy a ten-dollar bill. Jimmy was stunned. Ten scoots! Ten lousy scoots! Of course he didn't expect a full share. But if they'd been busted he'd have been busted too. And for what? For this he drove that miserable car with that miserable slithery clutch? For ten stinkin bucks? But he saw Greg grinning again, and he thought of how he'd grinned at Small just before he pulled down on him, and he remembered those eyes, how they'd changed. Cold. Blue. Staring. So he said nothing, and put the ten dollars in his pocket.

'Take this too, Jimmy,' said Small, offering the paper bag with the half bottle of whiskey inside.

'Thanks,' said Jimmy, turning and walking into the hotel, with a last futile hope that they'd call him back and give him some real dough, maybe a hundred at least, which he deserved. But he heard the car clunking into gear and whine as it slipped down the street.

Linda was in the lobby. She too was just coming home.

'Hi, Jimmy.' She smiled. He liked the way she smiled openly, like a kid. It made him feel a little better.

'Hi, baby.'

'What's in the bag, Jimmy?'

'Now, you're gonna have to come to my room to find that out.'

'Wonder if I dare to?' She giggled.

'Oh, baby, I had a rough night,' Jimmy said, and his smile faded and he felt exhausted and defeated and put his arm around her waist. They climbed the stairs together.

Linda looked as though she understood and said, 'Okay, I'm coming to see what's in the bag. And you better let me see whatever I wanna see.'

'Okay, baby,' said Jimmy and kissed her on the cheek, and it was a struggle to climb the darkened staircase. He was dog tired. When they got in the room he drank deeply until his throat and stomach burned and his eyes watered. She took a coffee cup and poured in half an ounce of whiskey and half a cup of water and sat on the bed sipping demurely.

'Don't like to drink?'

'Not very much.' She smiled.

'You're a sorta sweet baby,' Jimmy said, putting down the bottle and stroking her red hair which was now snarled and stiff, and lint dotted.

'You're a sweet guy, Jimmy,' she said, putting down the cup.

'Know what, Linda? I got a feelin for you. Like, I was thinkin. All night I was thinkin. I got a feelin for that chick. That's what I was thinkin, is all.'

'Kiss me, Jimmy, but don't hurt me,' Linda said, lying back on the bed and closing her eyes.

All the time he was getting her clothes off, Jimmy was thinking about what she said. What the hell's she talking about? Hurt her, my ass, he thought. Her box ain't no rose blossom. Then he remembered: Shit! That's what the Roman broad said to the big gladiator dude when he poured her the pork in that orgy movie they saw. Shit!

The lovemaking wasn't too good, he decided, after they were finished, lying naked under the covers. It wasn't even as good as last time. It was those two fools that did it. Making him so he couldn't think of anything else except how they cheated him, used him, took advantage of him.

'What's the matter, honey?' Linda asked, cuddling closer to him, rubbing her face on his chest.

That's another thing that Roman cunt did in that movie, Jimmy thought. 'Nothin,' he said.

'You're so quiet.'

'I'm tired, honey. You took it all outta me.'

'Glad to.' She giggled.

Then they were quiet. He smoked and looked out the window. It wasn't smoggy and he could see stars over the city. He tried to think about the stars. But instead he thought of how he was going to start boosting tomorrow. That's the way to make it, of course. Shoplifting was his game. Fuck this robbin and gettin shot at and gettin cheated by a drunk and a turkey-necked freak who was at least half off his bean and . . .

'Jimmy.'

'Huh?'

'We're sure having a tough time, Jimmy. Pat and me. We just ain't making it. We're two weeks behind in our rent and all. I just don't know what's gonna happen if I don't get some bread, Jimmy.'

'Well, you just keep bein my sweet lil baby, and I'll take you with me when I get outta this dump. Now you better run on down to your apartment cause Pat'll be worried, and anyway, I gotta lay here and think out my plans. Tomorrow I go to work on my own and that's when the bread starts comin in.'

'What're you gonna do, Jimmy?' asked Linda breathlessly, and she slipped her dress over her head. 'Can you tell me?'

'Naw, I can't tell you, sweetie. Like, what you don't know won't hurt you. Dig?'

'Oh yeah, Jimmy,' she said, carrying her underthings and shoes and gathering up her purse and cosmetics. Then she leaned over and kissed him long on the mouth, wrinkled up her nose, stuck her tongue out, smiled, and opened the door.

'Later, baby. Catch you later.'

'Jimmy,' she said before turning.

'Yeah?'

'It don't matter to me, but I was wondering. You're so light and all. Are you maybe only part Negro? Maybe just half and half or something?'

'Baby, I don't know,' he said in a tired voice. 'I honest to God don't know, and that's righteous.'

'Well it don't matter to me anyway, honey. You know that.'

'Yeah, I know that.'

'Bye, lover.'

'Bye bye.'

Jimmy lay in the darkness until 4:00 A.M. before he went to sleep. He slept only when he finally tired of watching a sliver of golden light in the hallway under the door. His heart beat fast and he watched it until his eyes fell shut.

Jimmy had a large and filling restaurant breakfast and three cups of coffee the next morning. By nine o'clock he was hustling down Hill Street heading for the shine stand with no clear-cut idea of what he was going to say to Small when he got there. The department store boosting could come later, but first he had to see

Small. There would have to be some kind of showdown and he would have to be paid something. Ten bucks! For risking his life! Now that Small was sober he would have to agree.

And then Jimmy had another thought. Would he *be* sober?

'Say, Youngblood,' said the grizzled black man, leaning against the wall of the stand. 'Lookin for cousin Billy?'

'Yeah.'

'Ain't seed him.'

'Listen, I know where that sucker be layin up,' said Jimmy, consciously changing inflections and dialect depending on whom he was with. One way with Linda last night, he thought. Or with his parole officer. Or with that fuckin square Greg Powell. Another way with this dude. Nigger talk for this dude.

'I don't give a fuck where dat sucker be,' said the man, drinking out of a paper sack.

Jimmy knew the man to be Billy's real cousin. Or were they play cousins? Probably real cousins, he thought. Whole fuckin family's probably drunks.

'Listen, brother,' said Jimmy. 'Billy be layin up wid dat white dude, you know? And dig, I was with em, and Billy got him some coin. And baby, I mean *coin*!'

'Say what?'

'Yeah. And he owe me some. And I needs me a ride.'

'Le's go,' said the drunken cousin and in ten minutes they were turning off the Harbor Freeway, heading toward Sixty-fifth. Jimmy had the cousin drive past the house and told him to park around the corner and wait.

'Listen, I ain't sure which pad,' Jimmy lied. 'You wait here and I be comin back after I knocks on a few doors.'

'Righteous.' The cousin winked, slumping back in the seat and sucking on the bottle inside the brown paper bag.

Maxine opened the door for Jimmy, who thought her friendly smile made her looks almost passable.

'Lookin for Small,' Jimmy said.

'Come on in,' Maxine said. 'Have some breakfast.'

'I'll have some coffee,' Jimmy said and saw a tangle of blankets on the floor and Small snoring on his back with his knees up in the air, elbows bent, and hands thrown back beside his head, palms up.

'Looks like a dyin cockroach, don't he?' Jimmy mused as he

stood over Small, who was buzzing and drooling saliva over his chin.

Max shoot her head as Small cut a window-shaking, boozy fart, and Jimmy followed her into the combination bedroom-dinette where Greg was having coffee.

'Hi, Jim.' Greg smiled. 'Glad to see you. Have some breakfast. How'd you get here?'

'Had a guy bring me,' said Jimmy. 'Cousin of Small's. Had him park around the corner. Didn't think you'd want too many dudes knowin where you stay.'

'Good thinking, Jim,' said Greg as they heard the knock at the door.

Maxine peeked out and whispered, 'Greg, it's some old colored guy. Looks drunk.'

'That's Billy's cousin,' Jimmy said into his coffee cup, feeling Greg glare at him.

'All right goddamnit, let him in,' said Greg. 'Might as well invite the rest of the neighborhood while we're at it.'

'This is Greg,' Jimmy said when the cousin staggered in. By then Small was snorting and groaning as he raised up on his elbow.

'Billy got to make my car payment for me,' the cousin explained. 'Billy know he got to make it today.'

'Well get him the hell up and go on and make it,' Greg said, tucking his shirt in his pants, and Jimmy wondered why he was so damn skinny. After all, he was making good scores and yet Greg was no more than a hundred and forty, about Jimmy's weight. But Jimmy had an excuse, he was starving, hustling, broke. He guessed he'd dropped ten pounds in the week he'd been out. In fact, Greg was five feet nine or ten, just about Jimmy's height. But the similarity ended there, Jimmy thought. He's a nut and I ain't.

'What kinda car you got?' asked Greg as Small struggled into his pants, groaning and taking a drink out of the cousin's bottle.

'T-Bird.'

'Yeah?' said Greg, apparently interested.

'Wanna drive it and go with us?' the cousin asked, looking at his bottle which Small was now draining.

'Sure,' Greg said, and Jimmy shook his head as the cousin continued making small talk about the car. Kissin the white man's ass all over the place just to get another fuckin bottle, Jimmy thought disgustedly.

As the four men left the apartment Jimmy got up his courage and tried to put a little steel in his voice as he motioned Greg aside. But it came out a plea and not a command. 'I'm really on my ass, Greg. I just gotta have a little bread. I mean, I think I should have a little somethin anyway. Like, you know what I mean, Greg?'

'Sure, Jim,' Greg said. 'Here. Here's ten bucks. Go around the corner to the liquor store and get some more Schenley's and wait here until I get back. I got plans, Jim. Big plans. And Small's such a drunk, he just ain't reliable. Have patience, Jim.'

'Sure, Greg.' Jimmy smiled. 'Like, I'll wait for you.'

Maxine was glad that Jimmy was going to stay and wait for Greg to return. 'Gets lonely here when Greg goes out,' she explained. 'You know what, Jimmy? I'm gonna fix you a real home-cooked meal today. How'd you like that?'

'Oh yeah,' said Jimmy. 'I ain't had one in years.'

'Okay.' She smiled, pouring him a tall cool whiskey and soda. 'But first, let's go shopping.'

Jimmy drove her in the station wagon to a market and while she did the grocery shopping, he bought the Schenley's, and by the time they were back at the house he was hot and tired and grateful when she fixed him another drink. Then she changed into a pair of shorts despite her stomach, and a flimsy halter, and sat down on the couch next to him. She turned and smiled and he could see she wore nothing beneath the red shorts which were unbuttoned at the top and pulled apart in the seam.

'You know, Jimmy, I didn't mean to get Small in trouble with Greg about the roll of dollar bills. Nor even about him patting my fanny. But he's so drunk all the time and clumsy. He really smacked me on the butt. I didn't say nothing to Greg, but it really hurt. Left a bruise.'

Lot to bruise, ain't there? he thought, getting more excited than he should by this dumpy little pregnant broad with her big white belly and homely kisser. Maybe it was those swollen titties. But now, he could see they'd hang like a cow's bags if she dropped the halter. No, he guessed it was because she was *his* woman. I'd love to fuck his woman and kick those fish eyes shut if he opened his mouth. And I would. I'd fix that fuckin giraffe. If he didn't have that gun.

'You know, Jimmy, you sure got a bold eye.' She smiled, fingering a ferocious pimple on her forehead.

'Sorry if I was starin',' said Jimmy. 'But you know, like, I just got outta the joint, and after five years, you know?'

'I don't mind.' Max laughed. 'In fact when a woman is all swole up and ugly like this she appreciates a man's approving look. Especially a handsome man. You're a handsome man, Jimmy.'

'It's downright painful for me,' Jimmy smiled.

'Oh, you're naughty.' Max laughed and curled her legs under and dipped forward, letting the halter slip.

Jimmy had only intended to play with her a bit, but now things were getting out of hand. He was aroused like he hadn't been since he got out. And the more he thought of Greg, and the fear of Greg which he hated to admit, the more excited he became.

Suddenly Max got up and waddled into the kitchen to make fresh drinks. When she returned she was looking directly at the bulge in his pants and when he reached for the drink she reached past his hand and set the cold glass right on his crotch.

'Wow!' said Jimmy.

'Did I hurt you, Jimmy?' asked Max.

'Naw, but turnabout is fair play,' said Jimmy in a husky voice, running his hand up and down her thigh.

'Oh what the hell,' Max said, as she unzipped the shorts, and Jimmy was panting before she got them off.

It was the best one, all right. Better than cute little Linda. Better a thousand times than that five-dollar whore. Better than anything he'd had since he could remember. And she was a dog.

'The curves kinda turned to lumps,' he said later. 'Soon as I got her naked.' But no matter. The fear of Greg made it good.

'Did I hurt you, Max?' he asked, patting her stomach gingerly.

'No.' She shrugged. 'I've had kids before. I'm used to it.'

Then Greg was at the door. But they were dressed. Jimmy's heart raced because he forgot about the tell-tale smell. He was sniffing the air like a birddog when Greg and Billy entered. Greg was all business. He didn't notice.

'Come on, Jimmy,' Greg said after a quick whiskey and soda. 'Let's drop you off at the hotel. But first, we'll take Billy to his pad.'

'So long, Max,' said Jimmy. 'Thanks for the coffee and hospitality.'

'Anytime, Jimmy,' said Maxine, with her usual friendly smile, and they left.

After dropping Billy, Greg and Jimmy didn't speak much. Then Greg said, 'Jimmy, I gotta dump Billy. He's no good to me the way he is. Maybe I can use him again when he slows down on the drinking but right now I just gotta dump him.'

'I can understand that.' Jimmy nodded, knowing what was coming next.

'I need a driver, Jimmy, and you fit the bill. Now you and me could make plenty, but we gotta handle it right. Billy and me are just living too high. You gotta hold back a nest egg. You can't save anything with old Bill, drinking the way he does.'

'You're right, Greg.'

'You and me can go all the way, Jim. I'd like you to be part of our little family.' Greg face swiveled toward Jimmy on the long pole of a neck and he smiled, the upper teeth protruding and hanging over the lower lip, the cheeks so hollow they drew the chin to a point so that his face was like a spade upside down.

'I'll get hold of you tomorrow, Jim,' Greg said as they pulled up in front of the hotel. 'Hang loose, partner.'

'Yeah, I'll sure be glad to make some bread. I sure need me some bread and all. You know, Greg?'

'See you, Jim.' Greg waved and the wagon whined and slipped from the curb.

'Catch you later, you pencil-neck motherfucker,' Jimmy whispered under his breath, while he smiled and waved.

All that talk. All that big talk and what did he get out of it? Not even a sawbuck to tide him over a day. Nothing. It was only two in the afternoon when he went in the hotel and he knew that the odor of disinfectant from the morning mopping would still be in his room, so he found a comfortable chair in the lobby, one which was not too greasy, had some nap left, and no broken springs. He sat and drank coffee from paper cups and smoked and thought about his future. That is to say, thought of the remainder of this week which was as far in the future as it was reasonable to plan. Life had taught him that. Think about a week ahead. He'd be way past the majority of street hustlers, who can only go a day at a time. Who could plan farther than a week? Not him. Not *anyone* he'd ever known. So he planned for the rest of the week.

The most important thing to consider of course was what Gregory Powell represented to him.

I should cut this bastard loose, right now. Right now, Jimmy thought. And then get down to business. Like, I'm a booster, a boss booster. And look at me. Fuckin around with this jiveass fool and not thinkin about my real business. Look at me! Shoes practically wore through. Dirty pants with the creases gone and flappin around my legs. Losin weight and hungry. Not even a change of clothes. Shoot, I shoulda had fifty suits boosted by now. Why ain't I out on Wilshire, rippin through those stores on the Miracle Mile? Maybe teamed up with another good thief. By this time we coulda had a hundred suits, is all.

And then Jimmy thought of the work that went into shoplifting a hundred suits. How many trips in and out of the store carrying only what he could conceal on his person. And he thought of hustling around with the stolen suits to sell them for less than a third of their retail value, if he were lucky. Sometimes the risk was greater on the sale than it was when he stole them, what with cops ready to jump on a guy carrying new suits. And one of his customers might get busted with a suit and snitch him off. Yes, the risks were great. It made him tired to think of the work and the risk.

Robbery was essentially a more dangerous game, that couldn't be denied. But for whom? For Greg. Jimmy was just a wheel man. Number two as he had always been. There was no real risk. Then Jimmy thought of the liquor store proprietor firing blindly in the night at the station wagon as they crawled from the curb with that miserable slimy clutch that Greg said he had fixed to avoid quick getaways.

But Jimmy's hands became clammy when he thought of that gunfire in the night and the fantasy he had of a .38 slug cracking through the back of his skull and him dying on the steering wheel as that lousy engine raced and that clutch slipped and slid him to oblivion.

But it was stupid to think about that. There was no real risk to the wheel man. If Greg should ever run into any trouble inside – a foolish gun-toting proprietor, or a police stakeout – and if shooting started, then it would be adios, turkey neck, because Jimmy would slide off in that wagon.

Hell, only two or three big scores with Greg and he could make enough for a good transportation car, pay the fee for the union dues, get a good job as a painter. Even some clothes, and some good restaurant meals. Maybe some records, and a record player, and a good radio, a TV, and some *women*. Like, why bother with Linda? There was lots of it around. And a decent pad someplace. Maybe an apartment with some chick who had a hustle of her own, or maybe even a square job of her own. They could split the rent and play house. But now he was dreaming and he knew it. Getting past the one week at a time.

So he stopped and thought of the coming week. Today was March 4. He would work with Greg this week. Maybe until after the weekend. Then he would have enough bread to do it all. He would cut Greg loose after this weekend.

For the rest of the afternoon and evening, Jimmy hung around at a Fifth Street bar, drinking and smoking until midnight, watching the winos mooch drinks. He wasn't interested in Linda or any woman that night. He was still thinking, and when Jimmy returned to his room he cursed the darkness. He was sick to death of dark and dingy rooms. He wanted white blazing lamplight and gold leaf wallpaper like he'd seen in pictures, and a snow-white carpet. But then he remembered a farm worker's camp he'd once lived in near Bakersfield when he was following crops. He remembered greasy clay, being wet to his knees in the mud. The mud was everywhere, cold slime you could not escape, where cockroaches crunched under foot like gravel. This was better. A dark hotel room was infinitely better.

Jimmy was shaving the next morning when the phone rang and made him jump and cut his lip.

'Hello, old partner,' a familiar voice said.

'Come on up, Greg,' said Jimmy. 'I'm shavin.' And in a few minutes Greg was sitting in the only chiar in the room with his feet up on the bed watching Jimmy brush his teeth.

'This is sure a crummy little room, Jim,' said Greg cheerily and Jimmy started getting irked already. 'But you won't have to live in places like this for long. I got good news, Jimmy. I cut Small loose so it's just you and me. And furthermore, I got a job in mind that's gonna be worth twenty-five grand. And we're gonna case it today.'

'Twenty-five *grand*?'

'And it's easy as pie.' Greg laughed. 'Come on, Jimmy, finish up and let's get going.'

In fifteen minutes they were once more in the vicinity of Small's shine stand, and Greg was explaining animatedly as they stood in a parking lot in the hot sunshine.

'There's a colored man here who takes care of this lot, Jimmy. He's crippled and he also takes care of the two across the street. He watches to make sure nobody steals the cars and all that. Now we know where *he* is and he's no problem. So let's go through the alley and see if there's a back entrance to one of the department stores that'll take us through to Broadway if we run in this way.'

Jimmy was still unsure of the nature of the robbery, but he would let Greg act like a general planning a battle if that's what he wanted. They found a public entrance at the rear of a large five-and-dime. Greg walked ahead with his swaggering stride. They passed a counter displaying cheap imitation leather briefcases.

'We need one of these, Jim,' Greg said. 'I want a big briefcase that'll open real easy. This guy carries money loose.'

'Uh, Greg. Who the hell *is* this guy?'

'A collector, Jimmy.' Greg smiled, paying for the briefcase. 'He collects the take from about one hundred parking lots.'

'A hundred parkin lots!' Jimmy whistled, now really impressed. 'I can dig it, Greg. A *hundred* parkin lots!'

Greg chuckled and Jimmy followed him out onto Broadway and around the block to Small's shine stand.

'I'll hit him when he drives into the lot beside Billy's shine stand, Jim. He drives in back of Small's at about eleven-thirty in the morning in a little coupe, dark blue, with a dent in the fender. Now, I want you to park on Broadway. Put it in one of those loading zones. Or double park it if you have to, but here's where you're finally gonna pick me up. Got it?'

'Yeah, but what's the first part?'

'Well, the first part is this: you park in a lot over on Hill where you can watch Billy's. You get your ticket but park the car yourself. Sit in it and watch the shine stand. When the guy pulls in I'll signal to you. See, I'll be sitting like I'm getting a shine. I'll wear my hat, the felt with the little feather. And when he pulls in, I'll take the hat off and run my hand through my hair. Then you pull out down to Seventh, turn and come around to Broadway. In a few minutes I'll be there with the dough. See, all I gotta do is stick a gun in his face, take the money, make him get down on the floor, and come through the alley and in the back door of the store and out the front. Got it?'

'Yeah.'

'Twenty-five grand. At least.'

'Yeah.'

'We're gonna do it now.'

'What?'

'Now. Today. It's almost eleven-thirty.'

'Jesus. I thought we was only gonna case it today.'

'No sense screwing around.'

'Jesus.'

'Just think of it, Jim. Twenty-five grand.'

'Jesus.'

Then they returned to the car, got the gun, put it in the briefcase, and Greg put on the hat and sunglasses and swaggered across the street toward the shine stand, turning once to wave at Jimmy.

It was very hot in the car in the parking lot. It shouldn't be this hot in March, Jimmy thought. And there's lots of traffic. And it's smoggy. And thinking of his half of twenty-five thousand didn't help. Because he couldn't think of twenty-five thousand. It was much too much to think about. It was at last just a number. So he thought about getting a contractor's license and going into business for himself. Painting houses, offices, meeting sweet little office girls while on painting jobs in air-conditioned buildings that had piped-in music. He sat for half an hour watching Greg, but Greg never removed the hat. Jimmy waited, sweated in the heat. At one-thirty Greg swaggered back across the street carrying the briefcase.

'Guess the guy changed his routine,' Greg said, getting in the car. 'Or maybe he came earlier today.'

And with that went Jimmy's momentary dream of a contractor's license and painting in air-conditioned offices with piped-in music. And sweet little secretaries.

'Let's go home and have Max fix us dinner,' Greg said. And Jimmy drove dejectedly to Greg's apartment.

When they got there Jimmy was even more depressed to see that Max didn't respond to his wink when Greg went in the other room. She showed no sign that there was anything between them.

Jimmy's vanity was hurt. He wondered if he had done an adequate job yesterday. Jesus, maybe he couldn't even do *that* very good anymore.

'No luck today?' Max asked sympathetically, kissing Greg on the cheek as she put a Schenley's and Seven-Up in front of him, and another in front of Jimmy.

'Nothing,' said Greg.

'How about a nice dinner?' Maxine asked, stroking Greg's hair and Jimmy winked again, but still she ignored him. Aw fuck it, he thought, and didn't wink again.

'Yeah, we could use something to eat. A nice dinner sounds good.'

Several drinks later the dinner was steaming in front of them. They were TV dinners. Jimmy tore the tinfoil and ate hungrily.

'This tastes real good, Max.' Greg smiled at her and Max beamed happily. 'Jim, why don't you stay the night?'

'Well, I dunno, Greg. My P.O. might call to check up on me.'

'Hell, phone the hotel. See if there were any messages.'

'Okay, I'll do it,' said Jimmy. And he walked down to the corner gas station to find a phone to call his hotel. When he came back the fantasies were already being woven and after a few more drinks, he eagerly took part.

'Listen, Jim,' Greg said as they sat on the couch drinking. 'I got plans for us after we get the big dough. We can't go on robbing forever.'

'You're right, Greg.'

'I used to work as a mechanic in Oceanside, and I figure I could open up a little garage or something, but it's gonna take about ten thousand.'

'Dig, I learned how to paint real good in the joint and I could get a contractor's license and paint houses, and that. And maybe you and me could go into painting, and all.'

'Well, could be, but I don't know anything about painting. Could be we could go in together, five grand from you and five from me. Half a garage and half a painting business.'

'An automobile garage that paints houses? Don't sound like it makes too much sense to me, Greg. But I could be wrong.' Jimmy finished his drink and poured another.

'Well anyway, we'll do something together. We're gonna be great partners.'

'We'll do somethin together, Greg,' Jimmy agreed. And he was right.

The next day was spent shopping with Maxine in Hollywood pawnshops. They were cheerful that day and Max was happy to be buying things: a radio, a record player, a ring for Greg, a watch for herself.

'This setting is loose,' Greg said to the pawnbroker after he'd been wearing the ring for a few minutes.

The pawnbroker examined it with the glass and said, 'No, it's just the way it's made. But you bought a lot of things. I'll give you a bargain on the record player and radio.' Greg seemed satisfied and forgot the ring setting.

'How about a watch, Jim?' asked Greg expansively. 'How about one of those?' He pointed to a tray full of watches with a cardboard sign saying: 'Your pick, $14.95.'

'I don't need a watch,' said Jimmy.

'Sure you do,' said Greg, picking one. 'Take this one. It's a gift. It's a good watch. I want you to have it.'

'Yeah, I think it is,' said Jimmy shyly. 'It's a pretty good watch. A pretty good watch for a cheap watch.'

'We'll take it,' Greg said to the pawnbroker, and Jimmy was truly grateful for the gift. His first gift since the small things he'd gotten as a child. 'I really like the watch, Greg,' he said softly.

The rest of the day was spent at Greg's apartment enjoying the purchases, drinking Schenley's and Seven-Up, eating a TV dinner, and listening to Greg recalling cons he knew from prison whom he thought Jimmy might know or might have heard of. And he *did* remember names. He had one hell of a memory, Jimmy thought.

After talking of boxing experiences in Vacaville and describing how his nose was broken in a grudge match, Greg said, 'Well, it's eight o'clock, Jim. Ready to go to work?'

It was the same outfit for Greg: felt hat, trenchcoat, dark shirt and pants, boots, the mole and the hairline moustache, but he had added one thing – a fast-draw holster. It was an old holster he had cut away below the cylinder section, and it could be strapped on the inside of the belt.

'Watch how this half-breed holster works, Jim,' said Greg, and reached for his waist, fumbling twice before getting the gun out.

'That wasn't too awful fast, Greg,' Jimmy said gravely.

'Well, it hung up on the trenchcoat,' said Greg. 'But the thing I really practice besides drawing and shooting is the look. I do it in front of a mirror.'

He drew again and stared at an imaginary victim, and Jimmy chuckled and smoked and watched as Greg said, 'Give me the goddamn money.' Then the imaginary victim apparently wouldn't comply because Greg's hand whitened at the knuckles and the gun trembled for a few seconds and the look started to come. Flat. Icy. With the glint. Jimmy started to go cold watching it happen. Then Greg laughed, put the gun in his half-breed and it was gone.

Baby, you and me are partin company real soon, Jimmy thought as he snuffed out the cigarette. His hand was shaking but Greg was giving Max a long drawn-out kiss and didn't notice.

They hit the first market at closing time. Jimmy's mind was racing as he waited for Greg. That fuckin trenchcoat. Jesus, the trenchcoat is too much for such a hot night. And there ain't hardly any white guys wear hats in L.A. Jesus. He's got stickup written all over him. If I worked at that market I'd hit the fuckin alarm button the minute he came in the door. Jesus.

Jimmy thought of Greg drawing that gun and cocking the hammer in the same motion. 'That's how it's done, Jim. Let them hear the click when you cock it. And point it right where they live.'

That cocked gun, thought Jimmy. The clock. And the look in his eyes. Jesus.

'Just relax, Jimmy,' Greg said as he got out of the car. 'And when you see me coming, don't panic. I'll be covering ground fast. I got this step I practiced. What you might call a hop, skip and jump that to a normal eye looks like I'm walking fast.'

'Yeah, Greg,' said Jim nervously. 'You thought of everything. Yeah.'

'I'm gonna take both checkout stands, Jim. Two tills full of dough.'

'Yeah.'

After he was gone, Jimmy noticed the old car parked next to the driveway with the Mexican and a flock of kids. What if they were there when he came out? Jesus, this could turn into real trouble. What if there was a shootout?

But in a few minutes he saw Greg moving fast across the parking lot with a skip and a jump every few yards, and to Jimmy he looked like a hungry kangaroo. He was craning his neck for cops as the car moaned and sighed and lurched away sliding.

'Turn right, Jim,' said Greg, jumping in the moving car. 'Turn your goddamn lights on! That's it, take it easy, turn onto the freeway there at the on-ramp. Where the hell're you going, Jim?'

'Sorry,' said Jimmy, clenching his teeth to control the quiver in his chin. Sweating. Cold. Light-headed. Panicked.

'What's the matter with you?'

'Nothin. I got bad eyes. I should wear my glasses.'

'Well turn around and go back.'

'Okay.'

'Get on the freeway and go to the Long Beach sign.'

'Okay.'

They were speeding down the freeway, the clutch finally catching up with the engine. Jimmy was studying the road, blinking into headlights, watching the off-ramp signs. He drove past the off-ramp.

'Jimmy, goddamnit, you did it again!'

'Well don't yell at me, Greg. It's my eyes.'

'Okay, take the next turnoff. We'll do another market. Let's see, we got that Safeway there on Whittier, so next we go to Long Beach and take another, then we return to the vicinity of the first job and take another one. Or we do a liquor store or something.'

'Why back in the vicinity of the first job, Greg? Shouldn't we hunt for virgin territory? I mean ain't that risky?'

'Naw, I got a theory that it's confusing to the cops. It always works for me. So we'll come back.'

'How much did we get?'

'I think we got a couple hundred, Jimmy.'

'Okay, Greg. That's okay,' said Jimmy tensely. Without willing it, he was pressing hard, and the station wagon was racing past eighty, then ninety, and ninety-five.

But Gret didn't see. He was counting the money and chattering excitedly. 'You shoulda seen that broad, Jim. I scared the shit out of her. I mean right *out of her*. I bet she can't talk yet. God, that was funny.' Then he looked up. 'Damn it, Jimmy, we're almost in downtown Long Beach. You're missing the off-ramps again!'

Jimmy cut the wheel to the right, bouncing over the divider curb, skidding sideways, screeching rubber as he stood on the brakes.

'You damned near dumped it, Jim!' Greg shouted as the car straightened up and slowed down on the off-ramp. In a few moments they were in the parking lot of the all-night market. 'Now this time no screwing up when we make our getaway.'

'Okay, Greg,' said Jimmy. 'I'm okay now.'

It was a short wait while Greg shoved the gun in the face of the man at the checkstand, stared at him, and whispered, 'Hurry up, punk,' in a way that almost paralyzed his victim. Then Greg was skipping and jumping across the parking lot.

Jimmy strained his eyes in the darkness for a passing police car, and then he was pulling on the headlights and dropping into low gear and trying to remember to drive off slowly just as Greg was getting in. It went smoothly this time. It was perfect. Greg said, 'I got both tills!'

They were on the street and Jimmy had not panicked.

'Turn right here,' said Greg, 'and just take it easy. I cased this one good. You can make a right-hand turn on the red light. That's it. Now go a block and make another turn.'

It *was* perfect. As Jimmy drove he grew confident.

'Make another right,' Greg said, 'and we'll be heading back toward the freeway.'

'Jumpin fuckin Jesus!' Jimmy yelped. 'We're at a dead end!'

'Now don't panic, for chrissake,' said Greg. 'Just make a U-turn.'

'Greg, the cops might be at the store by now!' said Jimmy.

'Just go back the way we came and don't panic,' said Greg, but Jimmy could hear the tremor in Greg's voice which was an octave higher now. Jimmy felt the panic grow as he instinctively stood on the accelerator and made the engine race too fast.

'Stop the car,' said Greg.

'What for?'

'Stop the car, I lost my hat!'

'Fuck the hat!'

'Jimmy, there's writing in the hat. They can identify me! Stop the car!'

So Jimmy pulled over in the light from a plate glass window and discovered to his horror that he was directly in front of the market they had just robbed!

'Greg, we gotta go!'

'Jim, did I have the hat on when I went in the store?'

'Greg, we gotta go!'

But Greg ignored him and was looking under the seat, beside the seat, over into the back seat saying, 'If we don't find it I'm gonna have to go back in that store.'

'Jumpin fuckin Jesus!' said Jimmy Smith.

And then, 'I got it, Jim! It was wedged down between the seats. I got it!'

Before the second 'I got it,' Jimmy was pressing on the accelerator and the station wagon was moaning down the street into the traffic, gradually picking up speed, heading toward the freeway.

It had been Gregory Powell's nineteenth robbery, the fourth he had committed with Jimmy Smith.

The last either of them would ever commit.

'I think we got in the neighborhood of a grand, Jim,' Greg said as they turned off the Harbor Freeway in the direction of the apartment. 'I think we done enough for tonight.'

Jimmy felt himself go limp when Greg said it. He knew he couldn't stand another job tonight. If ever. At last they were parked in front of the apartment. They were safe.

'Listen, Jim,' Greg said as they got out of the car. 'Tell you what. Let's play a trick on Max. We'll leave the money on the floor in the back seat and tell her we didn't score. And she'll be happy because she doesn't really want me to pull jobs, you know. And then you come out and get the dough and bring it in.'

'Okay,' Jimmy said listlessly. Drained.

'Watch how happy Max is when she hears we didn't do any robbing tonight.'

'I'll watch,' said Jimmy, and then they told her. But he missed the elation that Greg said would be there. In fact, he thought he saw her sigh disgustedly.

'Oh Jim, go out and bring in my trenchcoat,' Greg said. Jimmy nodded and shuffled out to the car to retrieve the paper bag. He decided to look under the seat, and sure enough, there was a five-collar bill there. Jimmy picked it up and put it on his shoe, then decided Greg had put it there purposely to test him, so he took it out and threw it in the bag with the rest. But now Greg knew how much there was and Jimmy did not. He's gettin ready to screw me again, thought Jimmy. Yeah, I just oughtta hold back twenty bucks right now. There's probably more than eight hundred here.

I'd like to fix that bastard. I'd like to tear that head off and turn it around and put it back on upside down on that skinny handle of a neck. Yeah. And then I'd like to dig up the fuckin concrete with that fuckin head! Yeah. That's what Jimmy thought. But he said nothing. And he did not hold back the twenty.

Greg grabbed the bag and pushed Max into the bedroom and dumped the bag onto the bed in a shower of bills. Then Jimmy saw on Maxine the expression Greg had predicted earlier. It was joy. Uncomplicated. Childlike. She jumped on the bed and played with the money, stacking it and restacking it. Greg laughed and swaggered into the kitchen for a Schenley's and Seven-Up.

'Let's let my little cashier count the money, Jim,' Greg said, undressing and walking into the kitchen in his undershorts. 'Let's have us a drink.'

They left Max counting aloud, eyes shining as she sat on the bed, fondling the bills.

When they came back in she had it arranged in stacks by denomination. 'A twenty-dollar bill is mine, Jim. I used it to fake paying for the Schenley's when I went through the checkstand. I take that and we can split the rest.'

'Sure,' Jimmy said.

Max counted a thousand and forty dollars after Greg took his twenty.

'What say we give our little banker the odd forty bucks, Jim?'

'Sure,' Jimmy said. 'Like, it'll pay for the good meals Max gave me. And the good drinks.' And the good nooky, Jimmy thought, and giggled, and winked at Max. But there still was no sign of recognition. Aw fuck it, he thought.

'Max, count out a hundred for Jimmy and me. Us two guys're going out tonight. How about a little bowling, Jim? A little fun to relax us. Or maybe you want to spend some money on a girl. That's okay with me, but I'll just have to make it back home if you do.'

'Yeah, a hundred in my pocket would be a groove,' said Jimmy as Max counted it out.

'If anything unlucky should happen and we ever get busted for something, why, Max'll bail us out and have us on the street in no time. Max is like money in the bank, ain't you, my little banker?' With that Greg smiled and kissed her on a somewhat bulbous nose.

'Let's take the ones and fives, Greg,' said Jimmy eagerly, thinking it would make for a fatter-looking flash roll. Jimmy was carefully rolling the large wad of ones and fives when Greg dressed to go out. Oh shit, Jimmy thought when he saw him in his tight pants and a string tie. Jesus, I gotta teach him how to dress, but I got to be cool about it so he don't get tight-jawed.

'You and me're just almost exactly the same height and build, Jimmy. Wear anything of mine you want,' Greg said, waving toward his closet.

'Thanks, Greg,' Jimmy said, selecting a sport shirt and a jacket.

'How about some pants, Jim?'

'No thanks,' said Jimmy. 'You wear them a little tight and I'm bow legged, and all.'

Jimmy managed to squeeze into a pair of Greg's shoes, and tight as they were, they were infinitely better than the hot dogs, which he vowed to get rid of tomorrow. Then they were rattling along the Harbor Freeway heading for Jimmy's hotel. Greg once again impressed Jimmy with some fancy driving and gear shifting, bad clutch and all.

At the hotel Jimmy found a card in his box from his P.O. saying he'd been by and would see Jimmy tomorrow, March 7, at the parole center. Jimmy was nervous when he went back outside to the waiting station wagon.

As they drove west on Eleventh Street Greg said, 'It's getting late but I gotta stop and see a friend. You'll notice, Jimmy, that I'm loyal to all my friends. That's a quality I admire, loyalty.'

'Righteous. Me too.'

'And a fink is someone I loathe, you know? I'd kill me any one of my friends who ever ratted on me.'

'Don't blame you, Greg. I damn near beat a dude to death one time who rolled over on me,' Jimmy lied. 'I found out he snitched me off and I broke his arm and two ribs.' Then Jimmy thought that didn't sound too out of line and added, 'I broke his jaw too.' He was going to add a leg, but thought it would be too much and Greg wouldn't believe him.

'I imagine you can handle your dukes, Jim.'

'I boxed a little in the joint, and of course I had a million street fights, and that.'

'I boxed in Vacaville,' said Greg. 'Pretty good welterweight if I do say so.'

'Yeah, you said.'

'Anyway, this friend I'm taking you to see is the first one I lived with when I came to California. She helped me out when I was broke and was good to me. Now I wanna see if she needs anything.'

'Okay,' said Jimmy always anxious to meet a strange broad, but then thinking any one of Greg's broads would no doubt be a strange one. And she was.

'Oh, looky here,' said the drag queen when the door opened. He was a tall black queen in a tight blouse and women's slacks, eyebrows arched, hair upswept. He had his hands on his hips, grinning wide, lips red and wet, shaking his head slowly, saying, 'Looky looky here.' He threw his arms around Greg, hugging him close. 'Good to see you, baby, good to see you.'

Jimmy Smith was to tell of it in detail at a later time: 'As soon as we got in the pad it was plain he was havin some fun. On the couch was some young white guy about twenty years old. He was in his shorts and nothin else. The freak had the little apartment all gussied up with colored lights and had crammed couches and easy chairs to match all into the one room. The place was fixed up for easy livin or layin around doin the things homos did. After goin over and kissin the pigeon on the couch, the freak went on into the kitchen and fixed up drinks. I turned down the drink sayin I just got outta the joint and wasn't able to handle too many drinks. I just wasn't all that hungry for a drink from the freak's glasses, not that it was dirty, and that. I just got a complex against those kinds of things. In the joint I seen some of the dirty rotten things freaks do on the sex side, so when I'm around them the pictures come to mind and I sorta choke up.

'It dawned on me that Greg seemed to have only black people for his special friends and that he dug on freaks. Later I connected the two together. It turned out Greg was a half-ass homo his own self, one that dug on both men and women. When he was a kid in some juvenile place some black guy musta been his jocker. One thing sure, when we ran together he never let me know he was a homo. In fact, he impressed me as bein a fairly tough guy, just a little off his nut, is all.

'We only stayed at the sissy's pad for about fifteen minutes and cut out. I think he was glad to see us leave so he could get back to work on the young pigeon. Before we left, Greg told the freak if he needed any bread, just to say the word, the sky's the limit for a friend. The freak turned him down, sayin everything was all right. Like, I'm sure he didn't want his victim to know if he was broke. Most of the freaks that hustle young guys know that without money they're outta luck.

'We left the sissy and went out on Normandy and Adams to the hot dog stand where all the brothers hang out at night. A few nice-lookin broads was hangin around lookin to turn some tricks, but they give me the ol freeze out. They could see Greg hangin in the background and thought maybe he was the heat, bein a paddy and dressed in that square style he wore. One of the whores was a little bolder but when she heard Greg's square little remarks she froze up too. She woulda gave up some action if Greg talked like a hip white guy.'

Hollywood was the next stop that night. Greg drove Jimmy up and down the boulevard and finally stopped at a strip joint on Santa Monica Boulevard where the cocktail waitresses wore tights and cat ears and long cat tails which drunken customers could pull when they wanted a drink.

'This is *all* right,' said Jimmy as Greg got them a good table, ordered two whiskeys and soda, and giggled as the kitten took off with tail flying.

'Looky that horny ol buzzard,' said Jimmy, pointing to a florid bald man who had one cocktail waitress by the cat tail, and was stroking it, and wouldn't release her.

'Let go, honey, and I'll come back to you,' she cooed, patting him on the head.

This was to be the biggest night in the life of Jimmy Smith. The future looked bright indeed, and he was ready for some real action, but the nightclub didn't look like what he had in mind. As he was to describe it:

'I kept my eye on the other tables and saw that most everybody was elderly or middle aged guys. They all at one time or other had their hands on the waitresses bottoms or was pullin at the long skinny cat tails. Most of the girls were good at avoidin the sudden hand holds, but some got fondled regular.

'After the stage show started, it was somethin to see. Each broad got announced like some famous movie star. The names was all male and the broads kinda matched the certain star in some offhand way. Like, Miss Sammy Davis Jr was black and a little short slender thing. But it wasn't really much of a comparison to the star because the little broad had teats like bowlin balls. A couple of the dancers did a lightweight all-the-way-off thing, which ended with the G-string gone, but still some very small thing was blockin the view. One broad had a special act with the teat thing. Like, she would rotate one tassel one way and the other teat and tassel would rotate in the opposite direction. It reminded me of the old thumb twiddlin game which I could never do.'

As they sat and drank, and Jimmy gazed with a permanent leer at the stage, Greg talked. Jimmy nodded politely every few seconds, looking blankly at Greg, hearing only snatches of his words which Greg spat like bullets. 'Yeah, Jim, I wish I was big like my kid brother. He's about six one, and two hundred pounds.'

'Uh huh,' said Jimmy, watching the little pussycat serving the table off to the left. 'She's rubbin it right in that old fart's face,' said Jimmy. 'Right in his goddamn face!'

'My dad's a big guy, Jim. Not too tall, but big, you know? I wish I was big-boned like him and my brother. But I was a hell of a football player in junior high anyway. Probably coulda made a hell of a running back someday. You ever play high school ball?'

'Uh huh.'

'You did?'

'I mean, uh uh.'

'Well, I sure did. And Christ, you should hear me with the tenor sax. I can play just about any instrument. Tell me the scale and I'll figure it out. I used to sing quite a bit. I ever till you that? Harmonize too. I got perfect pitch. When we get a real stake I wouldn't mind living in a small town again. I mean growing up in a town like I did, a town of ten thousand, that's the way to grow up. You always live in a big city, Jim?'

'Uh huh.'

'In a small town you get to do all kinds of things. I used to paint, ski, skate. Moreover, I had a band. You name it, I did it. Don't you have any hobbies? Anything you like to do?'

'Pussy.'

'I mean besides that. Hell, I could tell you about bed artistry. I've had some very strange sexual experiences in my time, Jim. Maybe someday when it's the right time I'll tell you about them.'

'Uh huh.'

'You boxed. You said you did some boxing.'

'Uh, huh. In the joint.'

'I was a hell of a welterweight at Vacaville. A hell of a fighter. Furthermore, I think I could've made it on the outside as a boxer, but I had this brain operation. Can't afford to get hit in the head.'

And Jimmy Smith started getting a headache, and stopped turning toward Greg, and gave up nodding politely, and Jesus, he thought, how much longer can I take this fuckin paddy cornpone with his fuckin family, and his fuckin moreovers, furthermores, alsos, wherefores. And when Greg got up to go the men's room, Jimmy tossed down the drink and scowled at the empty seat and whispered: 'Don't gimme any more moreovers or furtherfores or whereovers or any of that other HONKY BULLSHIT!' But when Greg returned Jimmy nodded politely as Greg talked of Cadillac, Michigan, and the girls he had conquered with the bedroom virtuosity he learned in strange mysterious places.

'Want another drink, honey?' asked the cocktail waitress who was coming by the tables every four or five minutes.

'Naw,' said Jimmy. 'I want some real action, and know what? Nothin on that stage does half so much for me as you, kitten. Like, what time you get off work?'

'Sorry, baby.' She smiled brightly. 'It's against house rules to date customers. Besides, I got a boyfriend.'

'Maybe this'll change your mind,' said Greg, throwing his roll of bills on the table.

Without dropping her smile the waitress said, 'See you fellows again, I hope.' She swished her tail at them, and wriggled away.

'Fuck it,' said Jimmy.

They went back to the car, but Jimmy was not through yet. 'Greg, Small told me about this pad on the west side. Two extra fine chicks work outta this here pad. Cost us maybe twenty, twenty-five bucks apiece, but Small says they're worth it.'

'Let's go.' Greg shrugged, making a U-turn on Santa Monica, slipping the car up on two wheels and burning rubber all over the asphalt.

Fifteen minutes later they were standing on the porch of a darkened apartment, striking matches and peering through dusty windows.

'There ain't a stick of furniture in that place, Jimmy.'

'No, there ain't.'

'There's nobody living here, for chrissake,' said Greg.

'Guess it's one of them floating whorehouses,' said Jimmy.

'Let's go home,' said Greg.

And so ended the biggest night of Jimmy Smith's life, when the future was glowing and he was full of good whiskey. When, for the first time in his life, he felt like *new money*. He spent the rest of the night on Greg's living room floor and woke up with a hangover.

It was the moaning which woke him. 'What the hell,' Jimmy growled, feeling the pain in the front of his skull as he sat up. Then he heard a muffled scream and a laugh and more moans.

Greg hadn't bothered to close the bedroom door and now Jimmy was awake, had to urinate, and was getting angry waiting for the bedsprings to stop squeaking. Finally, it was quiet for a moment and he cleared his throat, pulled his pants on, made sure they could hear him moving around, and walked through the bedroom on his way to the toilet.

Jimmy tried not to look as he passed through, but Jesus, they were lying there naked, and what could he do but glance down. Max paid him no attention, and just lay there. It was Greg who looked at Jimmy in an odd way, half smiling. It was strange the way he looked at Jimmy so intensely. Jimmy wondered about it at the time.

I didn't hear that bitch go in the bathroom, Jimmy thought angrily as Max fixed the ham and eggs. She better had washed her slimy mitts in the sink or I ain't eatin her dirty fuckin eggs. He thought about going to the sink to see if the soap was wet, but thought, what the hell, as the aroma of fried ham struck him. He downed four eggs and two large slices of ham.

Afterwards, Jimmy sat and smoked contentedly and thought about not waiting until the weekend to cut Greg loose. After all, he had over five hundred bucks coming right now. That was more than enough for a stake and a transportation car. He could join the union and be on a painting job by Monday morning. On the other hand if he hung around with Greg until the weekend as he planned, he might be in the morgue Monday morning. Or in jail.

'Jim,' said Greg, walking into the room, drying his body with a large bath towel. 'I got an idea. After we take you to the parole office for your naline test today, let's cut out. Look, you don't have to test again until next Thursday, right? Well, that's a whole week. Let's the three of us cut out to Las Vegas and Risco. I'm sick of these two-bit jobs. I heard that in Frisco you just walk in a bank with a note in your hand and walk out with the dough. They say a Frisco bank is nothing to knock over.'

'I dunno, Greg.'

'Listen, first we stop by Vegas and do some business. There's a car there I wanna buy. I saw it last month when I was up to see my sister. Actually it's in Boulder City, and it's a beauty. A '46 Ford coupe. It's the kind of car I always wanted to play around with. Whadda you say, partner?'

'I dunno, Greg. A trip outta state? I dunno. That's a bad violation, you know? I could go back to the joint just for that. Like, even if we got stopped for a traffic ticket, they might find out I'm a parolee and call my P.O.'

'They don't stop people for tickets in Nevada,' said Greg. 'They need the sucker trade and don't wanna piss off the tourists. We won't have any trouble.'

'You really think we could make a big sting in Frisco?'

'Jimmy, I'm telling you, it'll be easy as pie. And listen to what else I got lined up. Something I didn't tell you about. After we come back to L.A. I got a new bank at the Farmer's Market lined up. We'll need a shotgun and one more guy to go in with me. By then Billy'll be off the sauce and we can use him.'

'Well maybe,' said Jimmy, impressed that Greg had been planning so far ahead. Maybe this time he really knows what he's talkin about. Maybe he ain't so dumb after all, thought Jimmy. Imagine what he could do if he could get hold of five, ten grand. Maybe he really *does* know where the bear shit in the buckwheat.

'Let's do 'er,' said Jimmy, and he was stunned by the fire that flashed in Greg's blue eyes, at the wide triumphant smile, at the excited chuckle that burst forth. Jimmy watched the jerking Adam's apple and couldn't understand the jubilation, but then, Jimmy did not understand that Gregory Powell at last was the unchallenged head of his own little family.

So they prepared for their trip by going downtown to buy four recapped tires for the station waggon and clothes for Jimmy.

'No sense cheatin yourself with cheap clothes,' said Jimmy, buying a pair of tan, plain-toed, thirty-five-dollar shoes from the store at Fifth and Broadway, and a pair of thirty-dollar slacks, and a good sport shirt and socks, and underwear, and handkerchiefs.

'Jim, let's buy a leather jacket,' said Greg. 'They're just the thing for cool nights in Frisco.'

'Well, I *do* need a jacket, but really I ain't much for leather jackets, Greg.'

'Come on, Jim, let's both get one. Like partners.'

'Well, okay,' Jimmy acquiesced as Greg picked a black one for himself and a brown one for Jimmy. Then Greg decided that he should have a matching black leather snap-brim cap to coordinate with Jimmy's brown cap.

'Greg, I'm gonna run over to the YMCA and take a steam bath now because of the naline test this afternoon. Like, naline ain't supposed to show liquor in the system but I ain't taken no chance that my P.O. knows I been drinkin. I'm gonna sweat out all the booze.'

'Okay, I'll go with you.'

'Jesus, skinny as you are why'd you want a steam bath unless you *had* to have one? You'll disappear, you lose five pounds.'

'Might relax me a little,' said Greg.

'And when we was sittin there naked in the steam room,' Jimmy said later, 'I sorta felt that way again, like, uncomfortable. I thought he looked at me the way he did when he was in bed with Max. I thought he did, but I just forgot about it and soon we was back in the wagon with good tires and lots of clothes for me and headin for the Nalin Center.'

Max went with them that evening when they drove to the parole office. They parked in front, and Greg, as usual, decided what Jimmy should do.

'Give me the watch, Jimmy. Your P.O. might notice it. And I don't like it, you wearing those new shoes, but it can't be helped. He probably won't notice slacks and shirt. Don't carry a lot of bread in there in case he frisks you.'

'Shit, I ain't got none. I blew it all on these threads today.'

'Okay, I'll give you some spending money when you get back. Max and me'll go out to dinner while you're in there. Pick you up in an hour?'

''Bout two hours, maybe more.'

The line for testing was long, and it was more than an hour before Jimmy had his turn. While waiting on the benches in the hall he saw a bony woman fall off the bench, flattening her nose on the floor, crying out in surprise when she hit.

'Tryin to beat the naline,' said a tall black man sitting next to him.

'By gettin shit-faced?'

'Look, brother, they ain't supposed to violate you for bein drunk. And the naline ain't supposed to work if you *are* drunk, so I guess she figures she rather get a few days in a drunk tank than get violated for dope.'

'Her arms don't look too bad,' Jimmy observed as two parole officers dragged her up and propped her on the bench, blood smeared all over her face.

'Prob'ly ain't got a vein in her legs that ain't burned out,' said the tall black man.

After testing, Jimmy went upstairs for the 'grouping' and sat there smoking, talking freely to others who, like himself, seemed to be enjoying the effects of the naline.

'How about someone starting out,' said the discussion leader. 'Let's be honest and start out by telling the last time we fixed.'

With very little prodding Jimmy began the conversation because he was feeling like a fat mouth. And he told a dozen lies about nothing important and in general had an enjoyable time of it. At the end of the meeting, Jimmy was the first to leave, hoping to catch a little Mexican junkie he saw in one of the other groups. He wanted to offer her a ride home. He was rounding a corner in a hurry and crashed into Gregory Powell.

'What the fuck you doin here?' asked Jimmy, and then he saw the look. 'You carring a *piece*?' Jimmy whispered, not feeling it, but looking at Greg's eyes, or rather his expression, seeing the black leather jacket zipped up.

'I thought they had you, Jim,' said Greg. 'I thought maybe you been shooting dope without me knowing and maybe you tested dirty and maybe they had you. I was coming in to get you out.'

'Man, are you cr . . . Man, that wouldn't be too cool, would it? I mean takin off a fuckin parole office?'

'They don't have guns.' Greg grinned, the look vanishing quickly. 'Just handcuffs. I could've done it.'

'You're a real pal, Greg,' said Jimmy and smiled weakly.

They spent the remainder of the night loading the station wagon, all their possessions. It was a big load, making the old wagon creak down onto the frame. The back end was filled up to the window with clothes, food, small appliances, everything they owned. They drove through the night when it was cold on the desert, through Barstow and across the Mojave Desert to Boulder City. It was like most small desert towns, with burned lawns and sparsely planted trees swept clean of their leaves by the wind. At eight the next morning they were knocking at the door of a friend.

A tall young man about twenty years old opened the door sleepily, smiled warmly, and shook hands with Greg. He was a friend Greg had met while spending time in the small desert town when he and his mother were taking care of Greg's sister Lei Lani. Jimmy could see he was a complete square and knew nothing of Greg's business, so he kept a straight face when Greg told him he had been making some money boxing and introduced Jimmy as his sparring partner.

He had a pretty wife and a pretty new baby girl, both blond and plump. The house was tiny and almost without furniture and Jimmy liked them at once.

They were given breakfast and then the young man went to work at the naval base where they were doing experiments at the lake. Max stayed at the house with the sailor's young wife while Greg and Jimmy went to get the hot rod.

'I like her,' Greg said as he walked around the 1946 maroon Ford coupe. 'I'll give you a hundred and a quarter for this little hot rod.'

And they bought themselves a car. Then they pushed it to get it started, and drove to a gas station to get the wheels aligned and the battery charged. They decided to try her out and take the short run to Las Vegas, and one of the recently aligned wheels fell off at sixty miles an hour. They barely escaped with their lives.

'Jesus, Greg,' Jimmy said after they got the wheel back on and the lug nuts tightened. 'I only been knowin you a week and like, we seem to be havin too many close calls. I wonder?'

'You wonder what?'

'I jist wonder.'

'Forget it, we're just getting this family going. We're just getting our equipment lined up. Everything's gonna start going smooth, Jim.'

'I guess so, Greg.'

Why didn't the lug nuts come loose later, Jimmy was to wonder a thousand times. Why couldn't they have come loose on the Ridge Route *that* night, that fuckin Saturday night. But this was only Friday. Saturday was another day away.

After stopping every few minutes to check the lug nuts they managed to get the hot rod to Las Vegas. And after visiting the famous Golden Nugget casino Jimmy found himself trailing behind Greg into a Las Vegas pawnshop.

'We're gonna get you a piece, Jimmy,' Greg said, and surprisingly, even to himself, Jimmy blurted, 'Right, right, righteous. I want one.'

Later he thought about why he, who hated and feared guns, who had never fired a handgun, should be so anxious to get one. In a moment of self-analysis he understood. I'll be able to stand up to him then. That piece is gonna equalize this fuckin partnership, maybe even tip it a little my way.

'All you need in this town is a driver's license to get a piece,' said Greg and he picked out a .32 Spanish Star Echeverria automatic.

'Sixty-five dollars,' said the pawnbroker.

'Like it, Jimmy?'

'It's okay, Greg.'

So Greg bought the gun and a velvet-lined case and an extra clip.

After leaving the pawnshop they stopped at a hardware store for some .32 ammunition and .38's for the Colt, and Greg took the .32 out of the case and showed it to the proprietor.

'I just bought this. Whadda you think of it?'

'Nice gun.' The man shrugged. 'It's in pretty good shape. What did you give for it?'

'Sixty-five.'

'Too much. I could sell you a brand new four-inch .38 here for that price. Brand new. And a better gun.'

'Yeah?'

'I like the gun okay, Greg,' said Jimmy. 'A gun's a gun. What the hell.'

'No it's not, Jim,' said Greg. Then to the man, 'Tell you what, pick me out one and get her ready. I'm taking this one back.'

Now they were hotfooting it down the pavement squinting into the white Las Vegas sunshine, dodging the foot traffic, and Jimmy was saying, 'A gun's a gun, Greg.' But to no avail.

'Let me do the talking, Jim,' said Greg. 'I'll get our money back.'

'Greg, there ain't a pawnbroker born with a heart or a soul either one. And there ain't never been one who gave a guy back his bread once he gets his mitts on it. It ain't gonna work, Greg.'

'Watch me.'

And Jimmy watched, but it didn't work. Ten minutes later Greg's face was white with anger, and he was shouting, 'That fucking gun you sold me is no good. It's broken I tell you and here it is and I want my money back.'

'If you insist, I'll take the gun back,' the pawnbroker finally said, never losing his composure. 'But of course at a reduced price.'

'That fucking gun is broke,' Greg said hoarsely, and Jimmy watched the look come into Greg's face. Jimmy was sure Greg now believed the gun *was* broken, the way he said it. 'You kikes're all alike. I'll call a cop to take care of you.'

'Leave this store,' said the pawnbroker.

Jimmy was thankful, oh so thankful, that Greg was not carrying his own piece, but still they weren't safe, because what if Greg unzipped that case and loaded that automatic?

Once outside Greg said, 'Let's get a cop, Jimmy.'

'Jesus, Greg,' Jimmy pleaded. 'Dig, you can't mess with that Jew. Like, all pawnbrokers're snitches and he probably works with the cops. He's probably a good informant for them, and that. Jesus, Greg, it's *us* who's gonna get fucked over. The cop'll ask for our I.D., and what if they roust us into the station and find out I'm on parole from California?'

'I'm gonna call a cop.'

'Greg, I'm tellin you I been a thief all my life. I oughtta know. I mean I musta did business with a thousand pawnbrokers. Like, they deal in hot merchandise all the time and to protect themselves they snitch somebody off once in a while. And they take care of cops. Shoot, cops always buy stuff in pawnshops or get stuff free, and that. They're probably like fuckin brothers, the cops and that fuckin Jew. And somethin else, you never asked me, but I *like* the gun. You know? I really *like* the little automatic. It fits right in my hand so nice and all. Let me keep it. Please, Greg.'

Then the look started to pass and the color returned to Greg's face. 'Okay, Jim, but only for your sake. If it weren't for you I'd make that bastard give back the money. You can be sure I would.'

Driving back to Boulder City Jimmy once again made a decision about Greg. That does it, he thought. Fuck it. I gotta cut this maniac loose. I gotta shine him on. That's it. I ain't goin to Frisco with him. I ain't goin nowhere with him. That is fuckin *it*!

Just outside Boulder City Greg pulled the car into a remote junk yard in Henderson, Nevada. It was near a railroad track in the wasteland, and the sun was high and hot as they crunched across the sand and sagebrush for their first target practice together. Greg stood six to eight feet from a wrecked car and snapped off nine shots into the door of the car. He fired so suddenly it startled Jimmy, but the reports were not loud out there in the vast open desert.

'Your turn, Jim.'

'Why're we so close, Greg? I mean won't a shot bounce back and hit us maybe?'

'We're at combat distance, Jim. This is the kind of thing we talked about in the yard when I was in the joint. This is the combat distance cops learn to fire at. This is how close you'll be if you ever have to hit a guy. And see how the pattern I made would fit in the body of a man? You try it, Jim.'

Jimmy aimed and jerked the trigger and nothing happened.

Greg smiled and said, 'First you gotta jack one in the chamber, Jim. You gotta pull it back like this.'

Greg drew back the slide and released it. The oily metallic snap made Jimmy wince and now his palms were sweating. Then Jimmy aimed again and jerked the trigger, missing the entire car with half the shots, thinking only of the bullets bouncing back and striking him, thinking that it shouldn't jerk in his hand like this, and that maybe there was something wrong and it would blow up in his face.

'That's not too bad, Jim,' said Greg, and Jimmy was glad to give him the gun. Now it was over with.

'I coulda did better left-handed, Greg.'

'Yeah, but this automatic is for a right-handed person. Kicks the shells out to the right. You fire southpaw and you might get the ejected shells in your face. So fire it right-handed.'

'Okay.'

Then Greg shot at some cars and tin cans, once scraping up

sparks on a rock, and Jimmy instinctively stood behind him, fearing a ricochet.

Then Greg insisted that Jimmy fire another clip. He tried it with his left hand, but did no better. So they returned to the young couple's home in Boulder City and decided to have a night out in Las Vegas. Maybe this night would be the big one that Wednesday night should have been. Maybe Greg would break loose with a hundred bucks or so and Jimmy could get himself one of those fine little casino whores he'd heard so much about. Yeah, tonight he was gonna swing out and some chick was gonna give up some action!

So after changing clothes, washing up, and arranging for a baby-sitter, the five of them drove in the young man's car to Las Vegas. With Greg's promise to treat them to the best dinner they ever had they eagerly entered a large and elegant Las Vegas restaurant which advertised Italian cuisine.

Max and the young blond girl were as dressed up as they could be considering their meager wardrobes, and Greg and the young sailor both wore suits. Jimmy had on his only decent clothes, those he had bought yesterday. After ordering drinks and nibbling on the appetizers, Greg, playing the role of host, said, 'Damn, the service's lousy for such a fancy joint.' He signaled to a waiter.

'I don't wait on this section, sir.' The waiter smiled unctuously. 'But I'll tell the headwaiter.'

They chatted and waited a few more minutes and Jimmy could see Greg squirming impatiently. At last a milddle-aged hostess in an expensive off-the-shoulder gown drifted up to the table and with a friendly smile said, 'I'm sorry, sir, but we don't serve colored people in here.'

Jimmy nodded because he guessed it. If only I had a suit and tie, he thought. Lots of times I can pass. If only I had a suit and tie.

'I can't believe it!' said Greg, and Jimmy could almost *hear* the muscles tighten in his hollow cheeks, and watched his head swivel on the long neck, and saw the nostrils whiten. 'His money is as good as anyone's.'

'I'm sorry, sir,' said the hostess. 'I don't make the rules here. I'm sorry.'

'*I'm* sorry I picked this lousy joint, that's what *I'm* sorry about,' said Greg, bolting to his feet and storming out through the bar with the others following, quiet and embarrassed.

As they got to the car, Greg pulled Jimmy aside and whispered, 'We're gonna go home and get our pieces and come back here and knock this dive right over. Whadda you think of that?'

'No, Greg, let's just forget it. It don't mean nothin.'

'The hell it don't. They ain't gonna get away with that insult.'

'Listen, Greg,' Jimmy said. 'You can see all those wops in there are gangsters. I mean, like, everybody knows this fuckin town is full of Mafia. Shoot, we'd get burned down the minute we showed iron. Even if they thought we was packin heat, they'd just kill us and bury us out back in the desert.'

'Maybe you're right,' said Greg reluctantly, 'but I sure wish I could get even with those bastards.'

And at that moment, for the first time, perhaps for the *only* time, Jimmy Smith felt affection for Gregory Powell, a feeling he had not had very often in his life. As he put it: 'I felt kinda close to the nut.'

But the night was irretrievably lost. Jimmy told them he wasn't feeling too well and couldn't eat, and that was the truth, and that he would just as soon go to a movie alone if they would drop him off, and that was a lie. Finally, after many pleas and refusals he sat in the car while they ate dinner at another restaurant. After dinner, none of them would go on without him so they returned to the little house tired and depressed. At nine o'clock they were all asleep. So ended what was to have been another big night in the life of Jimmy Smith.

After the driving, and the desert, and worries, and frustrations, the sleep was incredibly short. In fact, it felt as though he had just closed his eyes when Max was shaking him awake. It was 2:00 A.M. and the house was dark and quiet.

'Greg wants to go now,' Max whispered. 'Get up, Jimmy.'

'Jesus. Okay,' said Jimmy, shaking himself awake. Max stood there to make sure he didn't lie back down, then she bent over him again. 'I said get up, Jimmy.'

He then truly opened his eyes and was staring right down the front of her nightgown which was hanging open. He reached over and squeezed one of her breasts thinking how much milk must be gurgling around in there.

Max whispered, 'Watch it, fellow. Don't start something you can't finish,' and ran from the room as he dressed.

Two o' fuckin clock in the morning, thought Jimmy, shaking

his head. Leave when it's cool he said. Drive to Frisco when it's cool. Two o'clock. He had his own way, this batty giraffe.

Now it vaguely occurred to Jimmy that they were actually doing it. Actually on their way to Frisco. And he hadn't shined them on. He was still with them.

They gathered together their things and crept out of the house before three o'clock leaving a note and fifty dollars on the table for the young couple who had been kind to them. It was Greg's idea and Jimmy approved. What the hell, he thought.

Then they were out on the road, with Jimmy driving the coupe and Greg and Max in the station wagon. They stopped at a diner for breakfast and Greg could contain himself no longer, he wanted to try the hot rod. So they traded cars, Jimmy with Max in the wagon, and Greg burning rubber off the tires and racing ahead.

When the little coupe was so far ahead he could no longer see it, Jimmy reached over to Max's leg and slid his hand all the way up.

'Kinda warm,' said Jimmy.

'It's gonna be if Greg catches you.' Max smiled, and Jimmy suddenly thought, I bet she even laid that drunken fool, Billy Small. I'll bet she did and then put that story on him that he patted her ass and stole the money and all. I bet she's a little bit of an evil bitch herself, he thought, and looked over, his hand still where it was. She was looking at the road and smiling.

He *wasn't* looking. The wagon veered off the pavement into the sand and slid sideways. Jimmy cut the wheel hard to the right and jammed on the brakes.

'Jesus,' he breathed afterwards.

'You keep your hands on the wheel,' said Max, and from then on he did.

They had been told that the short way to San Francisco would not be the best with the loaded wagon, that the pavement was badly in need of repair, so they had decided to go back to San Bernardino and take the Bakersfield route. At 9:00 A.M. that Saturday morning, March 9th, they were in Bakersfield, in the southern part of town at a place called the Hi-way Motel. To get there they drove through a Bakersfield tulie fog, driving blind for several miles before emerging in a forest of billboards and neon, trailer parks, drive-in restaurants, and movies.

This section of Bakersfield was one of the more unpleasant

parts of a town not known for its beauty. The city is in the southern end of the San Joaquin Valley, on dry and dusty flatland, with mountains too far in the distance. Even the southern valley is rich and fertile when water is piped to it, but when it isn't, the dust blows in powdery choking clouds and the sand whips across the highways and piles up in drifts, especially in this part of town which was left to the junkyards, and motorcycle racers, motels, and an occasional tent revival. The music is country, the drink is beer, the billboards are everywhere. This little motel was one of the cheapest on the highway, two rows of wood frame cottages with a bit of grass in the middle. They took rooms two and three.

They went to an automotive parts house, bought a fuel pump for the wagon, which Greg installed, and then went to sleep in their rooms.

Jimmy thought it was the radio which awakened him late that afternoon. He lay staring at the ceiling and realized that Greg and Max were at it again. The bed was creaking and he could hear it through the thin walls. That bitch sure is a moaner, he thought.

Jimmy got dressed and went outside to think. He came to perhaps his fifteenth decision to leave Gregory Powell, but this time he had the money, the car, the ability to do it. After buying the coupe they had divided the money and Greg had sixty dollars left plus whatever Max had. Greg could have the sixty scoots, the guns, everything. Except the coupe. *That* Jimmy needed and he didn't think Greg would come looking for him even though it was in Greg's name. And he didn't think Greg would call the cops and report it stolen. So that was it! He made up his mind! He got in the car and started it, backing up, grinding gravel noisily, squinting into the dusk. Then he turned and saw Greg standing in the driveway beside him, the door to the cabin open.

'Where're you going, Jim?'

'Nowheres, Greg, nowheres,' Jimmy smiled. 'Only over to Cottonwood Road. That's the black part of town, you know? Like, I sat in that room and heard you and Max goin at it and now I'm horny, is all. You know what I mean?' Jimmy nervously turned off the key to show him he wasn't up to stealing his car.

'Me and Max stirred you, huh, Jim?' Greg smiled. 'You know I'm kind of a virtuoso sexually.'

'A what?'

'That means I'm kind of a master at it. I've done a lot in my time, Jim, sexually that is. A lot of ways in a lot of places. Different things. I've learned great control. And technique. It's my control and technique that women seem to want so much.'

'Yeah, uh huh,' said Jimmy gravely.

'It's what keeps Max wanting me so much even though she's getting pretty far along and shouldn't be doing it now. I don't know how to say it without appearing to be bragging, but I seem to have the capacity to please women that they can't find in any other man. Max told me she couldn't even imagine herself with another man, ever again.'

'Uh huh,' said Jimmy.

'It can be a blessing and a curse, Jim. I mean it don't always make you happy to have to perform. I mean all the time, like women seem to demand of me.'

'Uh huh.'

'Come on inside, Jim.'

'But Greg, I jist gotta take a trip over to Cottonwood Road. You can come with me if you want to.'

'No, come on in. We decided to go to L.A.'

'To L.A.? I thought you . . . we was goin to Frisco.'

'Max wants to check on her kids before we go on to Frisco.'

'But they're with her folks, ain't they? They're in Oceanside, ain't they?'

'Yeah, but she wants to phone them.'

'She can phone them from here, Greg.'

'Yeah, but we need dough, Jim. We gotta pull another job in L.A. before we go to Frisco.'

'Jesus, like that's another two hours to L.A. and two hours back here. And Jesus, Greg, we been movin around a lot, and that.'

'Hell, that little hot rod'll make it in no time, Jim. Come on in.' And Greg laughed as Jimmy, hands in pockets, kicked gravel and entered the cabin.

It was after dark when the three of them sped up the Grapevine Highway over the Ridge Route which divides the San Joaquin and San Fernando valleys. Greg was anxious to show them what the coupe would do. The wagon was packed and parked at the motel and the rent was paid, and they'd decided that Max could visit the queen who lived on Eleventh Street or go to a movie while Greg and Jimmy attended to business.

Jimmy decided that he was leaving them in Los Angeles. But he was leaving *with* the coupe. He had it coming, at least that much. They could take a bus to Bakersfield because L.A. was a big town and they wouldn't be finding Jimmy Lee Smith no matter how tight-jawed Greg might be.

The only thing unusual that happened on the trip that night was that a man pulled up beside them and yelled, 'Hey buddy, your tail lights're out.' This was on Highway 99, south of Bakersfield near a turnoff called the Maricopa Highway, a cutoff that was to have great significance later that night.

'We can't afford to be getting stopped by cops for tail lights,' Greg explained, and they found a service station and discovered that the lighting wires were bare and shorting out against the frame. They patched them up and were on their way once again.

After reaching Los Angeles they drove straight for the apartment of the drag queen. Greg had decided they should rent a room there for Max to rest while he and Jimmy went to work.

By now Jimmy's stomach was in knots. The changing plans. The way things suddenly shifted from one minute to the next, so suddenly they were going off a hundred miles in an opposite direction from what they had intended. He'd been a footless drifter all of his adult life, roaming was not new to him, but Jesus, these people were too much. Fuckin gypsies, that's what they are, thought Jimmy.

The queen was not home. There were no rooms for rent either, so it was decided that Max should go to a movie for a couple of hours while Greg and Jimmy did what they had to do.

'Be careful, honey,' said Maxine, kissing Greg as they dropped her in front of the theater at Wilshire and Western.

'Tony'll be home by the time you're through with the movie and we're through with our work. We'll just stay at Tony's tonight and be on our way tomorrow.'

'Okay.' Maxine smiled. 'See you later, honey. See you later, Jimmy.'

You ain't gonna see me no more, you goofy little nympho, Jimmy thought, and he knew how he would do it now. The plan started working for Jimmy when Greg said, 'Let's head for a gas station so I can put on my makeup and get ready.'

Jimmy did not answer as they drove through the heavy traffic on this rather brisk Saturday night, heading in the general direction of Hollywood.

'There's a station. Get her gassed up,' said Greg, as he wheeled into the pumps. 'I'll just be a few minutes.'

Jimmy nodded, and then Greg was gone with his briefcase. Jimmy slid behind the wheel but the nozzle was already in the tank and the man had the hood up.

'That's okay,' said Jimmy frantically. 'It's okay. The fuckin oil's okay.'

But the man apparently did not hear.

'I said that's okay!' Jimmy shouted. 'It's okay!'

And the man looked up, smiled, and turned to wave at a customer driving by.

'Come on, come on,' Jimmy muttered, tapping on the steering wheel, and finally the hood was slammed shut.

'That's enough gas,' said Jimmy. 'That's enough. Just stop it there. I'll take what you got in it.'

'Oh, okay. In a hurry?'

'Yeah, yeah, yeah,' said Jimmy, and his heart sank as he saw Greg leaving the restroom and walking hurriedly toward the car.

Jimmy moved back over to the passenger side and thought, I'm *supposed* to pull one more job with him. It's just gotta be. It's just meant to be. And they drove toward Hollywood. Through the night. To their destiny.

It was good the way you could train the twisted juniper, thought the gardener. It's a living thing and yet it doesn't resist kindness and accepts your training. This one had been sculptured dramatically, and its largest branch was convoluted up and out from the balcony of the house, accepting the urging of the man who directed it.

You could guide a tree's behavior, but not a man's, thought the gardener. A tree just lived and breathed and harmed no one, least of all itself. It was for instance infinitely easier to control this juniper than it had ever been to control himself when he was committing his crimes. He thought of the day he stole his biggest thing – a sewing machine.

Until then his biggest theft had been a saber saw. The sewing machine was stolen during the last stop of that day. After that one he would be finished and could then go home. He had almost bypassed the stop. He had driven through the parking lot of the shopping center, but when he reached the exit he turned the car and went back. He had to. He hadn't made every stop of the day. He knew he could not begin to rest or sleep that night if he failed to make this stop.

In many ways the portable sewing machine had been the easiest thing of all. It was so large, looked so believable in his arms that it may as well have been invisible. When he got to the parking lot with it he felt something very much like disappointment. The heart-splitting fear and dread and excitement had vanished quickly. It had been such an easy victory it was hardly worthwhile. He had told his wife he had bought it for her birthday.

As the gardener crawled on his knees among the roses, jerking the little weeds and tossing them off onto the driveway, he realized the headache was almost gone. His chest was hurting a little, but he could live with the chest pains and the stomachaches and diarrhea and the other things. The headaches were hardest though. Maybe he could sleep tonight despite the trial tomorrow, despite the fact that he might have to tell about that night once again. He'd told it so many times over the years in so many courtrooms, what did one more time matter? He wasn't so afraid of it anymore. He was much more afraid of the series of crimes he had committed.

But he still dreamed of it, could feel the cold night wind in his face, could smell the onions in the field.

SIX

'A typical Hollywood Saturday night,' said Jimmy Smith, the tension festering in his guts, as he looked at rows of cars jammed up for blocks. 'And everybody's too law-abidin or too scared of cops or too fuckin lackadaisical to even toot their horns or swear at the guy in front.'

Jimmy thought of the Spanish automatic in his belt and what if the miserable thing went off by itself and shot his dick off? And yeah, that was somethin else to worry about. What if I killed myself with my own goddamn gun by squirmin around the wrong way in the seat? Bang! Off goes the cock and there I am, sprawled there dyin in the street. Bleedin to death! And he considered putting the gun in the glove compartment to get it out of his belt.

'Goddamnit,' Greg said. 'I'm getting out of this traffic. We'll head back toward downtown until we spot a liquor store to knock off.'

'But I thought you wanted to take off this market out here in Hollywood?'

'Too goddamn much traffic, Jim. We made the wrong turn off the freeway. We'll find something on the way downtown.'

As Ian Campbell drove the Plymouth into the alley near Gower Avenue on his ninth night with Karl, March 9th, 1963, he spotted some mauve-colored flowers in a window box, and rolling up his window against the night chill said, 'The sweet peas and azaleas are starting to bloom. That must mean spring is here. I keep warning mine of the Ides of March.'

Karl Hettinger grinned in the dark at his big-shouldered improbable partner who talked quaintly of flowers and bagpipes. Then Karl realized he had never heard a set of pipes firsthand.

'I'd like to hear you play those bagpipes sometime, Ian,' he said as the little maroon Ford passed by the alley westbound on Carlos.

Gregory Powell was heading north on Gower when he decided to circle the block to the west. He turned on Carlos Avenue, saw the short street called Vista Del Mar straight ahead, and mistakenly thought Carlos Avenue dead-ended there.

'Hearing me play the pipes can definitely be arranged,' Ian chuckled. 'No one else wants to listen to me. My wife and kids and friends run away screaming when they just see me blow up the bag. I wait for unsuspecting people like you to ask me.'

Ian slowed in the alley, flashing his two-cell light toward some shadows in an apartment house doorway, but it was just two bony cats slinking through the alleys, prowling hungrily.

The little maroon Ford made a turn and was coming back their way.

It was now 10:00 P.M. and the unmarked Plymouth police car known as Six-Z-Four was emerging from the alley onto Carlos when the coupe crossed their headlight beam and they saw the two gaunt young men with their leather jackets and snap-brim leather caps in their little car with Nevada plates.

They would have aroused the suspicion of almost any policeman in Hollywood that night. It was patently obvious that they were not ordinary out-of-town tourists cruising the boulevard. The caps were rare enough, but with matching leather jackets, they were almost absurdly suspicious, even contrived. It was as though they'd just driven off the Columbia Pictures lot farther south on Gower, two extras from a Depression-era gangster film, caricatures, Katzenjammer Kids.

But still, Ian and Karl had to look for something more tangible, something to tell the court for probable cause in case they came up with an arrest. They could not, or would not, depend upon their own ability to articulate a well-grounded suspicion, nor the court's ability to understand the several intangibles which go into the decision to stop and frisk and interrogate. So they looked for and immediately found something else: the tried and true 'rear plate illumination.'

Even if the Ford's license plate lights had not been out, it is doubtful that Ian Campbell and Karl Hettinger would have let this car go its way. The little Ford looked 'too good,' which in police jargon means it looked too bad, too suspicious, a 'good shake.' It had to be stopped and a reason found to search.

*

132

The little Ford had but to turn left on Vista Del Mar and it could have proceeded south to Hollywood Boulevard and never have been stopped by Six-Z-Four that night, but Greg decided on a U-turn, and on their ninth night together, the partners made their last wrong turn on Los Angeles streets.

'Fuckin dead ends,' Jimmy grumbled when they turned around. 'We always seem to be runnin into dead ends.'

'We should check these two,' said Ian as the little Ford stopped for the red light at Gower.

'All right. When do you wanna take them?'

'Right now,' said Ian, who pulled up behind the coupe, turned on his red light, and tooted the horn.

Gregory Powell glanced into the rear-view mirror, tightened his grip on the steering wheel, and said: 'Cops!'

As the coupe turned the corner into Gower and stopped, Karl saw the heads move a bit closer.

'Let's be careful,' said Karl.

Jimmy *felt* the red light before seeing it, felt the heat from the red light searing the back of his neck, and he was whispering, 'I knew it. I *knew* it,' even as he unzipped the brown leather jacket Greg had bought him, removed the .32 automatic Greg had bought him, and gingerly dropped it on the floor, kicking it across the car with his new thirty-five-dollar shoes. Greg's eyes were glued to the mirror and the kick was subtle but sharp enough so that the gun ended up very close to Greg's left foot where Jimmy wanted it. It was far enough from Jimmy so that he could swear that Greg had just picked him up hitchhiking and that he knew nothing of the two guns in Greg's possession. Weren't they in Greg's name? And in case that story didn't work he was sure he could come up with others.

'Just take it easy, it may just be a ticket. Just sit tight,' Greg said, looking at Jimmy for an instant, and Jimmy tried to answer, wanted to say something sarcastic, but found himself unable to speak.

He could not take it easy, was in fact frantic, wanting as much distance as possible between himself and Gregory Powell when the cops found the gun on the floor at Greg's feet, and the one in Greg's belt. Who knows, this maniac might just try shooting his way out! Jimmy wouldn't put that past him, and he just wanted to show the cops he was only riding along with this guy, a hitchhiker, that's all.

I got nothin to hide, and I just gotta be cool, gotta be cool, he told himself. But he was all the way to the right, as far as he could sit in the little coupe, and still felt too close to Greg, felt at that moment like they were Siamese twins. And then he leaped out of the car and looked into the eyes of Karl Hettinger, who was flashing his light, advancing slowly on the sidewalk.

Jimmy came forward, fear bursting all over him, and Karl reached inside his sport coat, placed his right hand on the gun butt in the cross-draw holster, and said what he knew was obvious enough despite the unmarked car: 'Police.'

Jimmy Smith froze at the sound of the word and threw his hands in the air.

Karl's pulse bucked. He glanced inside the car at Greg and quickly back at Jimmy standing stock still on the sidewalk, hands high in the air, though Karl had neither drawn his gun nor told Jimmy to raise his hands, and Karl knew for certain. Any policeman would have known. Something. There was something. Narcotics perhaps. They looked like hypes, but Ian was on the street side of the car and couldn't see Jimmy's panic signals.

Jesus, what if he sees the gun? thought Jimmy. What if Greg starts shootin? Christ, I gotta get away from this maniac!

Karl's eyes were not close set, nor did the irises bleed into the pupil, but Jimmy was to forever remember Karl's eyes as being close set and glittering behind his plastic-rimmed glasses. Jimmy bore it as long as he could, about five seconds. Then he said, 'What's the trouble, officer?'

'Police officers,' said Ian to Greg, coming up on the driver's side, not bothering to show a badge, because it went without saying that these two would certainly know they were police. He wanted one hand free since the other held the flashlight.

'Oh, Lord, I know what I'm getting a ticket for this time,' said Greg, with only a faint hope that he could bluff the cop, knowing that plainclothes police don't write traffic tickets. Knowing that when you get stopped by them it's usually a frisk and questioning. He knew it the first instant he looked up at the big policeman, seeing his dark sport shirt buttoned at the throat, and his old gray flannel slacks, and his well-worn sports jacket, knowing they were on something other than normal uniform patrol or traffic details. He *knew* there would be no traffic ticket.

'Would you mind taking your license out of your wallet?' asked Ian.

'Sure.'

'How long have you been in town?' asked Ian, glancing at the license.

'We just got in today.'

'Would you mind stepping out of the car?' asked Ian, handing back the license.

Greg placed the driver's license in the left front pocket of his leather jacket and lifted and loosened his gun.

'What's this all about?'

'It's just routine.'

'Okay. Okay.' Greg smiled, shaking his head and sighing, seeing Ian open the door and step back, seeing that Ian held only a flashlight in his hand. Greg turned to his right to back out, then wheeled to his feet.

Ian was looking at the Colt in Greg's hand and stepping backward slowly, unbelieving. Then Greg was behind him, holding him at the back by a handful of jacket, dizzily remembering the things he had learned in the prison yards about police disarming movements. So he clutched the big policeman by the jacket, and if he felt him turn he could push away and step back, and . . .

Karl had been watching Jimmy, who was licking his lips, cotton-mouthed, stone still in the flashlight's glare, asking, 'What's the trouble, officer?' And then Karl saw Ian coming around the car, with the suspect walking behind *not* in front, and that was wrong, all wrong. And then Greg peeked from behind Ian's back and said, 'Take his piece,' to Jimmy Smith and fluids jetted through Karl's body and he jerked the six-inch service revolver from the cross-draw holster and pointed it toward the man who was almost completely hidden behind Karl's much larger partner.

'He's got a gun on me,' said Ian. 'Give him your gun.'

And then no one spoke and Karl pointed the gun toward the voice, but the voice had no body. It was like a dream. He was pointing his gun toward Ian, toward a glimpse of black cap and a patch of forehead showing around Ian's arm, and there was no sound but the car sounds, tires, cars humming past on Gower, and headlights bathing them in the beams every few seconds. But no cars stopped or even noticed and Karl found himself now pointing the gun at Jimmy Smith, who was like a statue, and then Karl aimed towards the voice again. It was so incredible! It couldn't happen like this. Back and forth went Karl's gun and he was crouched slightly as on the seven-yard line at the police combat range. But this wasn't the combat range. There was no sound except from passing cars.

Ian spoke again: 'He's got a gun in my back. Give him your gun.'

Then Karl looked at Ian, hesitated, and let the gun butt slide until he was holding it only with the thumb and index finger, the custom wooden grips smooth and slippery between his cold wet fingers. Then he held it up and Jimmy, dark eyes shining, walked toward him and took Karl's Colt revolver.

Jimmy Smith held the gun clumsily in both hands at chest level and raised it toward the street light on Gower and squinted with astigmatized vision, like a primitive seeing a gun for the first time. And it *did* seem to him like the first time. This was a cop's gun! It was also unreal to him.

For another moment then they were inert. All four of them. Four brains fully accelerated, four bodies becalmed. Staggered. Inertia for a long moment. Four young men bathed in the purple glow of the street light. Detachment on the faces. Total bewilderment. Two policemen facing that which all policemen firmly believe can never happen to them. Two small-time robbers, fathoms deep, holding the Man at bay. Four minds racing. Tumbling incoherent thoughts.

Perhaps the first one to move was Karl Hettinger. Hands upraised, he began moving the big five-cell flashlight, ever so subtly, in a tiny circle, the beam flashing into the street, striking the windshields of the cars which passed unconcerned every few seconds. Then Ian noticed, and hands upraised, did the same with his little two-cell. Then Greg saw what they were doing and said, 'Put those goddamn hands down.'

Jimmy Smith stopped holding the gun to the light, stopped staring at it in wonder, and began trying to fit a Colt service revolver with a six-inch barrel into a four-inch pocket. He turned, staring from one to the other until he heard Greg's command, then he shoved the gun into his belt. Perhaps without a command he would have remained there forever.

'Get over there,' Greg said, nodding toward the coupe, and hearing the voice, Jimmy wanted to obey. Then he realized Greg was talking to the cops, so he waited for his own orders.

Then it came. His chance. His final opportunity to order fate. Greg said, 'Jim, go back to the police car and park it closer to the curb so we won't draw any more heat. And turn out the lights.'

Jimmy nodded vacantly and Greg said to the policemen, 'Get in the car.'

'Where do you want me?' asked Ian, standing at the right side of the little maroon coupe.

'Behind the wheel,' said Greg, who was thinking, watching, examining both men, sizing them up. At first it was merely Ian's physical presence which guided Greg. He was a big man. Put the big man behind the wheel where he can be watched more closely. The little man in the back.

'Where do you want me?' asked Karl.

'In the back.'

Karl struggled with the seat trying to pull it forward, not realizing it was a one-piece backrest and would not move.

'It's stuck,' said Karl.

'Goddamnit, get in that car and I mean right now. Climb over the seat!'

Then Karl was inside behind the seat, sitting on the floor of the coupe, knees up to his chin. In the cramped space behind the only seat, on the metal floor of the car, flashlight in hand, pulse banging in his ears so that it was actually hard to hear for a moment.

Jimmy Smith was wrestling with the gears of the police car and with the emergency brake, but most of all with his courage.

'Won't move,' he mumbled aloud to himself. 'Got it in drive and it won't move!'

He fought with the Plymouth, stepping on the accelerator and killing the engine each time he was caught by the emergency brake. Jimmy Smith didn't know that emergency brakes on late model cars were no longer controlled by clumsy levers hanging down. He desperately yanked on the emergency foot brake but didn't know to tug the little chrome lever under the dash. He had been away too long.

'If I'd only knew about late model cars,' he was to say later. 'I coulda drove off in that police car. I coulda cut him loose right there. But I couldn't get that fuckin brake off.'

'Hurry up,' Greg yelled, and Jimmy gave up, got out of the car, looked toward Hollywood Boulevard, looked toward escape and made his last choice. He walked toward the Ford hopelessly.

'I couldn't get the brake off,' Jimmy said to Greg who was seated in the passenger side of the coupe, Colt pointed at Ian's belly, hammer cocked.

'Leave the goddamn thing. Leave it,' Greg said, sliding slowly close to Ian to make room for Jimmy in the front.

At that moment a carload of teenagers drove by, talking loudly and laughing. One glanced at Jimmy for a moment and Jimmy became aware of the big revolver under his leather jacket and then the teenagers drove on. Jimmy got in the coupe.

'Did you check the police car for our license number?' asked Greg. 'They probably wrote it down when they stopped us.'

'Yes,' Jimmy lied, wanting to get away, to get away now, to have one more chance to cut Greg loose. If he just had one more chance.

'Where's the other gun?' Greg asked Jimmy, drilling Ian with his eyes, keeping the Colt at his belly, watching Ian's hands on the steering wheel. Already the little car was starting to reek from the smell of fear and sweat from the four of them.

'Where's the other gun?' Greg repeated.

'What other gun?' Jimmy asked, thinking of the automatic, hoping Greg would not notice that Jimmy had kicked it under Greg's side of the seat. And then Jimmy added further confusion to the moment by adding, 'You mean the .45 automatic?'

And then Greg, not knowing that Jimmy was referring to the Spanish Star .32 automatic, felt panic, suddenly thinking there was still another cop's gun unaccounted for.

'Was this guy carrying a .45?'

'I dunno,' said Jimmy, totally bewildered now, not knowing how many guns there were, or where they were.

'Well look around the floor for the goddamn .45 then,' said Greg frantically.

'Gimme that flashlight,' Jimmy said to Karl, and with the bright five-cell light he found the .32 automatic on the floor just under the edge of the seat where he'd kicked it. Now he had two guns in his lap: his own automatic, and Karl's Colt service revolver.

'This is all the guns there is,' said Jimmy.

'Okay, all the guns are accounted for,' Greg said in exasperation. 'Now let's get outta here.' And to Ian, 'Do you know how to get on the freeway to Bakersfield? I want Highway 99.'

'Yes,' Ian answered. 'We can go up the street here on Gower and get on the Hollywood Freeway.'

'Well get going,' said Greg. 'Don't break any laws and don't go fast, because if you get us stopped you're both dead.'

Jimmy switched his glance from Karl in the back to his partner Gregory Powell, and rode most of the trip in an uncomfortable twisted position where he could occasionally look at Karl.

Greg's voice had lost its rasp and was coming back normal and confident. 'Son of a bitch, we couldn't be any hotter,' said Greg, and Jimmy thought he detected a bit of elation in his voice. 'I've already killed two people. I didn't wanna get in this business, but now that I'm in it, I gotta go all the way.'

Oh Jesus, Jimmy thought. Greg was breathing regularly now and saying crazy things, and sounding like some punk Jimmy would expect to see in an old movie and, oh, Jesus.

'Why did you guys stop us?' Greg asked.

'Because you had no lights on your license plate,' Ian said as he drove onto the ramp of the Hollywood Freeway.

Greg's gun hit Ian's ribs. 'Just a minute. Where're you taking us?'

'This is the way to the Hollywood Freeway. I'm going the right way,' said Ian steadily.

'It's the right way,' said Karl, peering up over the window ledge from his place on the metal floor, looking over the space back there – finding a hubcap, rags, a bumper jack and handle, cans – nothing that could be of much help against two men and four guns in a cramped and tiny car, with one man holding a cocked revolver in the driver's belly.

'We're going on the freeway to the Sepulveda off-ramp. And that'll take us to the Ridge Route,' Ian explained.

'Jimmy,' said Greg, 'your job is to look to the rear and cover that guy. And also to look for a tail.'

'Okay,' Jimmy mumbled, thinking: Thanks for telling them my name, you dumb . . .

'How often you guys check in on the radio?' asked Greg.

'About every hour,' Ian said.

'I figure that gives us a fifteen-minute head start,' said Greg, who would occasionally glance back at Karl. He and Jimmy were sitting twisted to the left, toward the two policemen. Greg said to Karl, 'Don't try anything funny back there, because I got it in your partner's ribs.'

'I won't,' said Karl. 'We've both got families. We just want to go home to our families.' And he pulled the corduroy sport jacket up around his chin because he was suddenly very cold.

'Just keep that in mind,' Greg said, and now Jimmy sensed that Greg was totally relaxed.

Jimmy hated him more than he ever had because he himself was breathing so hard he was hyperventilating, and his heart was hammering in his throat. From this time on, Jimmy could never think of his friend as Greg. It would be Powell from this moment, whenever he thought of him, whenever he would dream about him.

Ian said quietly, 'Don't get excited, but there's a radio car up ahead.' And everyone in the car went tense as Ian kept up the steady speed in the slower lane, approaching the police car which was stopped in front of them.

'It looks like a roadblock, Jimmy,' said Greg, voice razor thin. 'Get ready!'

'It looks like they're writing a ticket,' said Ian. 'That's all. I'm just going to drive by at an even speed.'

'Okay,' Greg whispered. 'Remember. Remember. If we get stopped . . . '

'Yes,' Ian said, and they passed the police car at the Sepulveda off-ramp and then they were on Sepulveda Boulevard making good time in the nighttime traffic, catching most lights green, and each man was beginning to think about what all this meant, and to make and reject his own plans.

'Can I give you some advice?' asked Ian after several minutes during which time no one had spoken. The wind rushing through the window chilled them all because they were still sweating freely, but the car was filled with the smell of fear on all of them, so the window remained partly open.

'Go ahead,' Greg said.

'You should take off those caps. Nobody around here wears them and we're liable to get stopped.'

Greg immediately took off his cap, but Jimmy ignored the advice. Fuck it, he thought. I ain't showin them my hair. And I ain't takin no free advice from a cop. At night, in this dark little car, if I just keep my mouth shut or talk like a white man when I have to, they ain't even gonna know my race. And if Powell don't run off with his fat mouth and tell them all he knows about me, well shit, I might get out of this yet, I just might.

And after a few more blocks of driving, Greg reached down on the floor with his free hand and picked up the Schenley's and began drinking.

'If you drink in a moving car the Highway Patrol might stop us,' Ian said, and Greg put the bottle down sourly.

Karl peered over the ledge seeing they were passing Van Owen. Already his legs were cramped up, and he longed to stretch out.

Jimmy stopped thinking and listened to the tires hum and the wind rush, and occasionally he blinked when an oncoming driver failed to dim his lights. Then Greg said, 'Do you guys have any money?'

'I've got ten dollars,' Ian said, characteristically knowing exactly how much money he had.

'I've got eight or nine,' said Karl.

'If you take our money it'll get you clear to San Francisco,' said Ian, with a faint hope that the gunmen might be tempted by the few dollars. Might drop them off now. Might run for it up the highway in the little Ford. Or might feel that the policemen believed they would run to San Francisco, and then head in the other direction. Might do anything, but might just release them. That hope faded quickly as Greg snorted and said, 'You know better than that.'

They were quiet for a few more miles, and Karl tried to see his watch in the darkness, but could not. His stomach was twisted and he was sweating so badly the watch was sliding down his wrist. But Karl was not idle. He was looking at them, listening to the voices, staring hard whenever one turned. He would have to describe the faces later, and the car, and the voices. He tried to get a better look at the guns, but could not, except occasionally when the darker man pointed one at him over the top of the seat, his eyes like berries. The sight blade looked to Karl like his own gun.

Karl watched the blond one chew his lower lip with a craggy overbite. Then Greg said: 'Here's the plan. We're gonna take you guys up on the Ridge Route, drive you out on a side road, drop you off, and make sure you have a long walk back to the highway.'

Now Karl felt the tension subside a bit. It was partly what Gregory Powell said, and partly the friendly tone of his voice. The voice had softened now with the barest trace of hometown middle America in it. 'With just a little bit of cornpone twang,' as Jimmy Smith was to put it.

So a more relaxed Karl said, 'You know, those guns are paid for out of our own pockets. Would you do us a favor and after you drop us off, unload them and heave them into the brush so we can get them after you leave?'

'We don't make that much money,' Ian added.

'Sure.' Greg smiled. 'I think we can do that.' And now, even Jimmy's breath was coming at regular intervals.

They were driving through Sepulveda Pass, out of the heavy traffic. The infrequent street lights made it utterly impossible to see what time it was, so Karl gave up. Once in a while he would glance up at Ian, whose hands did not change position on the wheel. Ian looked calm except that there was a trickle running down the right side of his neck disappearing under the collar.

Greg said, 'There's a lake up here.'

'It's a reservoir off to the left,' Karl said.

'You're damn right there is,' snapped Greg, turning, mouth like an iron bar, as though Karl had challenged him. Karl felt himself go tight again. The abrupt change in tone, for no reason, puzzled and upset him.

They could see the reservoir shimmer in the moonlight and then they were near the Mint Canyon turnoff. After that, Sepulveda turned into the Ridge Route.

'Give me your money,' Greg said, and Karl removed his wallet carefully, seeing Ian do the same. Karl opened his with both hands and held it up for Greg to see and took out his nine dollars.

'We have a hideout a few hours from here where we'll be safe,' Greg said. And this was more than Jimmy Smith could bear. For the very first time he rebuked Gregory Powell.

'Shut up. Don't tell these guys anything.' And the moment he said it he stiffened, but Greg didn't seem to notice what he'd said.

The tires hummed on the highway, and Karl looked up and could gauge by the frequency of the stars how far they were getting from the smoggy skies of Los Angeles.

'Where does that road go?' Greg asked suddenly.

Karl raised up and said, 'That's the Mint Canyon turnoff. It eventually swings right back into Highway 99 here.'

'I know it does. I was just checking,' Greg said sharply.

Then they were four or five miles up the Ridge Route and Karl looked out and felt an overwhelming sadness mingle with the fear and he said to no one in particular, 'I was fishing out here two days ago. With my wife. She's pregnant with our first.'

'Shut up!' Jimmy Smith said. 'Shut your mouth!'

'That's all right,' Greg said soothingly. 'Let him talk.' And then Greg leaned over to Jimmy and whispered, 'Let them talk. Don't make them nervous.'

Jimmy Smith was thinking: Fuck you and fuck the lake and I don't give a fuck if it's covered with fuckin fishes. And you and your knocked-up ol lady, I don't give a fuck about any of it.

Now as he watched Greg's head turn on its swivel from time to time, he became even more angry, hating *everything* about Powell, especially that rooster neck. Then Jimmy became aware that Karl was speaking to him.

'Can I change positions if I keep my hands in sight? My legs're going to sleep.'

'Stay where you are. I'm cramped too,' said Jimmy, sticking the gun muzzle an inch or so over the seat once again.

Greg leaned over and whispered something to Jimmy. Then Greg said aloud, 'We changed our plan. We're gonna hold you guys until we stop a family car. We need hostages. We're gonna stop a family car and when we get one we'll let you guys go, but let me give you a piece of advice. I know you guys got a job to do, but if you turn us in before we can get away we'll kill every member of the family. Understand?'

'Yes,' Karl said, realizing at once the absurdity of it, and feeling once again in danger.

Jimmy Smith was to tell later what had been whispered. 'He said to me, "Jimmy, I told you it was only a matter of time before it would come to this. It's either them or us. Remember the Lindberg Law?" And I got an awful cold feelin all of a sudden. *"Them or us. Them or us."* I couldn't get it outta my mind, what he said. But I didn't know for sure what he meant. I never knew of no Lindbergh Law exactly. I mean, I felt like it meant death. And I wondered about the way he said, "I told you it was only a matter of time." He *never* told me. He musta told somebody else, because he never told me. And he said, "*Remember* the Lindbergh Law." Just like we talked about it before. I never heard of no such thing. Like, he was talking to me about things I never talked to him about. I started getting a bad bad feelin then.'

Karl worried over the whispered conversation and he wished the car muffler wasn't noisy, and that he wasn't down on the metal floor just over the differential of the car with the sounds banging around him, reverberating. He hoped that Ian could hear something. He hated the whispered conversations. But the blond man talked reassuringly. He talked so much he interrupted Karl's thoughts. When Karl was getting very edgy and feeling desperate, the blond one would say something to reassure. It was the other gunman Karl wondered more about, the dark one. What was he thinking? He seemed more volatile. Would he be a threat at the end of their journey? And now the dark one turned and looked at Karl again. He didn't talk enough for Karl to know for sure, but he seemed to be a Mexican. If only he'd talk more, Karl thought.

Then Greg turned and in a quiet voice said to Jimmy: 'We oughtta pull a stickup to get some money. Do you think you can handle these guys? If you think so, I'll go into the next likely place and take it off.' Then he raised a bottle of Schenley's and took a deep swallow.

'Are you crazy?' Jimmy whispered, caution be damned. And Jimmy was to say later, 'Powell's head turned on his long neck like a bird and I knew I shouldn't a said that word. Jesus, not then. The gun in my hand, the automatic, was pointed right toward him. I tried to make it up by humorin him. I said real casual, "Well, maybe I oughtta pull a job, Greg. You watch the cops and I'll go in someplace in Gorman here and pull a job." And I held my breath hopin he'd see how fuckin insane it was and he took a little drink and said, "No." He shook his head and said, "No." And that was the end of that. Jesus.'

Now Karl suspected that Ian was sharing his bad vibrations. The blond one was erratic and the dark one had just said something about pulling a robbery.

Ian said casually, 'How're you fixed for gas?' and looked

hopefully toward the gas station at Gorman. But the blond one looked at the panel and said, 'We got plenty,' and smiled. Then Jimmy Smith took a long desperate pull from the bottle, turning nervously as he drank, to keep an eye on Karl.

Now they were past Gorman, almost on that part of the highway known as the Grapevine. They were at the top near old Fort Tejon, with a view of the great, bleak, lonely San Joaquin Valley. The car started the long descent.

But while they were closest to the clouds, Ian Campbell looked up, ducking his head because the roof was low. He looked up and Karl pushed his glasses higher on his nose and followed Ian's eyes. Up. Up. But there was nothing. Only the black sky. Vast in this immense valley. Stars flickering close and familiar, as they do at the top of the Tehachapi Range.

Now Karl felt this was where he belonged. Out here. Where he'd always wanted to be. In cultivated land where things grow. Where the air is so pure and brisk it hurts. And now perhaps, out here in the farmlands, near the earth, he was somehow safe. Perhaps this nightmare was a city dream. He looked at his partner, who was switching his gaze from the road every few seconds as though he had never seen a great sparkling sky before.

The abandoned felony car was not discovered until eleven o'clock. By midnight, several police supervisors were belatedly panic-stricken and plans were formulated to search the area. At fifteen minutes past midnight a command post was established at Carlos and Gower. All residences, apartments, buildings in the area were being systematically checked. Many were carefully searched for any sign of the missing officers. Motorcycle units were called in for traffic diversion. Press relations were established, and all nightwatch units were held over to assist in the search. No resident of that Hollywood neighborhood slept until later that night when the search was abruptly called off.

During the descent down the mountain, the blond gunman turned all the way around to look at Karl. Or rather his head did. The torso seemed to remain motionless as the head twisted on the axis. The face was so thin and taut it was skull-like, but the voice was pleasant and it lulled, and he said: 'Here's your money back, I won't be needing it.'

And now Karl could have wept for joy because he was at last certain they were safe. The gunman was giving back the money. He'd never do that if . . . And then his thought was interrupted by the dark one, who again pointed the gun at him and stared for a second or two. So Karl put the money in his sock and waited.

Then Greg said, 'You should have that gun on half cock, Jimmy,' and he reached toward Jimmy's automatic and Karl heard the sound of the slide working, and then the blond one said, 'It's on full cock. Ease down the hammer to a safe position.' And he said. 'There. Now it's ready to go. You don't have to keep pointing over the seat. If anything happens, just fire through the seat. You'll stop him quick enough.'

Then Ian's gun in Greg's belt began poking him in the lower abdomen so he pulled it out and put it on the seat between himself and Jimmy.

With the automatic cocked and ready, Jimmy opened the glove compartment and put Karl's gun inside.

Then Greg said to Ian, 'What kind of a shot are you?'

'I'm not very good,' Ian said.

'Yeah? How're you classified on the police force? Sharpshooter? What?'

'Just a marksman. Not very good.'

'You been in the army?'

'Marines,' said Ian.

'See any combat?'

'I was in Korea during the war.'

And now the wind was blowing the little car and it seemed to be an effort to keep it in the number two lane and talk to Greg at the same time.

'Well I'm an expert shot,' said Greg. 'In my business you gotta be good. I killed one man with a gun and another with my hands. And Jimmy here can tell you about how I shoot.'

You fuckin fool, thought Jimmy Smith. You fuckin *fool*.

But Ian did not respond. He continued wrestling the car through the wind which was whistling around them now, making it difficult for Karl to hear every word. Ian's silence seemed to anger Greg, who said, 'Pull in at the next rest stop. I'm gonna give you a chance to see if you can beat me on the draw. I'll give you back your gun and we'll have a little contest.'

Oh, please, Jimmy thought. Oh, please, man, don't say nothin wrong, don't do nothin wrong. And he looked down the steep grade to his right and wondered, what if something happened now at seventy miles per hour. If something . . .

But Ian did not reply, and Greg nodded and smiled toward Jimmy who sighed sharply and took another drink.

Jimmy was to say, 'That big cop talked intelligent all the way, not like some dummies you meet on the police force, and now I respected him because he was sayin and doin just the right things for the fuckin crazy man. And I jumped in and said, "You're right, man, right not to fight him. Cause my partner here is some shot. Like, I seen him make a tin can jump like a country girl at her first dance."'

Jimmy smiled at Greg, who returned the grin and seemed satisfied.

Karl saw Greg's head move toward Jimmy again, and he strained forward, but heard only the muffler's roar. Jimmy Smith was to say later, 'Powell told me, he said, "Jimmy, remember, if you have to shoot a policeman, save one bullet for you." He said that, and my asshole slammed shut again.'

When they reached the valley floor, Jimmy Smith for the first time believed they were going to make it. *Had* made it. They were less than an hour from the motel and the station wagon – a *cold* car. They had kidnapped two cops and were walking away from it. It was hard to control his exultation. 'I know where there's a dirt road,' he said to Greg. 'I used to work around the Bakersfield area.' And they were at the Maricopa turnoff, where they had been earlier in the day with the shorted-out tail lights.

'Turn off here,' Greg said, and Ian did, and the road curved west over the highway.

'Do you want me to go straight?' asked Ian, now headed west.

'Yeah, if you turn it'll take us back to 99.'

And Jimmy was straining his eyes for a place, anyplace. A dirt road. Just a place. And then there were some dusty tire tracks out of the fields to the south. The tire tracks meant a road, and suddenly Greg said, 'Pull over and turn around.'

Ian eased the little Ford to the side and made a U-turn on the lonely quiet highway and Karl's heart began pounding with new vigor, as did Ian's, as did Jimmy Smith's and Gregory Powell's, as each approached his destiny. Then Ian made a right turn on the dirt road heading south toward Wheeler Ridge, south on the long dusty dirt road.

Karl strained upward, tried to see, but as far as he could see was black loneliness and quiet. The smell of sweat and whiskey was suddenly unendurable. Then Greg said: 'That's the farmhouse you're going to hike back to. To make your call.'

But it wasn't that far. It wasn't that far! Were they going to handcuff them? It wasn't that far to be logical!

They had driven ninety miles. They had four-tenths of a mile to go down the dirt road, down toward the ridge.

'I see it,' Karl said. 'I see the farmhouse.' That sight, that farmhouse in the middle of these lonely fields with its little lights, gave him hope. Somehow here, next to the earth, with *his* kind of people, on *his* kind of land, somehow he would be safe.

'We gotta tie these guys up,' said Jimmy.

'No, if we tie them up they'll be here all night long and freeze to death,' said Greg.

'Well, we oughtta tie them up just a little bit, you know, so they'll work themselves loose.'

'No,' Greg said vacantly. 'We'll just let them go.'

And Jimmy was to confess at a later time, 'Jesus. It dawned on me then. It really hit me!'

And then they came to a dirt road which crossed the one they'd been driving on.

'Do we turn here?' asked Jimmy Smith, and his voice was shaking and he rubbed his mouth with nicotine-smudged fingers, face numb, head feeling oppressively heavy.

'No, just keep going,' said Greg.

'It looks too soft. We'll get stuck,' Karl said, feeling that he could not bear it. Not another moment. Wondering if his legs would hold him when he stood. They were cramped, and weak from the fear.

Then they came to another dirt road which crossed this one and Ian again slowed to a crawl. There was a large ditch in front of them in which a huge, recently installed gas line was lying, waiting to be buried.

'This is the spot,' Greg said. 'Turn left and then turn around.'

While Ian made a careful U-turn in the soft dirt, Greg said, 'You get this car stuck and you're both dead.'

'Don't chatter the wheels,' Karl said to his partner. 'You might get the rear wheels stuck.' But Ian managed the turn and was now driving back toward the ditch, the little car pointing the way they had come.

'Stop the car,' Greg said. 'Turn off the lights. This is where we're going to let you go.'

Jimmy Smith turned and looked at Karl Hettinger, who was sitting knees up, hands hanging limply over his knees.

'I just started lookin at him then. I don't know why. He seemed to have big hands, the way they was hangin over his knees, and I noticed that, and I noticed the other officer was a pretty big man because I couldn't see around him out the side window on the driver's side. And I don't know, I just noticed little things about them ever since I got the bad feeling. Ever since Greg said he didn't wanna tie them up. And now I was breathin funny again. Couldn't get enough air, seemed like. And I could hear wind screechin through. Screamin like. I hated that goddamn wind. And I just saw the big ditch there where they was fixin to bury that big silver pipe. All the dirt was piled up beside it and I thought, I hope he don't drive off into that ditch. And then for the first time I started feelin cold. Cold!'

Then the car was stopped. The lights were turned off. Greg wanted the lights off.

'Get out, Jim,' Greg said.

Then a hand picked up Ian Campbell's Smith and Wesson which was on the seat between them.

Jimmy alighted and walked around the back of the car. Then Greg backed across the seat, still covering them with his Colt.

'Get out,' Greg said to Ian, and now Greg was standing beside the door on the passenger side. Ian got out and faced toward the rear, toward Jimmy Smith, and Ian put his hands in the air.

'Get out,' Greg said to Karl.

'Do you want me to climb over the seat?'

'What've you got in your hands?' Greg said unexpectedly.

'Nothing. See, I'm going to keep my hands in sight,' said Karl, crawling over the little seat in the darkness, using only his elbows.

'You're doing it just right,' said Greg.

Then Karl was standing beside the driver's door of the little coupe on Ian's left. He put his hands in the air and the two policemen faced Jimmy Smith, who stood several feet from the rear fender of the car, pointing his automatic at them, more in front of Ian than in front of Karl. It was just past midnight.

The silence rang in their ears during those seconds. There was no sound. No crickets. Only the wind howling.

But there was something. And then Ian recognized it. It was onion. They were between two sections of onions just beginning to sprout now in the month of March, already pungent. And it may have been the onion or the wind or something else, but Ian Campbell wiped his eyes quickly and then raised his hands high again. It may have been the landscape, so endlessly flat, so desolate at night, the earth so dusty beneath the crust, that the wind was blowing powdery choking dust balls all around them. The eye craved mountains, the heart shrank from the vast solitude, but the soaring ranges were far to the east and north, as far as the horizon. It looked to Karl like pictures of the moon, gray and desolate. Under a black sky like this, and except for the wind, the moon would be exactly like this, terrible in the darkness.

Now it seemed bitterly cold to Ian Campbell as he stood to the right of Karl Hettinger, and once, with their hands upraised, their fingers touched for an instant. Though Karl's fingers were icy, the touch of a hand of a friend helped calm Ian, and until they were safe at last, it would be better to think of other things, or anything, perhaps of the pipes, as he'd done all his life in troubled moments, as he'd done in Korea or in frightening moments as a policeman.

Though it was not the clearest of nights – now sporadically cloudy as puffy specters scudded past the moon – still one could *feel* the stars all around. It was on such nights many years ago that Ian loved to march by the tarpits. Perhaps Ian deliberately thought of the pibroch, of playing 'MacCrimmon Will Never Return' and vowed that when this was over, when he was home again, he *would* master the great pibroch.

The silence was broken by the sound of Gregory Powell's footsteps, scraping on the crust of earth as he rounded the back of the car in the darkness. Karl could not see his face in the moonlight, just the hair, short and dark blond, and the triangle of the face. Now he and Ian were looking at Gregory Powell, at the long neck and the shadowy face and the gun in his hand, the hand in deep shadows, and Gregory Powell walked over to Jimmy

Smith and leaned close and appeared to whisper, but no man except these two would ever learn the words he whispered, and then Gregory Powell moved to his right so that he was in front of Karl Hettinger. Jimmy Smith moved slightly to his right, just as he did on the desert that day in Henderson, Nevada, and he stood almost behind Gregory Powell. Jimmy held a gun in his right hand, the hand with L-O-V-E tattooed across the fingers. And though Gregory Powell was more directly in front of Karl Hettinger, he continued to look at Ian Campbell, at the *big* man, his gun pointed toward the policeman's feet.

Despite the cold wind, sweat poured down Ian's ribs and chest and burned his eyes, so he closed them, and perhaps it made it easier to imagine the pipes, to hear them wailing far back in the funnel of the wind. Perhaps it calmed him.

Jimmy Smith said later, 'The big cop seemed relaxed, you know?'

Gregory Powell said to Ian Campbell, 'We told you we were going to let you guys go, but have you ever heard of the Little Lindbergh Law?'

Ian said, 'Yes.'

And Gregory Powell raised his arm and shot him in the mouth.

For a few white-hot seconds the three watched him being lifted up by the blinding fireball and slammed down on his back, eyes open, watching the stars, moaning quietly, a long plaintive moan, and he was not dead nor even beginning to die during those seconds – only shocked, and half conscious. Perhaps his heart thundered in his ears almost drowning out the skirl of bagpipes. Perhaps he was confused because instead of tar he smelled onions at the last. He probably never saw the shadow in the leather jacket looming over him, and never really felt the four bullets flaming down into his chest.

Jimmy Smith was to say later: 'I can only remember his arm and hand. His *hand*! Each time a bullet hit him, his hand would jerk and jump up. Like he was grabbin for you. Like he was grabbin for your leg in the dark there! I'll never forget that arm.'

Jimmy Smith shouted in horror, 'He's still *movin*!'

But if Ian Campbell did not hear the voice, nor the staccato explosions, nor feel the four bullets bursting his heart, perhaps he believed that the bagpipes screamed and screamed.

Karl screamed and screamed. He didn't know that he screamed.

The scream tore from his throat a second after the first bullet crashed into Ian's mouth, when Karl's mind actually comprehended that they were killing Ian Campbell. When the two men heard him they turned and saw Karl screaming and running, stumbling and running a zigzag course down the road, terror leaping from his face white in the moonlight, yet with presence of mind to zigzag, change directions. Compulsively he turned and heard the explosions and saw the flashes – four oblong red and yellow fireballs blasting down into the chest of Ian Campbell. Then Gregory Powell ran after Karl and Karl heard more explosions and saw more fireballs, round this time, burning through the night, coming for him. Coming for *him*. He sobbed and cut to his left and his right, his coat flapping in the wind and heard shouts and dived head-first to his left through a row of tumbleweed packed hard against the barbed wire farm fence, ripping his hands and face on the wire. For a long second he was clawing through the wall of tumbleweed and was caught on the wire while he heard the padded footsteps running on the dirt road. Coming for him. As in a dream he was caught. Trapped. Weeping. Thrashing. And the footsteps were coming closer. He could hear the panting.

Then Karl was free. On his feet, stumbling over the torn trouser leg. He fell in the darkness still sobbing, then got up and balled up the ripped fabric and pulled it up to his thigh and ran, one bare leg flashing in the moonlight.

Jimmy Smith did not hear Gregory Powell running back toward him. He did not hear the distant echoes of the shots reverberating off the peaks of Wheeler Ridge. He no longer heard Karl's screams, nor the moan from Ian Campbell. He was only aware of pain, a burning sensation behind his eyes, and the first sound he heard was Greg's voice, breathless, frantic.

'What the hell is the *matter* with you? The son of a bitch is *getting away*!'

Jimmy nodded dumbly and Greg said, 'We gotta catch him! Where's the flashlight? Goddamnit, where's the big flashlight?'

'It's on the floor,' Jimmy said. 'It's on the floor of the car where I left it.'

Then Greg took Karl's flashlight and ran back to the barbed wire with the almost impenetrable tangle of tumbleweed blown against it. Greg got up on a fence post and flashed his light out into the field. Back and forth went the light while Karl lay burrowed into the ground not far from Jimmy Smith and the corpse of Ian Campbell.

Karl checked his watch and saw that it was 12:15 A.M., and he put his glasses in his coat pocket and waited. He was thinking now. He was weakened, dreadfully weakened by the shock. Unable to run without falling to the ground like a clumsy child. Still, his mind had not forsaken him and he was thinking rightly that the killers would believe he would run north toward the farmhouse. Toward the highway. So Karl had cut back south, doubled back toward the great ditch with the silver pipe, toward the coupe and the killers, toward the body of his partner. He was huddled there on the ground in the darkness and the cold, gasping erratically, fighting the shock and the agonizing stomach spasms. Fighting the terror. Waiting for his strength so that he could get up with a man's legs under him and run. But for now he just lay there watching the light sweep over the fields. Back and forth. Then it disappeared for a moment.

Jimmy handed Greg the automatic and sat in the driver's side of the car and Greg began the confounding job of reloading the weapons that had been fired. Jimmy handed Greg six rounds and the automatic, and Greg began fumbling and dropping shells and finally he got one revolver loaded and gave back two rounds to Jimmy Smith.

Jimmy sat in the dark and bullets and guns were clumsily exchanged by the two frantic men in the darkness, and finally Jimmy Smith said, 'I'm gonna drive down the road and stop and wait for you and work back.'

Greg decided to use the automatic since Jimmy could never be trusted to know how to fire it alone in the darkness and excitement. So Greg took only Karl's flashlight and the .32 automatic. 'All you gotta do is fire the .38, Jimmy. Just pull the trigger. And don't waste bullets.'

And Jimmy Smith drove north on the dirt road while Greg ran to the fence line choked by tumbleweeds and swept the beam across the field, back and forth. As Jimmy was driving off, he called for his last order from Gregory Powell: 'Do you want me to go down this way? And turn around?'

'Yeah,' said Greg.

While Greg passed the beam over the field, Karl Hettinger ran south and west and finally north, ran through the black night, falling often in the plowed ground, catching glimpses of light to the east and then of moving light to the south. He ran toward the ridge away from the lights.

Karl was afraid of the farmhouse to the north, the one which Gregory Powell had pointed to. He began running westerly, toward some other lights twinkling in the distance, another farmhouse. But distances were deceiving in the solitary blackness and his breath was ripping through his lungs and he was exhausted long before he began getting close to the lights. Then he heard a sound and for a moment he doubted his ears. He listened and reached for his glasses in his pocket and they were gone, lost. He peered through the darkness and saw it. A Caterpillar tractor!

Emmanuel McFadden had been working an especially long shift. He was supposed to work twelve hours from noon to midnight with a forty-five minute break at 6:00 P.M. It was a hard day and his khakis were sweat soaked. But somehow he was preoccupied when midnight came, and it was twenty past midnight when he unhooked the disc and drove the tractor back toward the barn so that it could be gassed up and taken over by his brother James. Sometimes he got caught up in the sound of the Cat's engine, and the hiss of the disc slicing furrows, and he would dream that it was *his* farm, and he would plow *his* earth gladly at midnight.

He thought of the little Arkansas farm, the forty acres his grandfather owned, and how it would be if he and James owned this farm here in the San Joaquin Valley, of how he would plant onions and garlic and cotton and cantaloupe and it wouldn't matter then, the twelve hours on the big Cat, nor the cold wet feet from placing the sprinklers, nor even the poor drafty barn where farm workers slept. None of it would matter if it was *their* land.

He was not a big man but he was young, twenty-five, and Mr Archie, the overseer, said he was a good worker, better than men twice his size. When he heard this, Emmanuel's mahogany face broke in two, in a prideful grin. His nose was scarred across the bridge and under the ball just over the lip, and it was off center, so that the face broke into two off-centered sections when he smiled. The dream had kept him working that night, past the hour when he could stop. He had thought several minutes before that he saw lights off to the east, and that was passing strange because who would be out there in those fields at this time of night except himself? When he looked at his watch and saw it was twenty past twelve he unhooked the disc, and was riding home, feeling the ache, feeling the exhaustion. Then he saw the apparition in the beam from the rear light. In the distance he saw it, and it was coming toward him, tottering crazily, ghostly weird, falling and crawling like an animal.

By now Karl had covered more than two miles at top speed, in the darkness, through the fields, with a body drained of energy from two hours of sustained fear and twenty-five minutes of overwhelming horror and shock. It was a terrible apparition to the tractor driver who saw Karl clearly now, struggling through the darkness like a drowning man, mouth agape, eyes round with terror, hands outstretched, lunging at the air as he staggered forward to catch the tractor.

'I was so scared. At first I thought it was a animal comin like that and I picked up a shovel I carried on the Cat and I was getting ready to hit it when he come up on me and I see it was a white man with his clothes half tore off. He's puffin and sayin, "Help me, help me." And I see a empty gun holster there whappin against him and I thought, Lord, he's fixin to shoot me off the Cat. And at first I jist try to ignore him. Pretend like I didn't see him. And he say, "Help me. I'm a PO-lice. Help me! They killed my partner and two men is comin to kill me!" I got scared. I jist can't hardly tell it, and I stopped and I took him on back of the Cat and turned it around and headed for the house at the Coberly West farmin camp, to those clapboard shacks where there was a phone. And he's tellin me, puffin and tellin me that he was kidnapped, and about the killin, and the two men.

'And when we get close to the farm I saw a light and I say, "It's them. It's them!"

'And the officer say, "Is there a car?" and I don't know if they was in a car, but I think they was on foot there. So I turn and I look and the lights is a ways off in the distance, but the lights is gettin closer and closer and the Cat can't go fast as we can run, so I say, "Let's go!" And we jumps off the Cat and leaves the motor goin and starts to run. Through the fields we runs. And it's dark and we jist runs. And runs. . . . '

And at that time, running for his life, Emmanuel McFadden turned back and saw that the policeman could not possibly keep up. That he was caving in, staggering like a drunken man, hardly able to breathe, falling. Emmanuel McFadden became angry. White men! This was white man's trouble. Police trouble. It had nothing to do with him. Nothing. And yet they were drawing him into it. Drawing him into a thing he could not understand. They were going to kill *him* now, those men, those two killers who had already murdered a policeman. And why should he die? He'd done nothing. Why should he die with this white man behind him, lurching across broken ground and falling in the dirt, fingers digging into the earth, retching. So Emmanuel McFadden ran away, far ahead, until he could no longer hear the wheezing rattling breath of the policeman behind him. Hoping he could outrun them. All of them. The policeman and the killers.

But when he got closer to the highway, to the Robert Mettler ranch, he stopped and looked at the house, so still and foreboding in the moonlight. He stopped and reconsidered, and lay down in the fields. It was bad. It was very bad. Because those lights he had seen near the place where he abandoned the tractor, those lights had circled the tractor and the killers now knew the policeman had a helper, and if they were car lights, it would mean the killers could easily have reached this same farmhouse by now. In fact, it was the logical place for them to come. There by the highway. This is where they would have to come. And there was an open field and a dirt road to cross before getting to the house. And if the killers were lurking in the dark . . .

So Emmanuel McFadden realized that he could not escape this policeman. That if he were to survive this night he would himself need a helper, one to watch and warn while the other one crept close to the house. But not this house. This looked like a trap. So he returned angry and scared to Karl Hettinger, who was still stumbling across the plowed ground.

'Look, man,' said Emmanuel McFadden, 'we gotta change directions. This ain't no good here. They could easy be there in the dark waitin. This is where I would wait if I was them. It jist ain't no good.'

'Whatever you say,' Karl gasped, sawing at the air, mouth hanging slack, eyes rolling back every few seconds, down on one knee, willing to follow anyone who could lead him away from his pursuers.

The farmhand made the decision and they headed due west

toward the Opal Fry ranch. Once they saw a flashlight off to the east, far off, and the farmhand thought they were safe and he could slow down and rest. Then he saw car lights coming north toward the dirt road where they could almost reach the Cat. 'Oh my Lord!' said Emmanuel McFadden. 'My Lord!' And they were up and running as hard as they were able. The car lights continued north and disappeared.

'I ain't never gonna forget that car. Never. It was a li'l ol round car. I heard that one killer say in court he never came lookin for us in the car, but it's a lie. I seen that car, li'l ol roundy car it was. I ain't never gonna forget that ol roundy car.'

They had to stop again for Karl to rest. It was so quiet, so still, that Karl's ragged breathing was panicking Emmanuel McFadden.

'You jist got to stop that loud breathin,' he said. 'They gonna hear you a mile away!'

'I ... I ... I ... can't,' Karl gasped, waves of pain breaking over him, and sounds in his ears like the memory of Ian Campbell's death moan. Karl's breathing actually got louder, and the farmhand looked fearfully through the darkness, knowing the sound would draw them in. He knew how they would come: like a brace of coyotes, quartering, circling a fleeing jack until they wear him down, trap him, and then walk in calmly, the jack fear-frozen.

'Man, I'm gonna run off and *leave* you if you don't stop makin so much noise.'

'I'm catch ... catching my breath now. I'm catching ... '

'You gotta go a little faster, too. You jist gotta.'

'What's your ... what's your name?'

'Emmanuel.'

'My name's Karl.'

'Can you get up now, Karl?'

'Yes.'

'We gotta do it together, Karl. We gotta stay together.'

'Let's go,' said Karl, and they were up and running again, running west, and then off fifty yards in front of them, a beam of light flashed slowly and they dived to the ground. Emmanuel McFadden wanted to put his hand over the white man's mouth. Now they were close. Now they were close to death!

'I can't ... can't go much more,' Karl whispered.

'Don't talk. Save it, man.'

'They're going to kill us.'

'Oh, no. Don't you be talkin like that. I don't wanna die. Don't be talkin crazy now. You jist lay down there. Lay in the furrows, Karl. Lay behind the tumbleweeds. We gotta stay together or we ain't got no chance. Neither one of us.'

And so they stayed together, the white man and the black man, strangers, but for a moment more dependent than either had ever been on another man. Hunted by a white man and a black man, each for the moment more dependent upon the other than either had ever been.

'And they was movin about,' Emmanuel McFadden would say. 'Movin about. I was layin on the ground. I saw the lights movin around the Cat. And I knew you could drive a car across those fields cause the gas company do it all the time. I saw the car, the same car. One car went into the mountains way back at midnight and one car came out. It was the same car. The lights went off the car when they went in and the lights came back on goin to the highway.'

And by this time, Karl Hettinger had run almost four miles, through plowed and unplowed ground, in the black of night, in shock.

They had another mile to go, and they ran it together this time, together through the darkness, their fear-filled sweat dropping to the earth, and finally, there in the distance, was the Opal Fry ranch.

Karl fell to his knees, dropped to all fours, head hung, not breathing now but rattling dangerously. 'I can't . . . can't . . . make it.'

'You can make it, Karl. You gotta make it. Cause if you don't make it, I won't make it.'

'They . . . k-killed . . . '

'Never mind about him, Karl. Jist try to quiet down and rest. Jist try to stop that loud buzzsaw in your chest. If you don't quiet down I'm gonna have to run off and leave you.'

'Go ahead, Emmanuel. I'm too . . . too . . . I can't go on.'

'Damn, Karl. I can't. We needs each other. But if you could jist breathe a little softer. You're gettin on my nerves real bad. Real bad, Karl.'

'I'm . . . trying to . . . be quiet, but . . . I . . . just keep thinking . . . about . . . about . . . '

'Karl!' Emmanuel whispered suddenly.

'What?'

'It's the car! See it to the north? The lights? It's comin down this way. Karl, it's comin!'

And they were up again and running full tilt to the patch of tumbleweeds fifty yards ahead. It was a pathetic sprint and at the end of it they fell to the earth and lay in the furrow and burrowed in. Karl sunk his knuckles deep into the ground, digging, holding on to the very earth itself, to keep from fainting.

They watched the car go sound toward Wheeler Ridge.

Then after three interminable minutes Karl raised up on his knees and said, 'I'm gonna try for the ranch, Emmanuel.'

'Okay, Karl. Okay. I'm gonna watch. And if you hear me yell, man, you come runnin back here, back to me in the dark.'

'If I make it across the fields. If I get in the house. If you see me in there, you come on in. You call my name. You call me. And then you say your name and I'll know it's safe to open the door. Okay, Emmanuel?'

'Okay, Karl, you go on now. Jist don't hurry across this last field. Just watch. Hear?'

'So long, Emmanuel.'

'See you later, Karl,' he said.

And less than a minute after Karl left him, Emmanuel McFadden felt it as he had not before. The fear. But now it was mingled with a dreadful awesome loneliness. Here in the night, in the fields, straining his eyes in the darkness, he felt like the last man on earth. Then he knew he was not. He gasped and felt his body go weak and he was not sure if he could get up and run. It was the car! It was coming south! Coming for *him*!

Down the road slowly it came, the same car which he had watched go north on the dirt road from the place where the officer had come. The same car which had gone west on the highway and had turned south down the other dirt road between them and the Opal Fry ranch. And now it was coming back south between Emmanuel and the ranch. Between Emmanuel and Karl. Between Emmanuel and safety. It was coming very slowly.

He knew he would die there alone. He who had done nothing. All *alone* like the fallen angel. And he knew he could not rise when it got near enough to hear the tires brutally grind on the hard surface earth. The fear had destroyed his will to rise, and he lay there waiting for it to stop. For the killer to see him and come forward with an implacable somber death step.

But the driver did not look to his right. Did not see the lonely figure huddling in the furrows, whimpering now because he was so alone. It kept going to the highway, and though Emmanuel McFadden and Karl Hettinger were still linked by their chain of need, Jimmy Smith and Gregory Powell were not. Jimmy Smith was now to sever that chain. But it was only a temporary break.

'I got a chance!' Jimmy Smith blurted as he sped east on the Maricopa Highway. 'I got a chance!'

The little Ford was doing ninety and Jimmy had to slam on the brakes when he reached Highway 99 and a decision. Los Angeles would be dangerous, there was the long drive over the Grapevine. The cop probably had reached a farmhouse by now and was already calling for help. They'd be watching 99. But maybe Powell *killed* the second cop. No, Powell, didn't kill him. Powell was too stupid to make anything right once he screwed it up as bad as this. The Grapevine was out. So where? Where? And then he thought of the 'hideout,' the cold car, Powell's station wagon. He drove toward Bakersfield.

En route he thought that maybe Powell will get *himself* killed. Sure, why not? The cop'll make his call and in fifteen minutes there'll be squad cars crawlin over every inch of that miserable farmland. Powell is nuts. He might try to shoot his way out. Sure.

Or maybe Powell will try to give himself up and the cops might shoot him anyway. Sure. Yeah. The fuckin cops'll be ready to kill anybody over this. Yeah. And maybe they'll just go ahead and dust him anyway. And then I made it for sure. That other cop won't know nothin about me. How could he? He knows I'm called Jimmy, that's all. And I kept my hat on and my mouth shut most of the time so he won't even know what race I am. 'I got a chance, baby, a hell of a chance!'

At that moment Karl Hettinger was banging on the door of the remote farmhouse with the shake roof and wood siding and grassy lawns. He was looking out into the darkness, ready to leap over the porch railing if he should see, hear, sense something out there in the dark. Then a light went on and the door opened and he was in the house of Jack Fry and the words were tumbling out. The stubby little rancher with the tousled red hair was blinking sleepily, and the ordinarily merry blue eyes were out of focus as he scratched his cheek and nodded, gradually coming awake as the young man talked. A sense of urgency and danger awakened him completely then, and he ran his hands around his suntanned lower face and over the high white forehead usually covered by a wide-brimmed hat. His wife and teenage son were standing there looking frightened and Jack Fry wondered about himself for letting a stranger in the house. But now it was all clear to him, all of it, and he said, 'I'll get my guns. You go ahead and make your call.'

At last, for the first time in three hours of terror, Karl Hettinger knew he was safe. He looked at the boy and the woman, and he realized how he must look to them with his pants half torn off, and the mud and blood caked on his face. He saw a cold half-empty cup of coffee on the table and he lunged for it and drank it down without thinking. It was a purely physical gesture, the crying out of a dehydrated body for moisture. The woman understood and went to work efficiently. She sat him down, got him water, and urged him to sip it. She made coffee and gave him a towel. Karl looked at them, knowing they were real, they were farm people, his own kind. He was truly safe! Then Jack Fry, armed with a shotgun, and his boy, with a .22 rifle, were dimming the lights and pulling the shades. They were ready for the killers from the city. They were more than ready.

They heard a scraping then, followed by a timid knock at the door. Karl dumped coffee on the plate and Jack Fry crept to the door, gingerly opened it, and Emmanuel McFadden shouted, 'Oh Lord!' and stared down the barrel of Jack Fry's gun.

'Good God!' said Karl. 'Emmanuel! I almost forgot you!'

And Emmanuel McFadden stepped gratefully into the rancher's home.

Deputy Berg of the Kern County Sheriff's Office then received the most startling call of his career.

'Sheriff's Office, Berg,' he said sleepily when he picked up the phone, and he heard Jack Fry say, 'I have a Los Angeles policeman here. He's in trouble. His partner's been shot.'

Deputy Berg became wide awake. 'Where is this? Wait a second there.' And he heard muffled voices.

'Hello,' Karl said. 'Is this the Sheriff's Office?'

'Yes. Yes it is.'

'This is Officer Hettinger, L.A.P.D. We were kidnapped from Hollywood Division, and my partner's been shot. Can you put out a description on two guys?'

'Sure I can,' said the excited deputy.

Karl gave what detectives later would admire as a remarkable description under the circumstances. After a drink and resting a moment, Karl was able to speak calmly and efficiently, when he once again assumed a policeman's role.

'Okay. They're driving a '41 Ford, maroon, Nevada plates. I don't know the number.'

'I have a report of a stolen vehicle in that area, so they probably changed cars by now,' the deputy said.

'All right. A male Caucasian, thirty-one, maybe younger. He's five nine, a hundred-fifty, light brown hair, blue eyes. He's wearing a dark leather jacket, waist length, dark pants. And his partner is a male Mexican, twenty-five, five ten, a hundred-fifty, black curly hair, brown eyes. He's wearing a brown leather jacket, waist length, dark trousers. They've got a .45 automatic and both of our .38 specials. Let's see, the Caucasian, he's got a light brown mole on his left earlobe. That's about it. They're talking about taking another car and going somewhere in this area, possibly northbound toward Fresno.'

At last someone had noticed the mole Gregory Powell had drawn on his earlobe.

After giving the ranch location Karl said, 'Will you call my people quick?'

'Sure can. But how badly is your partner hurt?'

'He's dead,' said Karl. 'My partner's dead.'

While Gregory Powell shone the light around the fields and ran up and down, unwilling to get too far from the dirt road, it occurred to him that Jimmy should be out in front of him. The headlights should be out there somewhere working back toward him. Then he realized it – Jimmy was gone. He wasn't coming back. Gregory Powell was alone. He panicked and began running back south on the road, looking over his shoulder for Jimmy, listening for the little Ford, hearing only the moan of the wind.

He was alone, all alone. Except for Ian Campbell.

He put the .32 automatic in his pocket and he ran to the body,

162

and began dragging it. Dragging the heavy body across the ground for no good reason. He came to the ditch and threw the body into it, down by the big silver gas pipe. If things had been different, they might have both been dumped into the ditch. And covered over. And soon the entire length would be dragged by bulldozers and no one would ever have found them. Not ever. But things were not going well for Gregory Powell. There were too many wrong turns. He spun around and began running. Running toward the Maricopa Highway. Away from the smell of onions.

At 12:40 A.M., Mrs Billie Riddick was sitting up watching a late movie on television. She was a farm wife living in the settlement of homes at the Clifford Mettler ranch. She thought she heard her husband's car start up. She listened then and heard it a second time and went to the window. She saw that the Plymouth *was* being started. It moved, but the engine died. Mrs Riddick went to the bedroom and woke her husband. He ran out the door in his underwear, but this was perhaps the luckiest night of his life. He was too late.

As Jimmy Smith pushed the coupe toward Bakersfield he was feeling better with each passing moment. What if Powell didn't get wasted by the cops? What if the snivelin little punk gave up? So what? He always brags about how he hates snitches, what he'd do to a fink. So maybe he'll keep his mouth shut. Maybe I'll *really* make it. . . .

Jimmy Smith, filled with hope, raced down the highway, actually convincing himself that the man he had abandoned in the onion field would not help the police identify him, would not inform on him. Then Jimmy saw that the gas gauge was on empty. He panicked, not wanting to run out of gas on the main road. He took a turnoff into the little town of Lamont, found what looked like a parking lot, and stopped Greg's hot rod.

Jimmy picked up Greg's trenchcoat and the guns from the floor of the car, wiping them, not leaving prints, and wrapped up the guns carefully, and shoved them under the seat, except for Karl Hettinger's gun, the only one which had not been fired. He took the paper bag from the glove compartment, found a ham sandwich and banana, and with the whiskey bottle in his pocket he was ready to hike.

Jimmy was walking quickly down the dark streets of the tiny

town heading for the grape vineyard he'd passed a block away when a dog spotted him. Within minutes it seemed to Jimmy that every dog in town was barking at him and lights were going on, and Jimmy was running toward the vineyard. As he scrambled through the vineyard fence he dropped his package. He stopped to retrieve it and dropped his gun. When he picked up the gun he dropped the whiskey bottle. Then he sat down in the sandy soil of the vineyard and sobbed and cursed his luck, and his life, and mostly Gregory Powell. As he sat there sniffing, he wiped his eyes and began eating the overripe smashed banana and the sandwich, and wondering if Karl's gun would blow up in his face now that sand was in the barrel. He took a drink of the whiskey but reminded himself that he needed a clear head so he poured out the whiskey, keeping the bottle in his pocket for water.

There was only one place to go, Bakersfield. So he began to walk, tripping over the vineyard wires every few feet. He had to bypass the farmhouses because of dogs, and now he was stumbling into muddy irrigation ditches in the darkness and soon Jimmy's pants were soaked and muddy to the knees. The new thirty-five-dollar shoes weighed five pounds each. So Jimmy sat again and sobbed, and cursed Powell and took off the wet socks and marched on with bare feet in the muddy shoes.

Greg was pushing the red and white Plymouth as hard as it would go. He had to beat Jimmy to Los Angeles. He knew what that punk would do, go straight to L.A., pick up Max and take the hundred and forty bucks she had. He had to get there in time. There would be a reckoning with Jimmy Youngblood. But he knew he could never make it in this stolen Plymouth. The farmer had probably reported it already. So he headed for the hideout. A change of clothes, switch to the station wagon, and he could drive to L.A. unmolested. It was almost 1:00 A.M. when he reached the hideout.

The first thing Greg thought of was a disguise. He decided to shave his head and started with the front at the hairline, but it was hard to shave the forehead there in the little room with his hands shaking from fear and frustration and rage at Jimmy Youngblood. Finally he threw the razor in the sink and changed clothes, discarding the leather jacket he knew the police would be looking for. In a few minutes he was ready to go and then it struck him –

the keys! Christ, the keys! Max had the keys to the wagon in her purse!

Then Greg crumbled. Panic set in, and Gregory Powell, who had stolen two dozen cars in his lifetime, and who could have hotwired a car in a few minutes, did not think of it. *Could* not. Frantically he fumbled for a dime to use to remove the screws from the license plates of the station wagon to switch with the plates of the stolen Plymouth.

Finally he was ready. He had on a light jacket with the spare clip to the automatic in his shirt pocket, a half box of shells in his coat pocket, and the .32 automatic in his belt. But he threw the .32 automatic under the front seat of the Plymouth as he was driving away.

California highway patrolmen Odom and Crist were cruising US Highway 466 on the east side of Bakersfield when they heard the stolen car broadcast at twenty minutes before one. They automatically jotted down the information: 1957 Plymouth four-door, white over red, and the license EOB 940. Though Mettler Station was several miles from their beat, they began watching passing cars more closely, but there were few cars to watch at this hour. They were bored and getting close to the end of their shift and made small talk to stay awake.

Twenty minutes later the communications operator re-broadcast the information on the stolen Plymouth. This time more was added: 'Above vehicle possibly occupied by two male suspects wanted for kidnapping and murder of the police officer.'

The sergeant, anticipating the reaction of young aggressive officers, came on the air and ordered all units to remain on their beats. But officers Odom and Crist looked at each other and Crist said, 'Screw the beat. Let's hump.' Odom drove toward Highway 99, across the Taft Highway to the new freeway, and south. He drove one hundred and forty miles per hour.

At that moment Gregory Powell drove south through the night, passing the road to Weed Patch. There were cows on both sides of the highway, and dairy smell, followed by fields of truck crops.

Greg was relatively calm. He knew exactly where he was, knew he was getting near the Maricopa Highway, near the onion field. He passed a desolate mile-long strip where nothing grew but scattered oak and eucalyptus, then the sign for Herring Road, and he figured he would soon be passing the Maricopa Highway.

Every tenth of a mile he strained his eyes off to the right, peering through the windbreakers, across the fields and the darkness. He thought there would be great lights out there, maybe a helicopter lighting the entire area. Maybe dogs and a posse. He was overrun with excitement and dread as he passed an open strip of highway through the grape vineyards. Gusts caught his car and whirlwinds spun through the sandy vineyards on both sides. Still he kept the Plymouth at a steady sixty miles per hour.

A few minutes to go. A few minutes to pass that place where the dirt road was off a mile to the right. A few miles and he would be safe. He believed that he only had to get past that road, past that onion field.

If only I knew another route, he thought. If only I didn't have to pass by so close to it, could steer away from it.

The whirlwinds danced on the sandy shoulders of the highway and powerful gusts sucked him to the right. Out there in the darkness, less than seven miles away, was the ditch where Ian Campbell lay.

Greg passed Mirage Station at a safe speed and officers Odom and Crist turned in their seats. They had already checked out another white and red Plymouth at Mirage.

'Wrong license,' said Odom, looking at OHA 459.

'Yeah,' said Crist dejectedly.

'Might as well check him out, though.'

'Okay,' said Crist, and he looked at the time and later testified it was 1:45 A.M. when they turned on the red light and the clear spotlight.

The left shoulder of the driver dipped a few inches and Gregory Powell discovered that the Plymouth had a deep well under the front seat. The .32 automatic had slid down and back. It was gone. He pulled the car over one mile north of the David-Copus turnoff.

Greg watched in the mirror as the two husky six-footers approached one on each side. He calmed himself, and he smiled before the one on the driver's side got up to him. He was smiling in a calm friendly way when Odom first looked at his face.

'Good evening. Could I see your license and registration?'

'I wasn't speeding, was I?' asked Greg, handing the license to Odom.

'No,' Odom said, examining the license and the registration slip Greg had given him, noting that the car was dirty but the plates shiny clean. 'This registration is for a Ford station wagon.'

'What? Let's see.' Greg examined the registration slip for a moment and shook his head, saying, 'It's just a goof-up on my part. I have two automobiles. I have a Ford station wagon my wife drives, and this Plymouth is mine. I registered both cars the same day and apparently just put the registration and the license on the wrong vehicle.' Greg clucked and shook his head.

To the officers it was believable. The new plates had just been issued in March and everyone was making errors. Odom and Crist looked at each other over the top of the car. Then on impulse Odom said, 'Would you step out of the car?'

'Sure,' said Greg. 'But what's the problem?'

'Someone driving a car like this just killed a policeman very close to here.'

Greg patiently shook his head and said in a tired voice, 'Surely you don't think it was me?'

In truth, at that moment, neither of the officers *did* think it was he. How could it be? Officer Crist was to say later, 'We were sure we had the wrong guy. I mean how could a guy who just killed a cop be that calm and indifferent? It was impossible.' *Surely you don't think it was me?*

Then Officer Odom shined his light under the seat and came out with a .32 automatic, and worked the slide, kicking the live round out of the chamber of the gun.

'Is this yours?' he asked. It was precisely 1:55 A.M.

'Sure it is. I have a permit,' said Greg, taking out the gun registration slip from Nevada.

'This isn't a gun permit,' said Odom. 'It's a gun registration card from where you bought it.'

'No? Well I sure thought it was, and anyway I thought I could carry a gun in the car. My wife has a .38 in her car.' Greg leaned back against the hood of the patrol car and folded his arms and shrugged again. Then Crist patted him down and found the extra rounds.

'Open the trunk,' Odom said.

'Sure,' Greg said, hoping the second key on the ring would fit, and it did, and the trunk popped open.

Crist watched closely as Odom leaned forward into the trunk. Greg seemed to be looking at the clamshell holster.

Then both officers saw the clumps of dirt in the trunk and the irrigation shovel and they knew this car didn't belong to any city man. This was a farmer's car for certain.

They found the other set of license plates, the ones that belonged on the stolen car.

'I'm a citizen, a taxpayer,' Greg said gruffly. 'It's degrading to be handcuffed like this. You're making a big mistake.'

Greg was seated in the back of the patrol car, and another Highway Patrol unit arrived as well as Chief Criminal Deputy Fote of the Kern County Sheriff's Office. Fote had been on his way to the murder scene when he heard the broadcast that the Highway Patrol had a car stopped. He came by and talked briefly with Gregory Powell and suggested they take him to the scene to let Hettinger look at him. Then another patrolman went up to the Plymouth and in a few minutes came back with something Crist and Odom had overlooked – a five-cell flashlight bearing the name 'Hettinger, L.A.P.D.' By now, Chief Fote was in his own car a mile down the highway, and for the first time the composure of Gregory Powell was shattered. He began to tremble violently as he looked at the hostile police faces surrounding him. His jaw was jerking and his voice went hoarse with fear. 'I'll talk to the big detective! Don't you take me away from here! Get the *big* man back here!'

Deputy Joe Hylton was astonished when he received the second radio call of the night. In those remote farmlands it was rare to receive a call at this hour, but now a *second* one coming on the heels of the stolen vehicle call at the Riddick home. And then when details were given to him over his radio, he floored the gas pedal and skidded into the driveway of the Opal Fry ranch.

'I was surprised by this Officer Hettinger. You could take one look and see what he'd been through and yet he took me straight back there, back where they'd shot his partner. The thing is that all this country looks the same to an outsider, and he'd been traveling maybe four, five miles through the dark on foot. He should've been completely lost. And we were coming from the opposite direction he came in from, still he found the dirt road right off and took me to the place. He had a farmer's instincts. I was amazed. Except that when we got there we didn't see the body at first. I looked around and then I found him in the ditch. He was lying there in the moonlight by the onion field. Of course I'd seen a lot of dead bodies as a sheriff's deputy, but this one struck home. The minute I saw that six-inch cross-draw holster on his belt it struck home. This was a brother officer.'

'I ain't ever gonna forget two things there, out there on that little dirt track that night,' said Emmanuel McFadden. 'Two things. First there was Karl. He almost started cryin when the deputy shined his light on the dead man. Karl and me, we jist back on off and let the deputy shine his light and write and make his calls, and Karl stood off there in the dark and held his head. Jist held it, you know, and said maybe it was all his fault and maybe he coulda did more. He was cryin and was awful nervous. I ain't never gonna forget him cryin so quiet like. I always says you can't care too much like that. You jist can't care *too* much.

'But the thing most passin strange that night was the dead officer. It made me come back, back over to the ditch in the dark and I stood there in the wind and looked down there while the deputy was shining his light around. I noticed the watch. He was layin on his stomach and he had his arm up. Looked jist like a livin man layin there looking at that watch, and his eyes was open a little and I say to myself he *is* lookin at that watch. Markin the time!'

Deputy Hylton's last official paragraph stated: 'Due to the fact that Officer Hettinger was very upset and the fact that he wished to accompany his partner to the hospital, the undersigned obtained very little information on his part about this offense.'

Deputy Joe Hylton recorded the time of the discovery of the body. It was 1:55 A.M., the precise moment of Gregory Powell's capture a few miles away.

Chief Fote was in the back of the Highway Patrol car with Gregory Powell and Greg was looking fearfully at the dozen uniformed officers surrounding the car, for by now the word had been passed that one killer was caught.

'Can I get a break if I talk, sir?' Greg asked the big man. 'Can I get a break?'

'I can't and won't offer you any deals. Anything you say may be used in court against you.'

'I'm aware of that, but I wanna talk to you. Only to *you*.'

'Why me?'

'I don't know. Maybe because you look like my father.'

'Well,' said the big detective, smiling slightly, 'I'm willing to listen to you – son.'

Karl Hettinger was taking his last ride with Ian Campbell. On the way to the hospital he was wishing for a miracle. Maybe he'd not really . . . maybe he's not . . . maybe . . . and he turned and looked in the rear of the ambulance from his seat in front and Ian didn't look dead to Karl. His eyes were open and they looked merely sad, deeply disappointed, grieving perhaps, not dead. But the ambulance attendant shook his head, and Karl looked at the blood on Ian's face from the hole under the nose in the dimple of the lip. The blood was leaking down both cheeks, running into the ears.

It was their second strange ride together this night. Only now Karl was in front and Ian was in the back.

SEVEN

Greg continued talking until 5:00 A.M. to Chief Fote and to other deputies of the Kern County Sheriff's Office.

'And Jimmy Youngblood was on the other side, so he got out first and he took the revolver and there was one laying on the seat beside him. So he handed me the automatic and told me to cover the driver. So the three of them were over there and I was still sitting in the car. I laid the automatic on the seat and walked around because I figured Jimmy would want me to drive. And when we got about . . . oh, I guess I was only about a foot or so from him, he said something to the officers. I didn't even catch it and he just started firing. And oh, there was the one officer hollering, and I was hollering. I hollered at Jimmy what the hell was he doing, and he was firing at the officer laying on the ground, and we both, the other officer and myself, took off running. Jimmy was firing at the officer ahead of me I guess, and I was running parallel with him, maybe fifteen feet to the side, and I jumped off to the side and crouched down in some tumbleweed. Jimmy hollered he was going to kill the son of a bitch and he jumped in the car and he took off after the officer that was running down the road.

'I was scared and pretty shook up, and I walked back up on the road and the officer was in the ditch and I felt his pulse. I found the flashlight back along the road and I turned around and started toward the highway. I found the automatic and I picked it up and started running toward the main highway.

'I just wanted to get the hell out of there, and I found this Plymouth and I took it and changed the plates on my wagon over to the Plymouth and hoped that I could get through to her. To my wife. And so that's it. I just took off and I got stopped. I never even fired a gun. I didn't even have any weapon when I walked around the back of that car.'

*

By three o'clock a Los Angeles homicide detective and his partner were driving north toward the Grapevine with the assurance of the chief of police that the detective was to relieve Kern County of all responsibility in the investigation, and that the evidence and suspect, Gregory Powell, would be turned over to him.

Usually the detective hated being awakened from sound sleep, especially on a Saturday night. Not so in this case, a police officer murder with a suspect still outstanding. The first execution murder he knew of in Los Angeles history. He didn't at all mind being called from his bed. Pierce Brooks was in fact exhilarated, stimulated – the traditional huntsman.

A fifteen-year officer, Pierce Brooks was already reputed to be the best homicide detective in the department, an honor usually accorded only much older men. But Sergeant Brooks sometimes *looked* old enough to inherit the title, even though his auburn hair was thick and not yet starting to gray, and his hazel eyes were a young man's eyes. It was his bearing which aged the forty-one-year-old detective, gave him a fatherly look. He was a round-shouldered shambling man whose slouching walk and baggy suits belied the body beneath. He'd played on all the department athletic squads before his transfer to homicide, and there was a hard body under the loose-fitting business suit.

When asked by strangers at parties what his police duties were, he'd reply, 'Catching felons.' And when asked what his hobbies were he'd reply, 'Catching felons.' The detective had led a diversified early life as a World War II naval officer, a blimp pilot for Howard Hughes (flying the dirigible with Jane Russell on one side and 'The Outlaw' on the other) and as a political science student with a taste for historical novels. Now, Pierce Brooks was a single-minded archetypal homicide detective.

Brooks glanced over at Glen Bates, his bearlike, gray-haired partner, as they drove silently down the Grapevine from Fort Tejon, the valley spreading below them. He and Bates had discussed tersely what they knew about the case and both had ridden without speaking for the past half hour. It was understood, though it had never been mentioned, that Bates, the older man, deferred to his younger partner, and in fact was glad to assume less exacting investigative duties connected with their cases because any investigation by Pierce Brooks made rigorous

demands. Brooks's imagination was admired, but his thoroughness was legend.

Brooks appeared to be lethargic, smoking listlessly as they rode,
occasionally glancing at his watch, but his mind was at work. He
was forming a mental picture from what he already knew of the
two killers, Powell and Youngblood, who would soon be
identified as Jimmy Smith, alias Jimmy Youngblood.

Powell had already talked freely to Kern County authorities so
the first step had already been taken for Brooks, but Brooks was
sure the initial confession was laced with lies and rationalizations
which Powell might feel obliged to defend stubbornly. And of
course with one suspect still at large, Powell no doubt had placed
all the blame on his partner. Brooks was sure without knowing
any of the details of the confession. It was as sure as nightfall.

He began thinking about the young officer Hettinger who had
been kidnapped, terrorized, who had barely escaped the fate of
Campbell. He wondered what condition Hettinger would be in.
Brooks could not think of a case in which a policeman had been
put through such prolonged terror. Policemen usually died cleaner
deaths: a sudden gunshot during a robbery attempt, a traffic
violator who suddenly turned out to be a gunman, a distraught
husband or wife who insanely shoots the arbitrator called to their
domestic quarrel. But not this. Not an *execution* in an onion field.
So he thought of what Hettinger would be like after being
kidnapped, witnessing the execution, being hunted. He couldn't
recall *any* murder victim in any of his cases being put through
such an ordeal. Then he thought of Harvey Glatman. His victims
had been sadistically terrorized for hours.

At one time Pierce Brooks had been involved in the cases of ten
separate residents of San Quentin's Death Row, such were the
kinds of Los Angeles murder cases assigned to him. Harvey
Glatman was the only man he'd ever seen executed. He'd
rationalized a hundred times to himself as to why he attended this
execution, deciding it was so that he would be a better homicide
detective because of it, that if he once saw a case all the way
through to the last gasp of life, he would be that much surer and
more thorough in his future investigations. He wondered if it
weren't merely the thing which drew most of the witnesses, a
ghoulish curiosity. One thing was certain, it did in fact make him
a fanatically thorough detective.

Then he thought of Harvey Glatman's victims, the woman kidnapped and bound, then photographed by the murderer, and strangled slowly. He thought of Glatman, the diffident little man who was so afraid of heights he wouldn't climb a ladder, who had a thousand dollars' worth of pornographic pictures in his home, who adored pictures of women in black lingerie, bound in ropes and chains. Who would dash to his television set with his camera and shoot a picture of the screen whenever a movie would show a woman bound, and who, while still a small boy, was discovered in his room with a heavy cord tied from his penis to a dresser, leaning back, groaning in pain and ecstasy.

Pierce Brooks tried to think of Glatman's victims, especially the pathetic lonelyhearts girl, and tried to remember the pictures Glatman had taken of them, tied and gagged, sure of their imminent strangulation, expressions on their faces ranging from hysteria to resignation, to absolute grief. But when they brought the killer into the gas chamber he didn't seem to recognize Brooks or anyone else at the observation windows. He seemed dazed and oblivious to it all, submitted pliably when they strapped him in the chair. And as the observers, jelly-kneed, reached for the supportive handrails, the cyanide was released. Pierce Brooks was to tell his colleagues that condemned men don't go peacefully to sleep in the gas chamber, as advertised. That on the contrary, Glatman died jerking, thrashing, gasping, strangled as piteously by the state as were his victims by him, though the motives were different. The punishment in his case had indeed fit the crime, and Pierce Brooks was indeed an even more diligent detective. But he never witnessed another execution.

'I'm still not against capital punishment,' he said later. 'But I've gotta be *real* sure of my cases from now on. *Way* past any reasonable doubt.'

By 5:00 A.M. Pierce Brooks was in Bakersfield sitting across a table from a young man with a red-blond crew cut in a short-sleeved shirt. The young man's husky voice was surprisingly steady. Other than for an occasional nervous touching of his eyelid, he seemed much like the other off-duty policemen still milling around the station.

'To recap what you've told me,' Brooks said, 'it was the one you now know as Gregory Powell who hit your partner in the face with one shot after making the Lindbergh statement?'

'Yes.'

'And you think he fired an automatic?'

'Yes.'

'And when you ran you looked over your left shoulder and saw oblong flashes?'

'Yes.'

'And round flashes?'

'Yes.'

'The oblong flashes were going down into your partner who was on the ground?'

'Yes. Yes.'

'The round flashes were coming toward you?'

'Yes.'

'And they were simultaneously fired by two different men?'

'Yes.'

'And you think Powell was firing at *you*, not down into the ground. In other words the round flashes were his?'

'Yes.'

'Take a break and then we'll go out to the onion field,' Pierce Brooks said, and he was more than excited now. It was as he hoped. *Both* suspects had shot the officer and this was important to Brooks, crucial in fact. For despite what the law said about principal defendants being equally responsible, he had been a policeman long enough to know that law as stated and law as applied are two different things. That despite the nature of the crime, the trigger man was almost always held more accountable than a passive partner. So he hoped very much that the evidence would put one of the murder weapons in the hands of Gregory Powell's partner. By now, a Kern County autopsy surgeon had made a perfunctory check and told Brooks there was one shot in the dead man's mouth and four in his chest, three of which had exited.

As soon as the sun rose, Pierce Brooks was in the onion field directing a ballistics expert to dust his brush lightly over the blood-stained earth where the blood began, before it tracked across the road to a ditch where a gas pipe lay.

On the second or third swipe with the brush the ballistics man recovered three .38 slugs which were later found to be from Ian Campbell's own gun.

Karl Hettinger was back again in the onion field, but now he was showing the strain and exhaustion – shivering, smelling the tender dew-wet onions, remembering how it smelled last night when he stood here in the dark, arms upraised, and accidentally touched his partner's hand. Pierce Brooks looked at him and said, 'Somebody better drive Hettinger back to Hollywood.'

But it would be afternoon before the police department was through with him and he was at last in his bed thinking the night had ended. He was later to wonder if that night would ever end.

On Sunday morning the Ian Campbell home was filled with people. It was a modest tract home with a floor plan indistinguishable from the others in the neighborhood. It was not large enough to accommodate the numerous uniformed policemen, family friends, and curious acquaintances who had by now heard of the killing and rushed to Chatsworth. Since it was so close to payday there was not enough money for coffee and food for all the people until Hollywood Station sent a policeman with some.

Wayne Ferber was there, his close-set eyes sunken and black against his paleness.

'I heard about it from Chrissie,' Wayne said. 'She called and said, "I want to tell you before you hear it on the news. Ian's been killed." Like that she said it. I couldn't believe it. I knew it must've been a car wreck. When I heard about the murder I thought that's *impossible*. It had to've been an act of God. Then I came to the house. All I could see were gray faces, blank stares. Adah, Art Petoyan. They didn't believe it. Neither did I. Chrissie believed it. She said she knew it the moment the phone rang in the middle of the night. She was able to tell me what happened. No sobs. Tears never fell. She talked with me and then took care of the little girls. I thought about how Ian had always been my best friend and how *he* never needed a best friend. He was completely self-contained. And then all I could do was sit there and wonder if his set and straight mouth would look different in death.'

Art Petoyan was in the bedroom sedating Adah and muttering clichés: 'I said things like, "God moves in mysterious ways," and all that crap which doesn't explain anything. All that crap which I'd said before and which disgusted me, and yet as a doctor all you can do is hope they have some kind of faith, some kind of inner strength. But this poor girl just wasn't like that unbelievable person in the other room. Chrissie Campbell was in there *serving* people. Serving them coffee and making sure the kids were okay, and God, I don't know what else. And of all of us, the close ones, she was the only one in the house who was dry-eyed. I offered to sedate her, but she said she didn't need it. And she didn't!

'After that for many months I treated Adah for her emotional problems, if you want to call it treatment. Like a charlatan, I gave her vitamin-B-complex injections, but she didn't know what they were and they served their purpose because she believed me when I said it would help her. I prescribed thorazine and talked to her when she needed me and finally she settled down into a kind of overwhelming loneliness. She didn't want to go back with her mother, but finally she *did* go back to northern California.

'Adah was no Chrissie Campbell. She couldn't raise those girls alone. I delivered them both. They were big beautiful babies. They looked just like him.'

Among the people at the house were the inevitable newsmen to interview the widow, but when they discovered it would be impossible they turned to Chrissie for the words that could be made both poignant and quotable. The reporters covering the hard news had pretty well gotten the *facts* of the killing and the capture of Powell. So until the other killer was caught, the Campbell family was all they had.

One of the reporters had been told that Campbell's mother was herself a widow who lived alone and that the officer had been an only child. So pencils were readied for a tearful litany of memories, prayers, maybe even a noble sentiment about a son not having died in vain, maybe even a comment on the ruthless senseless act of the killers. And of course a picture, if it could be arranged: a hanky pressed to a mother's face, pitiable, grief-etched.

Instead, Chrissie appeared before them completely composed. They saw an attractive woman with an erect posture, to them middle-aged, but actually past sixty. She said: 'I'm Chrissie Campbell.'

'Do you have a comment you'd care to make at this time, Mrs Campbell?' one reporter asked gently. 'Something that our readers . . . '

Chrissie's reply was buried in the middle of an article relating the events of the killing. The editor told the reporter that it had no value whatever as exploitation of grief or emotion or rage. When pressed as to why then he would boldface the remark and use it, he grinned and said: 'It's the most terse and sensible thing I've ever seen uttered by a victim of a tragedy. Not worth a nickel as news, but so goddamn true.'

Chrissie had merely said: 'There's no comment worth making. My boy is dead and anything anybody can say won't change it.' Then she had smiled politely, turned, and gone back to help with the children.

Karl Hettinger's house was full of waiting friends and family that Sunday, while Karl was still in Bakersfield. The story had been on television and radio by now, and they'd come from various parts of Los Angeles to be there. Among them was Bob Burke, Karl's ex-roommate, who was still in uniform.

Helen was nervously pouring coffee for everyone and running to the door whenever she heard a car. Little of the tumble of conversation was to remain with her, with one exception.

It was Karl's former roommate who said it. He was talking to several of Karl's other friends.

He shook his head and said, 'You can *always* do something. I just don't see giving up your gun to some crook under any circumstances. And even after that, you can do *something*. Karl should've . . . '

And then he saw Helen stop on her way to the kitchen and stare at him. 'No reflection on Karl,' he quickly added. 'I'm not trying to judge anybody.'

Helen went on to the kitchen, but suddenly her hands went clammy. But she dismissed it from her mind, not knowing that Burke, Karl's close friend, was only the first.

At noon that Sunday, Pierce Brooks was finished in the onion field and was facing another young man, just one year older than Karl Hettinger. Brooks, as always, watched carefully during the introduction. He saw flat blue eyes and a long neck and pale hollow cheeks. He saw hands which did not tremble when they met, and a jaw which remained firm. Pierce Brooks did not talk to Gregory Powell about the murder. Not a word. They drove him to Los Angeles speaking only when necessary. It wasn't until that afternoon in the interrogation room of Homicide Division that Brooks broached the subject.

'Greg,' Brooks began, 'would you tell us to the best of your recollection what happened last night, from the moment that the officers stopped you out in Hollywood? And start out first by telling us who you were with.'

'I was with Jimmy Youngblood in the 1946 Ford coupe . . . '

And after an hour of lies and truths and half truths Greg was excitedly relating the moment of death in the onion field very much as Pierce Brooks expected he would.

' . . . and Jimmy was standing out there, and I could hear him speaking to them, and I laid the gun down on the seat, and I walked around back to the car and I walked up to Jimmy and I was about maybe a foot and a half from him. I was going to ask him if he wanted me to drive, or if he was going to drive or what . . . and he fired, and the one policeman went down and the other started running and I hollered, "What in the hell are you doing?" And he kept firing at this one that was running. I turned around and started running, and he started shooting at the officer on the ground again. I don't know why. I ran about maybe forty to fifty feet, and I almost overtook the officer that was running. I dove off into the side because Jimmy was still shooting . . . and hell, I could hear the slugs hitting the ground. I was damned scared. I didn't know whether he was shooting at me for running or at the officer, you know, and then he took off.'

Pierce Brooks sat and nodded occasionally and Greg looked at the chestnut brown around the hazel irises of the tired patient eyes of the detective.

'You mean when you got out of the car in the onion field you weren't armed at all?'

'I was not armed at all,' Greg said evenly. 'I was so damned scared.'

'Was the .32 ever fired that night?'

'Not that I know of, no,' Greg said, telling for that instant the whole truth as he knew it. 'No, it was not. Definitely.'

But Brooks knew that it was.

Jimmy Smith's feet were bloody and he was sitting in a bed of sagebrush by an irrigation ditch with his feet in the water, drinking the irrigation water from the whiskey bottle he'd saved.

Jimmy slept all day, waking with a pounding heart when an occasional Sunday car passed down the lonely road.

Then Jimmy awoke, cold and shivering in the warm sunshine. He put on the socks, dry now, and began walking toward Highway 99 until he found a service station where he could wash and drink water. It was late afternoon and he was famished. He walked until the sun was down and then found a small well-lighted grocery market operated by two old women. He saw a newspaper stand but the Sunday morning paper was too late for the story. There was no murder headline and he was heartened. There was still time. It was just a matter of running. Running flat out. Where, he didn't know yet. Just running.

Jimmy bought a quart of milk, some candy bars, cookies and cigarettes. He looked at the cars parked in front and discarded the idea to rob the women, realizing if he was to do that and steal a car, they would know he was still here in the Bakersfield area. As it was they might think he was in Fresno or San Francisco. He needed another day, a room to sleep in, a bath, one more night to think about it. He couldn't afford to have them know where he was now, and he didn't think they'd find the car just yet. He should have one more night before they found it and maybe he could steal another car by then.

With a belly full of candy, milk, and cookies, much the same diet he'd had as a child shoplifting the Fort Worth markets, Jimmy Smith hiked all the way in to Bakersfield, to Lakeview Avenue, to a place he knew called Mom's Rooming House. Mom's was near Virginia Avenue and Lakeview. It had a lighted multicolored star atop which said 'Mom's Rooming House and Dormitory.' It looked like a private, one-story stucco residence with a chain link fence around it. The neighborhood was black ghetto: shine stands, a pool hall, liquor stores, a bail bondsman, bars on both sides, lots of trash on the street, lots of street corner loiterers. To Jimmy Smith, it looked like sanctuary and peace.

On Sunday evening, an autopsy surgeon employed by the Los Angeles County Coroner's Office was looking at the naked body on the stainless steel table in the glare of fluorescent light in a tiled room with rusty floor drains. On the forward end of the table was a hose. At the other end was a drainage tray. It was unusual to be doing an autopsy on Sunday evening but of course this was a very special case and would need to be done with great care, with no unnecessary mutilation of the head and face.

'File number 88883. The unembalmed body is that of a Caucasian male reported as 31 years of age, measuring 6'2" in length, weighing 195 pounds, with brown hair, hazel eyes, and medium complexion. No scars, tattoos, or identifying marks are observed.'

The police officer from Detective Headquarters witnessing the autopsy digested the technical findings of the autopsy surgeon for police department superiors.

Number 1 slug enters upper lip, shatters upper center incisor teeth, continues through hard palate and through tongue, and lodges in third cervical vertebra. Is a Colt .38 cal. special.

Number 2 slug enters chest at a downward angle, goes through left lung and exits left lower back. Through and through wound. No slug recovered.

Number 3 enters chest at downward angle, parallel to number 2, goes through left lung, breaks rib, and stops just beneath skin of lower back. Is a .38 cal. Smith and Wesson 200 grain.

Number 4 slug enters left upper chest, goes through heart, diaphragm, liver, spleen, kidney, and nicks the aorta. Exits lower left back. Through and through.

Number 5 slug enters right center chest, through right lung, through left lobe of liver, right adrenal gland, and vena cava. Exits lower back. Through and through.

An examination of the vital organs of the deceased indicates he was in excellent physical condition prior to being shot.

A probe was placed in each wound and photos taken to show the trajectory of the slugs.

The trajectory indicates that the weapon used to inflict wounds number 2, 3, 4, and 5 was held 34 inches from the top of the victim's head and 24 inches above his chest.

Autopsy surgeon states that wound number 1 would not be immediately fatal. However, either of wounds 2, 3, 4, or 5 would be instantly fatal.

It was a long and tedious autopsy, much work for the doctor because the remains could not be routinely mangled, but had to be presentable for the ostentatious police funeral sure to follow. It was after 10:00 P.M. before the doctor could remove his smock and finish his notes and leave the morgue.

It was some minutes later when the morgue attendant said, 'Well, that boy's done his all for the goddamn city. Let's call the mortuary and get him outta here. What's left of him.' And he called. His remains were released to the mortuary at 10:40 P.M.

It was 10:40 P.M. when the detective at Bakersfield police headquarters sat stock-still at his desk, the cigarette burning his finger unnoticed, his whisper almost as breathless as the one on the other end of the line. 'You sure?' he breathed. 'You sure? Goddamn! Okay. Okay. Sure. We'll take care of you. Sure. Don't worry. Be there in ten minutes! Goddamn!'

Mom's Rooming House seemed to Jimmy like it should have been a good place for a black man to hole up for a day or two. But thanks to police mug shots, Jimmy Smith was quickly becoming the most famous black man in Bakersfield. He checked in only ten minutes before the Bakersfield police received the hushed phone call.

Jimmy was relieved that Mom didn't give him a second glance when he registered, and Jimmy was happily surprised to find the room was clean. As soon as he closed the door he began stripping down, but despite the mud and stale sweat he was too exhausted to bathe that night. But he *had* to wash the pants and shoes and socks or they'd never be dry tomorrow.

Jimmy pulled himself up from the bed in the tiny room, cursed as he discovered there was not even a wash bowl, and trudged wearily down the hall with the soiled things, heading for the community bathroom.

As Jimmy walked down the poorly lit hall, he passed a tall, light-skinned man who looked at him strangely and kept walking toward the front. It was a small bathroom, barely large enough for the tub and stool. He scrubbed the clothes for perhaps five minutes and as he sat there on the toilet washing the socks he heard voices out front. Jimmy, wearing only his shorts, peeked out and the door came crashing in at the same moment. The back door was also shattered. Then Jimmy was hurled out into the hall and was against the wall, and strangely, he couldn't keep his eyes off the back door as one policeman was methodically hacking through, shouting, 'I cut my goddamn hand!'

Now Jimmy was ringed by uniformed police and detectives and his hands were jerked up behind his back. He was handcuffed and dropped to the floor on his stomach. One officer said, 'Move, you cop-killing bastard, and I'll blow your head off.'

Then Mom came in and said, 'Look at my tenant! Look at that nice young man layin there on the floor with only his underwear on!'

And a detective said, 'Listen, Mom, you get on back in your room and we'll handle this.'

Jimmy felt sure the tall, light-skinned man had recognized him and called the police. He was never to even wonder how so many uniformed policemen and detectives could have gotten there less than five minutes after the tall man saw him. Jimmy was grateful to hear the solicitous voice of Mom. She was a nice old lady like his Nana, he thought vaguely. If she wasn't here they'd kill me.

'Just move, you cop-killing son of a bitch,' another policeman whispered to him, 'and I'll take your head off.' Jimmy did not move and did not speak and thought how strange it was to hear the label 'cop killer' applied to him. He was only a thief, he thought, incoherently. Just a thief. Just a liar and a thief. Been one all my life. Just a sniveler and a crybaby and a thief. How can they say I'm that. *That.* They *gas* people for that.

Then Jimmy was jerked to his feet and pushed inside the room where a big detective was examining Karl Hettinger's gun and he said, 'Is this the gun you took from the policeman?'

'You got it wrong,' Jimmy whimpered. 'I didn't kill nobody.' Then he was being hustled out the door barefoot, with a blanket over his shoulders, walking through the broken glass, feeling nothing. He was turned over to two Los Angeles detectives who had been in Bakersfield and were summoned to Mom's.

'Powell forced me to come to Bakersfield with him and the officers,' said Jimmy to the Japanese detective. When the detective did not respond, Jimmy turned to the Caucasian and said, 'Powell shot that man in cold blood. It shocked me. I was scared. I didn't kill nobody.'

But neither detective looked as though he believed Jimmy Smith. On the way to the Bakersfield jail, Jimmy said, 'You just check those guns. You ain't gonna find my fingerprints on none of them. I didn't fire no gun. Not at all. You check, then you're gonna know I'm tellin the truth!'

EIGHT

At 8:40 that evening, while the body of Ian Campbell was being cut and sawed and disemboweled, Gregory Powell was taken from his cell and was again talked to by an exhausted Pierce Brooks, who knew the coupe had been found in the small town of Lamont with two revolvers, wiped and wrapped in a trenchcoat, stuffed under the front seat. Brooks doubted that there'd be any good prints on the guns, and even if there were they wouldn't prove much, with each man shuffling four guns around in the car after the shooting.

'Here're some cigarettes, Greg,' said Brooks when the young killer sat. 'Now, just before we shut down for the night, there's something that's confusing me. I need to know exactly what Jimmy did, and what position he was standing in at the time the shooting happened.'

'First, can I ask you, have you found out anything about Max?'

'Max is fine.'

'She got arrested, didn't she?'

'That's right.'

'What for?'

'For robbery.'

'Robbing who?'

'Max is under investigation to determine if she's been involved in any of your robberies, and if she hasn't she'll be released. You know darned well Max isn't in trouble, that she's all right.'

'I haven't seen her. I don't know this. I haven't seen her. You didn't tell me that she'd been arrested. You said you were going to level with me, Mr Brooks.'

'I *am* leveling with you.'

'Well you damn sure didn't.'

'Don't talk to me that way,' said Pierce Brooks. 'Just simmer down. How often does your temper flare off like that, young fellow?'

'Pretty frequently.'

'Pretty frequently. It flared off pretty frequently less than twenty-four hours ago too, didn't it? When you shot that policeman in the face.'

'I didn't shoot any policeman.'

'Yes you did. You know the other officer is alive, don't you?'

'Yes sir.'

'And he was there and he told us what happened.'

'I don't care what he said. I told you exactly the way it happened.'

'Who made the statement, "Have you ever heard of the Little Lindbergh Law?" just prior to the shooting?'

'I don't know.'

'You remember the statement being made?'

'I don't recall. We'd spoken about kidnapping when we were riding up there, but now I don't think there was anything mentioned about a Lindberg Law.'

'Do you know what the Little Lindbergh Law is?'

'No, not specifically. I have a general idea.'

'What's your general idea?'

'That kidnapping carries capital punishment.'

'The gun you had during the ride north was your Las Vegas Colt?'

'Yes.'

'This is also the gun you had when you walked around the car and fired at the officer?'

'No sir.'

'And then Jimmy fired into the officer as he was lying on the ground?'

'I didn't have the .38. I didn't have any weapon when I walked around that car.'

'He was shot with two different guns.'

'He was shot with two different guns?'

'That's right.'

'Well I didn't even know this.'

'All right then, why did you have to get out and face the officers? Why didn't you just drop them off and leave them there if your plan was to get a head start into Bakersfield?'

'I don't know. Because Jimmy was standing out here on the side, and I just walked around back and asked him if he wanted

185

me to drive or if he was going to drive, and there was . . . as far as I was concerned . . . well, we had even joked about the fact that it was freezing cold and everything.'

'Why did you have to ask Jimmy that when you do all the driving anyway?'

'I don't always drive. I bought the car for Jimmy. It was Jimmy's car. Just as soon as we got it registered it was gonna be in his name, and it was something that he was proud of, and I just . . . I got in the habit . . . making it a habit of always speaking to him, you know, like if he wanted me to drive or if he was going to drive.'

'You say you saw Jimmy shooting into the prone body of the officer?'

'I saw him shooting downward . . . I don't know . . . I think he was shooting . . . I *know* he was shooting into him.'

'I'm going to tell you now that he *did* shoot into the officer's body, but I'm going to tell you that before this happened, the officer was shot down with another gun, the gun you held all the way up.'

'It didn't happen like that, Mr Brooks. Look it, I know I've had the course anyway. You know, no matter how it goes. I've had the course and I realize this, you know? The officer just got shook up and doesn't remember right.'

'You say the automatic was not fired at the scene?'

'No, it was not.'

'What if I told you that it *was* and that we could prove it?'

'Well if you can prove it, then I'm wrong. But I say it wasn't.'

'Well, I'm telling you that it was, and now there are *three* guns fired. Are you going to tell me now that he fired all three guns?'

'I'll tell you this much, I'll tell you that .32 automatic was not fired at the scene.'

'You're losing your temper again.'

'No, I'm not losing my temper. It wasn't fired when I was present, and I don't believe it was fired, period. Despite anything you might say.'

'You *do* lose your temper pretty quickly, don't you?'

'Yes, I have a rather nasty temper.'

'Do you always shake like that when you get mad? Do you always look like that, like you're looking right now at me?'

'No, Mr Brooks. I can't help it. This is the way I've been all my life. Well, not all my life. I didn't used to have this bad a temper, but since I got out of Vacaville it's gotten worse and worse and that's the way it is.'

'All right. That's enough for tonight. It's 9:30.'

Before going back to the homicide squad room Pierce Brooks smoked a cigarette and paced the hallway and wondered what ballistics would uncover now that the little Ford coupe had been discovered. The .32 shell casing found at the blood spoor proved the automatic had been fired at least once, but preliminary indications were that .38 slugs had done all the killing – the overkilling. Brooks thought of the bullet in the face, enough in itself, and the four in the chest while he was writhing helplessly. They shot him to pieces. They killed him two or three times.

Then he thought of it again: *We told you guys we were going to let you go, but have you ever heard of the Little Lindbergh Law?* Pierce Brooks snuffed out the cigarette and smiled grimly. Losers, he thought. Small-time losers who couldn't do anything right. Who didn't even know that the Lindbergh Law applied only to kidnapping for ransom or kidnapping with great bodily harm. So at the moment Gregory Powell made the statement, he hadn't committed a capital offense. It was Powell's ignorance of the law which killed the young officer. Punks. Stupid, stupid punks, thought Pierce Brooks. Then he walked through the doorway into the squad room.

At ten o'clock Sunday night, after having been on duty twenty hours, and after having had only two hours of sleep before that tour of duty began, Pierce Brooks was home, stomach knotted and acid-full, nauseated from the cigarettes and coffee. But he would barely have time to eat and bathe. He would have *no* sleep. He received another telephone call. Jimmy Smith had been captured. This time Brooks and his partner were driven to Bakersfield in a detective car by a third detective, since neither of them was in any condition to drive, and they dozed fitfully as once again they crossed the Grapevine Highway.

At 3:30 A.M. on Monday, a bedraggled Pierce Brooks sized up Jimmy Lee Smith and decided to interrogate him there at Bakersfield police station. He seemed anxious to talk.

' . . . Now, wait a minute, before I get started, I wanna tell you that I'm not sayin this lookin for no help or nothin because I know there ain't none now. It's too late now. This is actually what happened. . . . '

And then Jimmy began a long tale full of lies, truths, half truths, and like Greg, he drew numerous diagrams for Brooks and was encouraged by the noncommittal nods of the detective, and thought that at least the detective didn't *dis*believe him. On went the story of Greg whispering strange things to him on the trip about a Little Lindbergh Law, and saving a bullet for himself. Pierce Brooks merely nodded occasionally in encouragement up to the moment Jimmy Smith was standing in the dirt road in the onion field and Greg came around the back of the car.

'I forget exactly what he said, somethin . . . he said somethin . . . Then he fired and he shot this officer. I don't know if he shot this officer once, twice. I think he shot him just one time. But anyway, I don't even move. I'm just petrified more or less. He starts to shoot at this officer who breaks and runs right down the road. Greg runs over there. He's runnin, but he's steady firin. Then Greg comes back and tells me, he says, "Let's get the car and catch him," you know? So I started comin toward the car. "Did you kill the officer?" I said. "Is he dead?" "Yeah, he's deader than a mackerel," he says. So he starts reloadin the guns and droppin shells. "Do you want me to drive around this way and turn around?" I say. He says, "Yeah." I took off and got down that road and went off and left him. Now that's just a fast of what happened.'

'You didn't fire any gun?'

'No sir.'

'You had the .32 automatic in your hand?'

'Yes sir.'

'When Greg stepped around the car just before he shot, which gun did he have?'

'I don't even know. When he comes around the car I stepped back three or four steps in order to give him room, see?'

'What did you give him room for?'

'I don't know. I just stepped back, you know?'

'Well why did you give him room? You saw him carrying a gun?'

'No, I . . . '

'You knew there was going to be a shooting.'

'No.'

'He whispered something to you.'

'He didn't.'

'You knew there was going to be a shooting then.'

'No sir.'

'You didn't want to get in the line of fire.'

'No sir.'

'Then why did you move back?'

'I don't know. I'd tell you if I did because it wouldn't make no difference.'

'Did you fire any shots at all?'

'I did not fire a shot. It don't make no difference. I want you to know I'm not lyin to you, you know?'

'That's what Greg said, that it doesn't make any difference and he has no reason to lie.'

'He did? Well, it don't make any difference because we're both gonna get gassed anyway. It don't matter, you know? This is the point I've been tryin to get over all night to these other officers. I wanna get the point over to them, to any man. I was scared to death. I mean, hey, I don't have the nerve to kill no man just cold-blooded, just outright. He shot that man down like a dog.'

'How many holdups have you pulled with Greg?'

'I haven't pulled none with him.'

'Are you telling me you have not been involved in any way either inside or out as driver on any robbery at all?'

'That's right, sir.'

'And are you also telling me you didn't fire a shot in that onion field?'

'I didn't fire a shot.'

'I'll finish by asking you one more question about what happened after Greg fired that first shot and the officer fell.'

'Yes?'

'Did he then fire some more shots at the fleeing officer?'

'I guess. I don't know how many.'

'Is that *all* the shots that were fired?'

'Yes.'

'How many times?'

'Three or four. Until the gun went click.'

'Did you fire even a single shot?'

'No.'

'Do you think you could ever kill, Jimmy?'

'No.'

'Jimmy, we believe you were involved in robberies with Greg. At least as a driver.'

'I can tell you don't believe me, Sergeant.'

'That's right, I don't believe you.'

'Please, Sergeant, believe me. I swear it on my mother's name. I want you to believe me.'

Pierce Brooks made a careful note that Jimmy Smith had given an account of the crime leaving out one conspicuous fact – the four shots fired into Ian Campbell's chest.

The Los Angeles papers were full of execution news on Monday, March 11. 'Moonlight Execution' was the headline, with pictures of the officers and the killers and Pierce Brooks.

Brooks drank coffee and munched toast that morning and read, and his newspaper told on the same front page of legal executions which were taking place in other parts of the world. In Paris, Jean Marie Bastien-Thiry was executed as a member of the political underground which merely *plotted* the murder of Charles de Gaulle. Brooks smiled crookedly at this and at another article telling of the execution in Leningrad of five men who were found guilty of black-marketing those goods their factory had produced above its quota. Death was the sentence in Russia for a crime which would hardly qualify as a high-grade California misdemeanor. Brooks wondered what his counterparts in the police forces of Paris or Leningrad would say to his fears that one of his two killers would evade the gas chamber for kidnapping and brutally executing a police officer. *We told you we were going to let you go . . .*

He was thinking such things because he had received a bit of distressing news from a reinterview with Karl Hettinger. Karl had changed his mind about a positive identification of Jimmy Smith as the one who stood over Ian Campbell and fired the four shots into the dying man.

'I'm just not sure, Sergeant Brooks,' Karl said, looking even more overwrought than he did that morning after the killing.

'But you're at least sure the shots were simultaneous with those being fired at you as you were running?'

'Well, I hate to say for sure. After all, I was wrong about Powell shooting Ian with the automatic. Now you tell me it was the Colt revolver.'

'Karl, that's a tiny detail. It was dark out there that night. Listen, your story's been right down the line and remarkable for what you've been through. I've been bragging about what a great witness you are. Don't lose your confidence because of a tiny discrepancy as to which gun you thought you saw in the dark in Powell's hand.'

'I'm not.'

'But on the other hand if you're just not sure about who fired the shots down into the body, well . . . '

'I'm just not, Sergeant. I was running. It was dark. Guns were flashing. I'd bet my life Smith fired down into him from where they were standing when I looked back and from the shots all being nearly together, but I guess I just feel it's remotely possible that Powell shot at me and then ran over and shot Ian four times in the chest and it only looked like Smith in the darkness.'

'*Was* it Smith, Karl?'

'I *know* it was Smith, but I can't say for sure in court because I couldn't see that well.'

Pierce Brooks, despite his disappointment, understood Karl Hettinger's unwillingness to say he was certain when he was not. It would be so easy to say it. Powell was with Hettinger precisely on this point. But Brooks understood and respected the young officer's honesty. He knew Karl would be criticized implicitly by others for reversing his earlier positive identification of Smith as the final executioner. But Karl would not be criticized by Pierce Brooks.

It had just been the year before that Brooks had investigated the Ronald Polk gang, a group of highway bandits who would pick up hitchhikers, rub pepper in their eyes, and rob or murder. Brooks suspected the gang of five murders. Only one could ultimately be proved. It involved the robbing of a young sailor who was picked up hitchhiking and, after failing to respond to caresses of the tall transvestite member of the group, was robbed and finally shot to death because he fought and screamed when his penis was being cut off. After the shooting the transvestite did a ritual dance around the bloody corpse as he waved the severed trophy in the air and finally put it in his mouth, laughing with a woman's voice.

Ronald Polk was, during the course of the trial, given a psychiatric exam. The psychiatrist concluded her evaluation: 'I would consider this individual incorrigible. He has much hostility in him for being poor and seems to have an unending reservoir of energy. This type of habitual criminal neither profits from experience nor punishment. He can only work against society and thereby derive power, and he will always be able to find followers whom he can impress with his intelligence and destructive drives. He will never be able to work within society. Diagnosis: sociopathic personality, antisocial type.' And then the French psychiatrist could not resist a Gallic quip which Brooks felt the defense had a right to pounce upon at the penalty trial of Polk: 'May I suggest in all sincerity that this individual be given a rightful place in lifesize format in Madame Tussauds Wax Museum.'

It was Pierce Brooks himself who helped the gang member he considered most dangerous to escape a murder charge, much to the disgust and consternation of other detectives. Brooks, troubled by a time sequence in the story of the accused killer concerning the murder in Tulare County, reopened the investigation in mid-trial and uncovered jail records to show that it would indeed have been nearly impossible for the suspect to have driven in time to the Tulare murder. Though he wanted the gang in the gas chamber more than any he had ever investigated, he did not regret his decision.

'This beast should be in the ground,' a Tulare County detective had said, outraged that Brooks's inquiry had resulted in a motion to dismiss the murder charge.

'I don't disagree,' said Brooks.

'He's what Death Row's all about, for chrissake!' the detective sputtered.

'I think you're right,' said Brooks. 'I've been to the row. I've put lots of people there who would just eat up average folks. There wouldn't even be any bones left.'

'Then *why* did you do this? Why screw up our case? Why?'

And later Brooks took some kidding when another gang member, a moronic, incredibly violent black man was asked by the psychiatrist if there wasn't somebody, some one person in his entire life whom he wanted to emulate, *besides* gang leader Ronald Polk. And the killer thought, and chewed his lip for a moment, and then smiled brightly, saying: 'Sure. There's *one* person. I'd like to be just like Sergeant Brooks!'

Pierce Brooks would never criticize Karl Hettinger for saying he simply wasn't sure. He understood what it was to have to be sure far past a reasonable doubt. So he sighed and accepted the minor setback and decided it just meant he'd have to work harder on Smith. He knew Jimmy Smith did not have Powell's tremendous ego which was producing a gush of information about the robberies, most of it remarkably accurate, indicating a good mind and an extraordinary memory for detail. Smith was rather without ego, a quiet-spoken, wily sneak thief, an ex-hype, a street hustler who survived by his wits. Brooks was painting a mental picture of Smith, of a leech who clings to aggressive movers like Powell, who is satisfied with crumbs. He knew Smith would trust no one, especially not a cop, white or black. Smith would respond neither to kind words nor harsh ones, was merely clinging to a glimmer of hope that if Brooks could not prove he fired the four shots, he would escape the gas chamber.

Pierce Brooks studied the records of both men for any clues to their personalities, especially Jimmy Smith's, and then he sat at his desk and examined the mug shot taken on the eleventh of March and compared it with one taken in July of 1950 when Smith was not yet twenty years old. In the first he saw a light-skinned, curly-haired mulatto youngster staring into the camera arrogantly, his eyes tough, the set of his mouth defiant. In the recent mug shot life had taken its toll and Pierce Brooks looked at a pleasant face, anything but tough, eyes soft and rather moist, the naturally strong jaw not set strongly, rather as though he was used to holding it loose, ready to break into a subservient smile. Smith's forehead was wrinkled. Brooks guessed his forehead was always wrinkled in expressions of distress, humility, servility. He was convinced that Smith was a whining coward and that was the key to him.

Brooks swore to himself he *would* confess. Jimmy Smith would follow Gregory Powell as he always had – right into that apple-green room in San Quentin.

On Monday night, with just one day's rest, Karl Hettinger was back on duty. He arrived at Hollywood Station an hour early to get it over with, the questions, the well-meaning questions, the insatiably curious questioning policemen he knew he must face. He hadn't had time to do much thinking about Saturday night and what it meant. He'd eaten, slept, awakened to talk to relatives and friends, and slept some more. He'd slept brokenly, an hour of deep exhausted sleep followed by an hour of hot, fitful, dream-laden sleep. Then another hour of merciful exhaustion.

Now, approaching the parking lot at Hollywood Station, he felt like a stranger, like he'd never been here before. He'd always liked this station. It was an old comfortable building. It *looked* like a station house. Now though, it looked different. He knew most of the uniformed policemen he saw leaving and entering, but *they* looked different. For a second as he pushed through the old swinging door he felt at home, and he wondered if it was his or Ian's turn to drive tonight. And then as he thought it, the blood jetted to his head and he felt dizzy and had to talk silently to himself until he was calm again.

One hour later he was *not* calm as he entered the patrol watch commander's office.

'Hettinger!' said the lieutenant. 'How's . . . '

'Lieutenant, I'd like to talk to the men at rollcall.'

'You would? Why?'

'I've just had thirteen guys ask me about Saturday night. I wanna talk to the rollcall. I wanna tell everyone about it at one time and get it over with. I just don't wanna keep telling it.'

Karl *did* talk to the rollcall. In fact, his superiors thought it would be a good idea for as many policemen as possible to hear about the kidnapping so he talked to rollcalls again and again. On the very first rollcall discussion that night at Hollywood Station, to the uniformed officers of the nightwatch, an older sergeant was the first to ask a question when Karl finished his recitation of the events from the moment of kidnap to the rescue at the farmhouse.

'Question,' said the sergeant, his foot up on a chair, the three chevrons looking bold on the blue sleeve.

'Yes?' said Karl.

'The purpose of this talk is to help other policemen, so the things that happened to you and Campbell don't happen again. Now let's hear your opinion about how you guys fouled up. The things each of you did wrong. Or what you *didn't* do and should've done.'

Karl stared at the sergeant and his mouth went dry and then he looked back at the faces in the room and some of them were looking at the sergeant and some at Karl and some were looking away. He answered something but did not remember what he answered, and then after a few more questions the watch commander broke up the meeting.

That first night he was back on duty, that Monday night, Karl and his new partner stopped a truck as their first contact of the evening. It looked very like one which had just been reported stolen in the vicinity. But the occupants were not auto thieves. A woman was driving and the passenger was her daughter. Karl's partner had a laugh with the woman when he explained why they had been stopped. When Karl walked back to the police car he wanted to grab the fender for support. He actually stopped walking and pretended to be checking his flashlight because he was afraid his legs would cave if he took another step. He clenched his teeth to keep his jaw from trembling, but there was nothing he could do about his legs. Only walk and hope they held him up. They did, and his new partner never seemed to notice. It was the second longest night of Karl Hettinger's life.

On the day before Ian Campbell's funeral, Pierce Brooks decided it was time for a confrontation between Gregory Powell, Jimmy Smith, and Karl Hettinger. By now, with the preliminary ballistics work done, he had a pretty good idea of what had happened when the shooting started.

Powell had fired one into Campbell's mouth. Hettinger ran and Powell emptied the gun at him. He fired three, had one misfire, and clicked on the empty cylinder. That accounted for the five he carried in the gun. At almost the same moment Powell was firing, Smith cranked off a round from the automatic at Hettinger. That took care of the shell casing they found at the scene. Then the ambidextrous Smith, with Campbell's Smith and Wesson in his other hand, stepped forward and blasted four more into Campbell's prone body. In the panic and confusion, Powell sat in the car with the heap of guns and reloaded Hettinger's Colt which one of them took from the glove compartment. In the dark he actually reloaded the only one of the four guns which hadn't been fired at all. Powell took out the rounds, fumbled around, and put two of them back in along with four of his own 158 grain ammunition. The other guns weren't reloaded and told the story.

*

Pierce Brooks took Greg out of his cell that day and into an interrogation room. Before bringing the others in, Brooks said, 'Sit down, Greg. I've got a little news for you.'

'What's that?'

'Jimmy Smith has been arrested and is in the building right now. He's told us a different story than the one you've told us.'

'Well' – Greg shrugged – 'as long as you've got Jimmy I may as well tell you, I popped off the first cap and Jimmy popped the caps into the officer after he was down.'

And that was that. He'd said it so casually it was anticlimactic. Almost disappointing. The fight was over as far as Gregory Powell was concerned. Brooks glanced at his partner and said, 'All right, let's tape it.'

Before Jimmy Smith's arrival, Greg softened his spontaneous declaration of a moment ago: 'Well, everything was just exactly like I stated it before except for one thing. When I walked around back of the car and walked up to Jimmy, I don't know exactly how it happened, but my gun went off and I hit the officer and he went down. And he was still moving and the minute it happened, I knew, well, there's nothing else to do but go ahead and try to get the other one too, you know, and so I started shooting at the other one. And he was running, and I ran off just about even with the other officer, and while I was shooting at him Jimmy said, "Hey this son of a bitch is still alive," and started popping caps into him.'

'All right now, Greg, when you got out of the car, which gun did you have?'

'I had my .38.'

Brooks took the four-inch Colt from a briefcase and held it up. Greg smiled and said, 'That's my baby.'

'All right, do you know positively which gun Jimmy had?'

'Yes. He had the .38, the police .38 that I had previously had in my waistband.'

'Would this be . . . '

'The driver's. He put the other officer's gun in the glove compartment so I handed him this one. He didn't know anything about automatics.'

'After the first shot at Officer Campbell do you remember how many times you fired at the officer running down the road?'

'Yes, I fired until my gun was empty. I carried five in the gun, always keeping the hammer on an empty cylinder.'

'After Jimmy handed you the .32 automatic, did you take one shot with that .32 at the officer?'

'I don't think so.'

'The .32 was fired, Greg.'

'I didn't fire it, no. I don't think Jimmy did. Maybe he did, but I don't think so. I didn't fire it because I didn't have it until afterwards when he handed it to me.'

'And now, if we bring Jimmy Smith up here, will you tell him the same story to his face as you told us?'

'Definitely, but there's an awful lot of hostility towards Jimmy.'

'We'll keep you separated here.'

'I have an awful temper.'

'All right, let me warn you, we're not going to permit any altercation. We'd like you to conduct yourself as a gentleman.'

Before Jimmy arrived, Brooks left and spoke to Karl Hettinger, who waited in the squad room.

'Okay, Karl, there's one thing I'd like you to do. When we all meet in here, you go ahead and tell your story the way you did originally, that you saw Smith fire the four shots. It'll just be a form of accusatory statement here in the interrogation. I think he might go ahead and cop out then. Only leave out the Lindbergh statement. I want Powell to talk. I'm willing to let him save face and rationalize. That statement might frustrate him so much he'll clam up.'

'I *hope* he cops out.' Karl sighed. 'I'd just like to get it all over with.'

At 11:00 A.M. Jimmy Smith was brought inside the room and he and Gregory Powell sat on opposite ends of the table. They looked at each other with hate and accusation, each feeling victimized by the other.

Gregory Powell was the first to tell his story, omitting the Lindbergh statement which Pierce Brooks let pass because it was so devastating, so brutal, it could upset him if it were dwelled upon, and Jimmy Smith had verified it separately. We'll save that one for the jury, Brooks thought.

' . . . I walked around back of the car,' said Greg, 'and shot the officer, and he fell to the ground and the other officer hollered and started running. I started shooting at him and Jimmy said, "That son of a bitch isn't dead," and started firing into the officer that was lying on the ground.'

'All right, Jimmy, you've heard Gregory's statement?'

'Yes sir.'

'Do you want to tell me what happened?'

And Jimmy said, 'Yes sir. We stepped around the car like I told you and he fired and shot the officer. I had the automatic in my hand, just like I told you I did, and I hadn't fired at nobody. I didn't . . . '

'Greg said he handed you the driver's gun when you got out of the car in the onion field.'

'No. I left the .38 layin there. This is the one the officer out in Hollywood, that he . . . the one that he handed to me, the little officer handed it to me. I still had that one.'

'Jimmy, you put that gun in the glove compartment,' said Greg.

Pierce Brooks then said for the benefit of the tape, 'The time is now 11:05 and Officer Hettinger has just entered the room.'

Then to Karl, 'This man sitting to my right – Gregory Powell – what did you see him do when the first shot was fired?'

'Powell was holding a weapon in his hand,' said Karl very deliberately. 'He raised it. He fired at Officer Campbell, struck him, knocked him down. I then spun and ran northbound along the same dirt road we had come up in the car. I looked back and saw Smith standing over Officer Campbell. He was firing a gun into his body. Then I saw Powell fire at me twice and didn't look back and heard two more shots. I assume they were fired at me.'

'I don't know how to say it,' said Jimmy. 'I swear I don't . . . I . . . '

'Before you say it,' said Brooks, 'let's ask Gregory here, what the officer told us, is that what happened?'

'That's substantially the truth except as I walked around the back of the car, without saying anything, as I was walking up, the gun went off the first time.'

'Now, do you want to tell us your story, Jimmy?'

'Anything you want me to say,' said Jimmy.

'I want you to tell us what happened. You've heard Powell and you've heard the officer.'

'Yes sir. I did not fire into that officer's body.'

'All right,' said Brooks. 'How could Greg possibly have gone over to the place you mentioned and shot at this officer and at the same time fired into Officer Campbell? The trajectory and the physical evidence bear out the story of Gregory Powell and this officer.'

'Yes sir.'

'It does not bear out your story.'

'Yes sir. But I didn't shoot him. I'm positive.'

'You say that Powell fired one shot at the officer. The officer fell down. Then he fired more shots at the fleeing officer until you heard a click.'

'Yes sir.'

'Then after the shooting was over, he came over to you and you got in the car for the reloading. How did the four bullets get into Campbell's chest? And how did one shot get fired from the .32 automatic? The empty casing was found by the blood spot.'

Then Jimmy began a rambling statement that took them back to the Hollywood streets, and Pierce Brooks tapped his fingers impatiently, but did not interrupt because one never knew when an innocuous statement could end in confession. Then Jimmy finished by saying, 'I don't know if he is firin at this officer or how. I said he's firin at this officer because to me it's the way it looks. It's all hazy to me. If you want me to admit, sir, I will.'

Pierce Brooks stiffened, because a remark such as that usually signals a seeking for rationale, a push, a man who wants to tell, doesn't know how, needs help. But of course he could not encourage him or the confession would be inadmissible in court.

'We want you to tell the truth,' said Brooks, and then he slumped disgustedly when Jimmy said, 'I swear by my mother I didn't shoot him. This don't mean nothin to you, but I'm gonna get the same he gets, so why deny it?'

'Do you recall, Officer,' said Brooks, 'when Powell ordered you out of the car did he have just one gun in his hand?'

'I only saw one gun in his hand,' said Karl.

'All right, Jimmy, I want you to explain to me how there were four shots fired into the officer's body after Powell had fired all of the ammunition in his .38 and you drove off with the other guns.'

'Yes sir, when we got back to the car . . . '

'You are not answering my question, Jimmy.'

And so it went, and after it was over Jimmy turned as they were leaving the room and said to Brooks, 'Sergeant, I'm sorry my memory was hazy and that I lied to you before. But now I *do* remember Powell firin into the officer's body on the ground. And I know I fired one shot outta the .32. It bucked in my hand so I *musta* fired a wild shot somewhere, maybe at the officer runnin away.'

'He's a goddamn liar,' said Greg. 'You're just a lousy punk and too chicken to tell the truth!'

'Well you told your share of lies too,' said Jimmy.

'You're a goddamn liar.'

'Okay, bastard, I'm puttin a curse on the baby, on that unborn child that Max is carryin around. It's gonna be born dead, hear me?'

'That's enough of that kind of talk,' said Pierce Brooks and the two friends were led back to their separate cells.

Pierce Brooks, during the days to follow, talked individually many times to both of them, on virtually every crucial point, until one of them verified each event told by Karl Hettinger, including a remarkable recorded admission by Gregory Powell which he was later to deeply regret making.

'I *did* think of killing the officer, Mr Brooks, as we came down the Grapevine. The thought came to my mind if we hid their bodies in one of those canyons, they never would be found. I shied away from the thought to find some other alternative to turning them loose, and yet keep them tied down long enough for me to get back to L.A., get my wife and come back to Bakersfield to get my station wagon and get a running start.

'As I got out of the car I was still trying to think up some alternative, and didn't let the thought of killing them enter my mind. I thought of handcuffs, but realized that daylight was only four to five hours away, and there was no possibility of holding them long enough to do everything I needed to do.

'So without even considering anything for fear that I'd change my mind, without really facing the fact that this is what I intended to do, I deliberately kept my mind occupied with other thoughts as I walked around the back of the car, raised my gun, and shot the officer.

'I didn't consciously think that I had to kill. I didn't dwell upon it. I just raised the gun, fired at him, and immediately tried to hit the other officer, still without thinking consciously: this man I must kill also. Because if I let the thought enter my mind of what I was doing, I might've been confused and too scared to do what I knew or felt had to be done.'

Pierce Brooks, anticipating the defense at the trial, the only defense other than insanity, said, 'Greg, do you think by any chance, for any reason, that you fired the gun accidentally at the officer when you came around the back of the car?'

And Greg looked at the patient paternal tired eyes of the detective and said, 'I've handled the gun enough that I'm competent with it, and the chances of it going off accidentally are nonexistent.'

Also anticipating the defense that Greg's Colt had a 'hair trigger,' Brooks tested the gun and found that, cocked, it had a trigger pull of more than five pounds. A police revolver's authorized trigger pull was only between two and a half and three and a half pounds, making a hair trigger defense impossible.

'You've told me you're pretty good with a gun and a good shot. Now I know where the officer was hit. Just for kicks, you tell me where you were aiming when you squeezed off that shot.'

'I aimed for his heart.'

'Okay, the heart.'

'You can end it this way,' said Greg. 'The only other alternative was to give myself up or kill him. I think I thought of this previously and thought I would rather be dead myself than give up.'

'Greg, you told us a story and Jimmy also told us an identical story of somewhere in the desert you stopped to fire off a few rounds.'

'That's when we bought the gun. It was outside of Henderson, Nevada, about three miles, and it was about a mile on this side of a railroad pass, and we pulled off to the right of the highway, and we drove down to this sort of junkyard there out in the desert. We fired at an old car.'

After Greg related the story of his string of robberies to a robbery detective, exaggerating his daring only when his ego demanded, often glancing toward Pierce Brooks for approval, Brooks had him relate in detail how he and Jimmy practiced shooting in the desert.

Though Greg could not imagine why the detective was interested in their target practice, he was proud of his memory and happy to draw Brooks a detailed map and diagram of the

Henderson, Nevada, trash dump where he and Jimmy had practiced firing. At the trial, four months later, it would be used to destroy defense contentions that the shooting of Campbell was a spontaneous gesture by men who had never intended, nor were they prepared, to hurt another human being in any of their holdups.

Greg demonstrated for Brooks his understanding of the police combat stance which he had learned in the prison yard and which he tried to teach Jimmy Smith. Afterward, Pierce Brooks drove to Henderson, Nevada, and returned, more than satisfied, to write the following:

As a matter of interest, and to demonstrate Powell's memory, the following comparison is listed, indicating actual measured miles, etc., versus information related by Powell during the interview.

	Powell	Actual
Dump location from Henderson	3 miles	3.4 miles
Markers, in order from Henderson	Speed sign	Speed sign
	Rest Area	Quarry Road
	Quarry Road	Rest Area
	Road to dump	Road to dump
	Railroad pass	Railroad pass
Dump distance from R R pass	1 mile	1.4 miles
Name of quarry	White Rock	White Rock
Distance, highway back to dump	400–500 yards	330 yards
Fired .38 Colt	3 or 4 times	Recovered 4 casings
Fired .32 auto.	4 to 5 clips (32 to 40)	Recovered 33 casings

It is to be noted that all empty casings, except one, were recovered about 8 feet away from the old car used as a target. It is estimated that Officer Campbell was approximately 8 feet away from Gregory Powell when he fired the first fatal shot.

Respectfully,
P. R. Brooks, #5702, Homicide Division

The defense had all but lost the spontaneity theory as well as the contention that much of Powell's confessions were made by a man with a poor memory.

Then, after Greg had in detail discussed the murder and his string of robberies and the practicing in the desert, it was safe to broach the statement and try to secure a remorseless admission to having made it:

'Greg, there's one point that I think you've hedged on a little. When the four of you were standing there, the statement you made about the Little Lindbergh Law just before you fired ... '

'That's wrong, I said absolutely nothing, Mr Brooks. I heard the statement being made. I don't know who made the statement, whether it was Jimmy or them, but when I walked around the back of that car my mind was blank as I've told you.'

'All right,' said Pierce Brooks. 'Now, I want to ask you once again, would you object for any reason at all if I come and see you once in a while at the county jail?'

'No, I sure wouldn't.'

'I mean, does it ... does it bother you? Or do you want to sign it off here?'

And Gregory Powell's recorded answer was:

'No. I'd like to see you from time to time, Mr Brooks. You've been very fair and if I would ... if the circumstances were reversed, I don't think I could treat anyone more fairly than you've treated me, and I do feel a certain ... '

'Well, it's part of my job, Greg.'

' ... friendship, you know?'

'Greg ... '

'Well, above ... above the job. I know you're doing what you do as far as the job goes, but above that ... '

'Greg, I'm trying to ... '

'I feel you've shown me more than just a normal policeman doing his job. You've shown me consideration about that, and I appreciate it.'

'Well, I ... I've told you the things I could and would do for you. ... '

'And you've always got it done. Things you didn't have to do at all, not policeman things.'

And now Brooks, knowing the tape would be played for the

court, was embarrassed by Greg's declaration of affection and feared it might taint the confession.

'They weren't things that . . . in other words, I've never said to you that if you don't tell me the truth, I won't do these things for you?' said Brooks.

'Oh no, no.'

'And actually it's part of my job.'

'No, you've been rather good to me, Mr Brooks.'

Gregory Powell, at a later time, crestfallen that he endangered his life by telling so much to Pierce Brooks, sought to explain and understand his feeling toward the detective:

'He was remote as a mountaintop, but yet forgiving and understanding as the dearest person in the world might be if he lived up to everything you hoped. He was stern and disapproving, yet not angry, just rather sad that you were so weak.'

But if the detective had earned something approaching filial devotion from one killer of Ian Campbell, he was meeting stubborn, frustrating resistance from the other.

'Jimmy, your original statement was that you had not fired a shot. That Greg had done all the shooting.'

'Yes sir.'

'In fact, you said nothing about the officer even *being* shot after he'd fallen on the ground.'

'No sir.'

'Now, listen carefully. On one occasion while you were relating this story to the Bakersfield police, you said you left the car with the .38, *not* the automatic. Then you corrected yourself.'

'Well I can't remember sayin that. I swear I don't.'

'You said a .38 revolver.'

'I don't remember sayin it. I swear I don't.'

'And you said the first time you felt Greg was going to kill them was driving down the little dirt road.'

'But I didn't know it for sure. Greg said in my ear on the trip it's either them or us, you know? And about the Lindbergh Law and everything, and I kind of flinched, you know? And I tried to change the topic right away.'

'What do you remember Greg saying just before he shot Campbell?'

'He said . . . he was saying somethin about the Lindbergh . . .

but I . . . if I finished the conversation I'd be lyin on him. I don't
know what all he said.'

'Jimmy . . . '

'I know I'll never forget that officer's light-colored jacket
fallin . . . '

'Jimmy . . . '

'Fallin, you know?'

'Jimmy . . . '

'I heard a pow, you know, and then I see this officer look like
he'd been lifted up.'

'All right . . . '

'If I could a just thought of the handcuffs I coulda saved it all,
but I couldn't think . . . didn't think of the handcuffs.'

'All right. That's it for tonight, Jimmy.'

'Yes sir. I wonder if you recorded all this? Is it all recorded,
Sergeant?'

'Yes,' said Brooks. 'It's recorded.'

Pierce Brooks was not surprised when neither man at any time
showed the slightest hint of remorse or concern for the dead
policeman. The detective understood sociopathy, knew that the
thing ordinary people call by numerous names, such as conscience, is as absent in the sociopath as the central nervous system
in a shark. And since he didn't expect remorse and understood
that for them it does not exist, he never became incensed or
horrified no matter how foul were the deeds confessed to him over
the years. Therefore the detective knew no outrage. Like most
policemen, Brooks could easier forgive a cop killer than one who
frustrated every interrogation technique he used on him, who
spent hours upon hours talking to him, and told him nothing.
Who stubbornly resisted any attempts to cajole, ensnare, outwit.
And who, Brooks knew by now, would reveal a bit at a time, only
when he felt he must, after he first was fed police information and
discovered how much the detective knew for sure. Who simply
would not buy what a homicide detective is selling: a trip to the
gas chamber. This was not in the detective to forgive.

More than Gregory Powell, Jimmy Smith was what a man like
Brooks despised: an admitted liar and coward. Brooks reasoned
that after Smith snapped one quick shot at Hettinger with the
automatic he hated, he pulled the other gun, Campbell's own
revolver, turned, saw the policeman writhing on the ground, saw

the man helpless at his feet, and in panic and horror and even rage for that which he felt had victimized him his entire life, he stepped forward and jerked the trigger four times. Smith's story that he left Campbell's gun on the seat when he got out, and that both Powell and Hettinger did not see it there in plain sight when they slid out, just didn't make sense.

But most of all it was that Jimmy, during their first conversation, failed to mention the four chest shots at all. Though he remembered what happened before and after, it was as though the chest shots never happened. As though he blocked them out of his mind.

Pierce Brooks's contempt for Jimmy Smith was clear in a report he later wrote for his department superiors:

> It is hardly necessary to state that both men were homicidal, however, they are definitely of two different types. Powell, a boastful egomaniac, was a cool, treacherous, scheming, cold-blooded killer. Smith, con-wise and cunning, was more impetuous and cowardly. Thus, if possible, Smith could be considered a more dangerous killer than Gregory Powell. There was no question of their sanity, and no pleas of insanity were entered.

During his first night in county jail Jimmy Lee Smith had a fearful dream. In the dream the arm of Ian Campbell, the jacketed arm and hand, jerked toward Jimmy four times, once for each bullet that was fired into his chest. Jimmy told the detectives of the jerking of the jacket sleeve. And though he could almost draw a mental picture of Gregory Powell standing over Ian Campbell firing those shots, each time the image would slip and fade. He could not clearly see Powell, much as he wanted to, only the sleeve and the hand jerking.

For an instant Pierce Brooks was forced to consider the possibility that neither man was willfully lying about those four chest shots. And if that was the case, then of course he could never get Jimmy Smith to confess. Brooks would have to do it all with physical evidence.

But Brooks angrily rejected that interpretation. Like most policemen he'd seen too many criminals escape justice with con jobs done on well-meaning psychiatrists. Jimmy Smith was consciously, deliberately lying. No other interpretation was possible to the detective.

After a few dreams in the county jail, Jimmy Smith never dreamed of the murder again. 'Why should I?' he said. 'My conscience is clean, if this thing you call a conscience is for real, and not just somethin rich white people *say* is for real just to make guys like me feel like they oughtta punish themselves. If the truth be known, I don't think it really even *exists*. Not for anybody.'

At a much later time, when Greg and Jimmy were together again planning strategy to beat the gas chamber in any way possible – scheming, promising, hating, fighting – each would become firmly convinced of one fact. Each man would say almost identical words about the other: 'One thing's sure, there's no sense fighting him anymore. The sick bastard really *believes* he didn't fire those four shots.'

NINE

A police officer neighbor of Ian Campbell was one of the pallbearers. He later said, 'It was probably hard for the department to choose pallbearers for this thing because, really, Ian didn't have any close policeman friends. All his friends were civilians. I guess I was the nearest thing to a friend because I was a neighbor and had worked with him.

'A weird thing happened at the funeral home just before we buried Ian. I was standing there in the mortuary with the other pallbearers, all uniformed policemen picked by the department, and remembering things about Ian, like how the only time he had cigarettes was when we got some freebies at a cigarette shop. And how he was so quiet to work with, and a big guy but gentle for a policeman. And how he tried to teach me to play the bagpipe chanter, but how it scared hell out of my little kids.

'I didn't know this guy Karl Hettinger very well, but I was looking at him, thinking that they shouldn't have ordered him to be a pallbearer because he looked kind of worn out. Anyway, he must've heard Ian and I were neighbors or something, because he walked up to me in the funeral home and began talking to me in a pleading kind of voice. "I'm sorry. I'm awful sorry about Ian," he said. Looking me in the eye, teary, and then dropping his eyes and apologizing to me. Really, I wasn't *that* close to Ian. And he sure didn't owe *me* any apologies.

'Then he started blurting things out. "Maybe I should've done something more. Maybe I should've grabbed the wheel and wrecked the car! I'm so sorry."

'Now I was getting nervous with this spooky guy. I sure didn't want to hear all this. I just told him it was okay and don't worry about it. I was surprised when I heard he was back on duty. He didn't look like he was in any condition.'

It was Grog Tollefson, the psychology student, who had remembered Ian's casual remark about his bagpipe teacher's funeral. Grog suggested that the police department obtain a solo piper to play 'Fleurs of the Forest.' That was before he knew what the department was planning.

At the gravesite, Grog experienced emotions other than ordinary grief and fear for one's own mortality. It was disgust and anger at the meaningless panoply and pomp: One thousand people. Fifty motorcycles. Hundreds of uniforms. Flags. A firing squad. Police department brass hats who did not know Ian Campbell and cared no more for his death than did the rest of these ghouls. Grog was tall enough to view all of it, and his chalky ascetic face was damp and cold.

It's a goddamned three-ring circus, he thought, staged by professional police mourners who grieved professionally like hags at an Irish wake. And there was the cemetery with its imitations of *objets d'art*, and uniformed hostesses on duty, and a twenty-one-dollar stone placement charge. And the vampires from the press with their shutters clicking every few seconds at Adah, and at the firing squad, and at the uniformed pallbearers. And the uniformed police chaplain, himself a cop, with his platitudes which could be mouthed at the funeral of every cop from here to hell and gone. Grog observed that the police department spends a good bit of money and takes damn good care of its dead. He wondered how they do with their living.

Art Petoyan had an observation about a living policeman: 'The thing I most remember was Chrissie and that Officer Hettinger. She holding his hand in hers. I was close enough to see. *She* seemed to be reassuring *him*. I thought it was strange.'

Karl could not remember what he said to Chrissie that day. Chrissie was never to forget. It was the one thing which was to remain with her among the thousands of fleeting impressions of that mob scene. It wasn't the words so much as the disoriented expression on the face of Karl Hettinger: pain drawn, eyes pleading, confused. He offered his hand and said four words: 'I loved your boy.'

Grog was sweating now, seething at all these strangers who never knew that dreamer, that professional Scotsman with his inexplicable attachment to those damned pipes. What the hell *was* he? *Who* was Ian, now lying there in that police uniform? Is that who Ian Campbell really was, a *cop*? And what the hell were those bagpipes really? What the hell *were* they that they were so much a part of his life?

And in college, his endless, unanswerable questions: 'Well, do you think Karl was wrong then when he said . . . ' And listening, always *listening*, damn, it was enough to get on your nerves. Why wasn't he opinionated like I am? Why did I always feel he was in control and I was out of control? Why did I like him so much? Him and his bloody music, his Bach, his Stravinsky. Did he *really* love his music? Was it just *her* who made him *think* he loved his music? And *her*: Why is she so correct, so courteous, so cordial that she scares me? Why is she sitting there now so sphinx-like? She's not some fanatical religionist. She's not tranquilized. So how is she doing it? Look at Adah, demolished, one step from hysteria, and Adah sedated by that Armenian doctor. It was easy at this moment to rage at all of them.

Grog knew that years before, Chrissie had brought her husband's body here and that she had given this plot of earth to Adah for Ian's burial beside his father. There were two flat stone markers: William Campbell, M.D. (1898–1944), and Ian James Campbell (1931–1963). And next to Ian an unused plot, conspicuous in the crowded graveyard. And Grog imagined how it must have been when Chrissie offered Ian's plot to Adah and how she must have been inside: not breathing, heart hammering, until Adah said, 'Yes. All right. I guess so. I don't care.' And how Chrissie must have closed her eyes, and secretly: 'Thank you. Thank you. Thank you.'

The only still-unused plot of ground was not beside Dr Campbell. It was beside Ian. Grog never had to ask. He knew who owned that unused plot of ground. And who would one day lie next to him.

So you're not fooling me with that relentless impassive expression, thought Grog, his shirt soaking against his chest. I know who he was to you, your young lord. Your sun, yes, s-u-n. And what's left now? Adah has the kids and she's young. She'll marry again. But you, what've *you* got? What is this Gaelic, Calvinistic, monstrous ethic that's holding you up? I don't understand it. I don't. And what . . .

The wail. Clear and piercing. Eerily distant at first. But then, like cold wind slashing through the crowd, cascading down on those who really knew him.

Art Petoyan said: 'It was like a *déjà vu* experience. I thought of Piper Major Aitken's funeral with Ian standing beside me. I knew it was coming but still I started trembling when I heard it. Back, back, up on the hillside he was. That solo piper. That solitary piper. Playing that ancient plaintive dirge for clansmen killed in battle, "Fleurs of the Forest." And I was shivering all over, and I noticed the hairs on the back of my hands were moving, swaying. I looked at *her*. She sat there erect, looking straight ahead. Adah of course was totally destroyed even before the piper started. She was almost collapsed across her brother's arm. But though I was worrying about Adah's condition I took my eyes away. I had to look over at the *other* chair, at *her*, sitting there, straight, impassive, motionless.'

Wayne Ferber was standing at the grave looking at Ian's wristwatch which Adah had given him. He was thinking of Ian's killers. 'I thought how the Ian I'd always remember couldn't even have believed there *were* people like that. Then I saw the wreath shaped like an anchor. It was sent by an old limping sailor, who looked like Popeye. I thought of us as children, playing in Hancock Park. Of how he was. Never frivolous. How even his kid games had to have an end, a conclusion, a point. He'd insist on it. And now I thought: *Is* there a point to it, Ian? Maybe you know at last. Then I heard those pipes. . . . '

Grog Tollefson's eyes were raw and burning, and he looked around and saw the effect of the pipes on those who really knew.

Well, Ian, he thought finally. Well, my friend, we got to them with that one. We got a *little* of you into this mad carnival.

Now his cheeks were wet and he found himself swallowing hard. He stole a look, one last secret look, at Chrissie Campbell. Still she was motionless. Without expression. Staring. Straight ahead.

TEN

'Outrage and horror' was the phrase most often used by police spokesmen. And it hardly described the police climate. For this was 1963, before the revolutionary assassinations of policemen. Despite their cynicism, American policemen are Americans. Perhaps a *gendarme* or *poliziòtto* would only have been deeply angered by the gratuitous act in the onion field. But an American policeman was horrified. There had always been rules in the game. One had to have a good reason to kill a cop, such as eluding capture. Smith and Powell had already won that night. In killing Ian Campbell they had scoffed at fair play, scorned the rules of the game. This was the thing an American policeman could not bear.

The young red-faced vice officer at Wilshire Station had been a policeman less than three years, but he had learned certain fundamental truths about policemen. Policemen thoroughly believed that no man-caused calamity happens by chance, that there is always a step that should have been taken, would have been taken, if the sufferer had been alert, cautious, brave, aggressive – in short, if he'd been like a prototype policeman. They saw themselves as the most dynamic of men, the ones who could take positive action in any of life's bizarre and paralyzing moments.

To suppose that a policeman's vicious murder was inescapable from the moment that little Ford made the wrong turn on Carlos Avenue was inimical to the very essence of the concept of dynamic man.

The ranking Los Angeles Police Department officer who went to Bakersfield the night of the killing was Inspector John W. Powers.

John Powers was greatly admired by Pierce Brooks and indeed by most policemen. Some twenty years before, when he was a young detective, he had been involved in a sensational shootout wherein he was wounded and had earned the nickname 'Two Gun

Johnny.' He'd lived up to that name. Even now, when he was a police administrator whose most hazardous task was driving to his office on Los Angeles freeways, John Powers wore two guns on his lean hips. It did not matter that many of his administrative colleagues left their guns in lockers or desk drawers. John Powers never was without *both* of his guns under the coat of his business suit. He was said to be the Patton of the Los Angeles force. He was tall like the general, with white wavy hair and eyebrows like crow's wings. And Inspector Powers had the Patton charisma with the line policemen, would talk their language at rollcalls, would brief stakeout squads and robbery teams in regard to shooting. A good clean bandit-killing pleased him as it does most policemen. He was known as a cop's cop.

What John Powers said carried much weight with the street cop. He had been one of them somewhere back in the old days, they were sure of it. And they were sure he wouldn't kiss anybody's behind. He talked like a real man.

Just five days after the murder of Ian Campbell, John Powers drafted what would be called by many policemen the Hettinger Memorandum. Actually it was Patrol Bureau Memorandum Number 11. The subject was: 'Rollcall Training – Officer's Survival.' It was considered so urgent that no officer in uniform or plainclothes was excused from rollcalls, and the division commanders were instructed to assure that every man was apprised of it:

> The brutal gangland-style execution of Officer Ian James Campbell underlines a basic premise of law enforcement. You cannot make deals with vicious criminals, such as kidnappers, suspects who have seized hostages, or those who assault police officers with deadly weapons.
>
> Officer Campbell will not have died in vain if his death causes each member of the department objectively to evaluate his personal role as a policeman and the objectives of the department as a whole. . . . Just as the armed forces protect the nation from external enemies, local police departments protect their communities from internal criminals every bit as vicious as our enemies from without. The police are engaged in a hot war. There are no truces, and there is no hope of an armistice. The enemy abides by no rules of civilized warfare.

The individual officer, when taking his oath of office, enters a sacred trust to protect his community to the best of his ability, laying down his life if necessary.

All men return to dust. The manner of a man's living and dying is of paramount importance. Although some moderns have attempted to sap the strength and ideals of this nation by slogans such as, 'I'd rather be red than dead,' there are situations more intolerable than death.

John Powers's lesson number one of the rollcall lesson plan was read once by the red-faced vice officer, three times by Karl Hettinger:

Surrender is no guarantee of an officer's safety or the safety of others, including that of his partner and other brother officers. The decision to place these lives and his own in the hands of a depraved criminal is not one to be made lightly.

In lesson two, Powers became specific in his recommendations to officers who find themselves suddenly covered by a gunman. Some of the suggestions are to tell a nonexistent policeman behind the suspect not to shoot, hoping the suspect might turn around to look, or to pretend to faint to get near the suspect's feet and trip him, or to jab a pencil through the suspect's jugular vein.

Perhaps the entire memorandum is summed up in lesson four, where officers are advised that, 'If shot, all wounds are not fatal.' And that, 'A strong religious faith gives you calmness and strength in the face of deadly peril.' And, once again, that, 'Surrender is no guarantee of safety for anyone.'

Pierce Brooks had mixed emotions when Inspector Powers consulted him about the memorandum he was about to write. On the one hand, Brooks subscribed to the unwritten police commandment about not second-guessing a field situation where you were not present. On the other hand, he was too much policeman not to believe in the dynamic man concept. Campbell's death *had* to have been preventable. Powers was right. But in his final reports some weeks later he softened his appraisal of the officers' conduct. He couldn't go so far as the Powers memorandum and imply that Campbell and Hettinger were almost – he hated even to think the word – cowards. So he finally concluded that Karl had merely used poor judgment in surrendering his weapon, and that once surrendered, it was too late on the ride up to use any of the fancy tricks recommended in the memorandum. He wished Powers would have waited before releasing the order. Brooks by now had come to know Hettinger, knew that the murder had disturbed him, but not how much. Still, he thought that the Powers memorandum could cause the young policeman to feel *some* guilt. Then he dismissed the thought. He was too much policeman to believe very strongly in other than physical trauma. He had been too often frustrated by defense psychiatrists.

'I've read the order,' said the young red-faced vice officer to his sergeant. 'And personally, I don't like it. We've been telling robbery victims for years not to try something as stupid as drawing, or shooting it out with a guy who had the drop on you. Now we're throwing it out the window as far as policemen are concerned.'

'They call that typical police overreaction,' said a second vice cop, an older policeman who was reading a racing scratch sheet trying to pick a daily double.

'The department's writing general policy because of one specific isolated case,' the young vice cop argued. 'It just doesn't make sense.'

'I'm not saying I disagree with you,' said the big sergeant to the younger officer. 'In fact, I more or less agree. But you have to understand what's happening in the department. Policemen are . . . are . . . '

'Outraged.'

'Yes, outraged. We've never had an officer taken to an onion field and tormented and needled at the very end and . . . '

'Look,' said the red-faced vice cop, 'I understand that. Christ, I feel the same, but I think the department's making a bad mistake with this Hettinger Memorandum. I'm gonna say so when we go the patrol rollcall tonight.'

'Now just a minute,' said the sergeant, 'the captain said the whole damn station has to go and *hear* this memorandum, that's all. Just listen to it. You're still free to do as you like in a combat situation. It's not handed down from a mountain.'

'Have you read it?'

'No.'

'It's on the watch commander's desk upstairs. Read it. It *was* handed down from a mountain.'

'Well, I don't see any sense of you or me popping off about it.'

'Listen to this,' said the young vice officer, his face not just ruddy now, but flaming, his voice cracking with emotion as he thought of standing at the crowded rollcall and daring to dispute the order of Inspector Powers. 'These are articles I clipped. It says that this is the fourth kidnap of policemen in Los Angeles County in the past four weeks.' Then the vice cop began heatedly reading:

On February 24, officers Albert Gustaldo and Loren Harvey spotted a woman parked alone in a car. She told them she was waiting for her husband. The officers remained nearby. Soon a man appeared, his arms buried in packages. The police ordered him to drop the packages. The man complied, but when packages fell, a sawed-off shotgun remained. He disarmed both officers. The officers were released unharmed.

On March 1, Whittier police officers Arthur Schroll and Richard Brunmier made a routine field check on an auto containing two men. The men drew guns, took the officers to an isolated section where road work was in progress and handcuffed them to a piece of heavy equipment. The men, both robbery suspects, escaped with the officers' revolvers.

On March 9, Inglewood police officers Arthur Franzman and Douglas Webb signaled a lone male driver to pull over to the curb in a routine traffic violation. The man got the drop on the officers, took their revolvers and made them drive to Inglewood Park Cemetery where he ordered them to lie down. The man later was identified as a bandit who held up a café and escaped with twenty-seven hundred dollars.

'So what's the point?' asked the sergeant.

'The point is that this kind of thing's been going on since time immemorial and right here in the L.A. area, and in fact on the *same* night Campbell and Hettinger were snatched. Now all of a sudden because one set of maniacs blows up a cop we're gonna say Campbell and Hettinger did it wrong and change our whole policy. Campbell and Hettinger must've known about these recent kidnappings. They must've figured *their* case was no different.

They were gonna be taken somewhere, maybe handcuffed, and that'd be it.'

'I think you better keep your articles in your pocket,' said the older vice cop. 'Once our leaders make up their minds there's no changing them. If they say Campbell and Hettinger done wrong then that's it. If they wanna tell us how to do it right, fine, I'll listen, then I'll do what I think best anyways. So who's gonna be hurt by their chickenshit special orders and rollcall training?'

'I was wondering about how Karl Hettinger might take it,' said the young vice cop.

The red-faced policeman had good reason to wonder. Karl Hettinger did read it. He read it again and again. He memorized paragraphs. He could after a time recite portions of the order to himself in the night. He would eventually come to know the memorandum better than its author, John Powers, ever did.

Ironically, the young vice officer later read an issue of *Official Detective Stories* magazine which featured an erroneous, lurid story of the killing of Ian Campbell, and also ran an article about a Salt Lake City police officer who, less than one month before Ian Campbell's murder, was disarmed by a gunman while answering a robbery alarm in a market. Then the gunman ordered the officer to signal a second officer, a sergeant, into the store pretending that all was well. After disarming the sergeant, the gunman ordered him to signal yet a *third* officer into the store, and disarmed that officer too, finally kidnapping the sergeant and a store employee. Both hostages were later released unharmed.

The young vice cop cursed and wished he'd also had *this* article that night in the crowded rollcall room. It was the perfect proof of his thesis: No less than *three* police officers in this single situation were suddenly and individually confronted with an armed bandit at point blank range. All three officers separately not only gave up their weapons, but cooperated fully with the gunman and urged the others to do so and submitted to kidnap. The suspect was caught a short while later and no one was hurt.

But then, the young vice officer tossed down the magazine and admitted ashamedly that it wouldn't have made any difference. Not after what the captain said.

At that rollcall, the station captain, an ancient veteran of forty years of police work, prefaced the reading of John Powers's order with a white-lipped remark:

'*Anybody* that gives up his gun to some punk is nothing but a *coward*. And if any of you men ever think of doing such a thing you'll answer to me. By God, I'll do my best to get you fired! You do like the order says. You take positive action! You're policemen! You trust in God!'

The red-faced, green-eyed vice cop had difficulty controlling his anger, for he was an emotional young man. But he was also accustomed to line authority, first in the marines and now on the police force. He would one day try to record what he knew about police life, but for now he seethed in silence. He kept his newspaper clippings in his pocket. He could not stand and dispute the captain. He lacked that kind of courage and he knew it.

Not all policemen though were as timorous as the young vice officer. Something very different happened at a crowded Central Division rollcall.

'The department's position is abundantly clear,' said the sergeant, at the conclusion of the reading. 'What's more it's logical and carefully deliberated and ... '

'Balls,' said a heavy voice from the rear of the room, from the old-timers' seats where rookies and slick-sleeved hot dogs dare not sit, where all the occupants of the chairs had at least three service stripes on the lower left sleeve, each stripe indicating five years of service.

The voice was familiar to everyone. It was often heard rumbling through concrete valleys near Main Street or Broadway, and you smiled if you were a policeman. But you trembled if you were a hype, a paddy hustler, or a confidence man.

The sergeant glanced self-consciously toward the voice. This was a challenge to his proclamation, his first challenge. He was a probationary sergeant, one of the youngest on the entire department. What's worse he *looked* young and knew it. He had tried smoking a pipe for a time, and even tried cigars. He had dreaded the moment at hand, when one of the crusty veterans would challenge something fundamental and true, something from which no good-natured wink or friendly grin would let him retreat with honor, something like the Powers memorandum.

'Balls,' the voice repeated and now he was staring into the eyes of the gray-haired, overweight beat cop whom he refused to admit he feared.

'Is there some disagreement?' the sergeant asked breezily, trying a knowing glance around the crowded rollcall room to show he was going to patronize the owner of the voice.

The beat cop had no reputation as a rollcall popoff. On the contrary, he was a quiet man, and this made the swollen rumbling voice more fearful when it commanded the denizens of the beat. He was a twenty-five-year policeman who preferred the one-man beat, a virtuosic beat cop, one of those who fades into police myth and legend, who rules his beat, and is frequently the very best or very worst that police work has to offer.

The sergeant lit a cigarette and hoped no one noticed the flame missed the tip three times.

'Balls!' said the beat cop again and stood up and this was no longer anything to take lightly. No one *ever* stood up at rollcall to make a point on anything. The sergeant felt the blood drain from his face. It smacked of mutiny. But so far all the old policeman said was 'Balls.'

Then the beat cop added, 'Sometimes I think there ain't two fuckin man-sized balls on anybody in this organization. At least there ain't none with hair on them.' He put his size thirteen hightop shoe on the chair and rested his elbow on his knee and locked eyeballs with the sergeant who had never noticed how tall he was before. He wasn't just a fat man. His belly was big, but looked hard. He was a goddamn big man, thought the sergeant.

'You disagree with some portion of the training?' the sergeant asked, keeping his voice even and talking slowly.

'I disagree with the whole damn thing.' He lit a cigar and there wasn't a sound in the room. Not a sound. A baby-faced cop in the front row absently let a portion of bubblegum pop through his lips.

'I been walkin a beat down here pretty near as long as some of you kiddies been on this earth,' the beat cop began, looking again right at the sergeant who dropped his eyes and began fiddling with something on his sleeve. 'I think I maybe made as many good felony busts as anybody on the job. I think I had my share of back-alley brawls, and I even been in a shootin or two.'

His voice, pervasive, enveloping, was trembling a little because he was not accustomed to making speeches. So he spoke with more force to control the trembling, and now he was growling.

'I'm tryin to say one thing here. In all my years there was one goddamn thing we was always sure of. One thing that was sacred, you might say. And that is that you're the boss out there. You know what you shoulda done at the time you did it. Police work is that kind of business and only the guy that's there knows what he shoulda done or shouldn't a done. Before this Campbell murder we all sort of agreed on that. You see, unless I'm reading this wrong, it seems like the department is faultin these kids Campbell and Hettinger. Now I don't know these boys, but I heard they were good coppers and I'm willin to accept that they did exactly the right thing out there the other night. Just because those psycho cocksuckers killed Campbell, that don't change nothin.

'Now I'm particularly pissed off about this order because once, a good many years ago, some asshole took my gun off me. He braced me and there I was point blank from this little prick and him with a .45 pointed right at my belly and not for one little minute did I even consider somethin as stupid as this crazy shit in this order. Sweet fuckin mother, can you imagine me rollin around on the ground like some big goddamn walrus trying to knock him down, or yellin, "Look out behind you, you little cundrum!" Or trying to grab that scrawny neck so I can shove a pencil through his crummy fuckin jugular? What the hell is goin on up there these days?'

The old beat cop paused and looked directly up, not at heaven, but at the true seat of power, the sixth floor where the chief sat, at that time with as much fearful authority as could be found outside the military service.

'Anyways,' the beat cop continued, 'I looked at that gun and at his mean little eyes and I knew as sure as there's shit in a goat that if I didn't do what he said, the coroner would be puttin in a special order for sawdust to fill this slop bucket.' He paused, patted his stomach, and puffed on the cigar. 'Anyways, I says, "Yes sir. Whatever you say, *sir*!" And I gave him that gun real careful, and if he wanted my Sam Browne he could have it, and if he wanted my fuckin pants and shit-stained skivvies he coulda had them too! But he didn't, and I was allowed to walk away with my life. One last thing is that if he had told me to get in a car and drive, I woulda done that too.'

This has gone far enough, the sergeant thought, and his anger was taking hold. He interrupted the monologue, saying, 'The point of this training is . . . '

'The point *I'm* tryin to make, Sergeant,' the beat man thundered, and the sergeant fell silent, his face draining again, 'is that you got to leave total fuckin authority with the cop on the street. You go tellin these young tigers here to draw against a brace and you're goin to be buryin some of these boys one day. Cause they might be stupid enough to believe you. Now I say normally you don't never draw against a fuckin guy that's got you cold, or that's got your partner cold. But *normally* you gotta be crazy to do such a thing. In any case, you fuckin well leave it up to the man on the spot.'

'Well, the department is of the opinion . . . '

'One last thing and I'll shut up,' said the beat cop. 'The guys that draft these orders don't have to live by them. They work in cushy offices and pinch plump little asses that bend over their desks every day to stir their coffee. Now this is a stupid and panicky order with no thought put into it. Everybody's shook up by this murder and they're panicked. Christ, it's all you see in the papers. This order is callin Hettinger and Campbell cowards no matter how you slice it. The thing I'm wonderin is this: Does this order make *me* a coward too? I'm wonderin if there's somebody in this room or even on that fuckin sixth floor who's got enough hangin between his legs to call *me* a coward too?'

'It's getting . . . getting . . . we're late. Let's relieve the watch,' said the sergeant, walking quickly from the room.

In a relatively short time after the order was issued, a Los Angeles policeman was faced by a berserk gunman in West Los Angeles. The policeman ignored the memorandum of John Powers and surrendered his gun to the suspect who held him for a time in a private residence. The kidnapped officer, when he had the chance, dived out a window and escaped unharmed.

The day after the memorandum was read, two nightwatch patrol officers in a station locker room were putting on an impromptu dramatic performance. One was dressed in his underwear and a Sam Browne belt. He held his right hand on his hip, the back of his left hand pressed to his forehead and he was saying, 'Oh, please don't shoot me. Oh, I'm gonna faint. Oh. Oh.'

Then he collapsed in a heap like a hairy ballet dancer, while his half-dressed partner stood imperiously over him holding a banana like a pistol, saying, 'Ah-ha! Just as I thought, all you cops're a bunch of chickenshit fruits!'

And as he leaned over, dangling his handcuffs, the hairy one in the underwear sprang to his feet, a number-two pencil like a dagger in his teeth, and jumped on the first one's back shouting, 'Gotcha! Gotcha, you prick! Where's your fuckin jugular? Huh? Huh? Huh?'

But the memorandum was given the chief's blessing and became part of the Los Angeles Police Department manual. All of the police department training schools would from this day on have a class on officer survival wherein an instructor would dissect the actions of Ian Campbell and Karl Hettinger which led to what was judged to be a preventable murder. Both the dead man and the survivor were implicitly tried by police edict and found wanting. There *had* to be blame placed. If you let yourself be killed it had better be by an act of God. And He did not kill by the gun. He killed by thunderbolt.

Working with the flowers was sometimes the best part of the working day. But on the other hand it could be the worst part. It depended on how hard the gardener felt he must work. Sometimes he would feel the fear come and he'd have to work like a field hand, and sweat, and hope it passed. Sometimes, if he felt better, he could afford to do a more relaxing and artistic kind of work, and tend to the flowers.

He didn't think he'd bother with flowers today, not the way the day was going. He looked at the flowers, spied some foxglove. There was something about the foxglove. The flowers hung long and tubular and purple. They dangled limply from the stems. They looked sick, neuter, disgusting. Young foxgloves looked gelded. He started getting afraid. For no reason at all.

He thought about when he had almost reached the end of his days as a thief, when he stole the fishing plugs. It didn't seem particularly different from his other thefts. At a later time someone tried to say it was different, but the gardener only half understood, half believed him.

He had driven to the Sears store in Glendale this particular time. That in itself was a little unusual. The store was a bit farther than he liked to go. Somehow though he had to go there that day. He had been stealing for a long time, almost a year. He had never been caught, never even come very close to getting caught even though he stole during peak shoplifting hours when the stores were sure to have watchers.

He felt very strange when he walked into that Sears store. Usually when he entered a store to commit crimes he didn't know what to steal, and just wandered until something struck his eye or he thought of something he might need. This time though, in the crowded store, in the middle of the day, he had seemed to know exactly what to steal. He didn't consciously decide. It was very strange. He just found himself walking straight for the sporting goods department. Straight to the counter where the fishing plugs would be. But they weren't there! They weren't in the same place they had been when a young boy and his friend had pilfered them so many years ago. When they had been caught and warned by the store clerk. The store had changed the counters.

He looked around frantically. Then he spotted them. Now the pulse ticked in his neck, his lightly freckled face was crimson. He stopped breathing when he approached the counter. He looked *around. He dipped into the tray of lead plugs. He stole them. He put them in his pocket. He looked around. His heart was cracking. He stole some more. He couldn't catch his breath. He walked slowly, deliberately from the store. There was no voice. There were no footsteps. There was nothing. He had escaped once more.*

ELEVEN

At their annual Christmas party the Los Angeles District Attorney's Office always selected an earnest young deputy district attorney to be the recipient of the Marshall Schulman Nasty Prosecutor's Award. The namesake of this honor was the prosecutor chosen to try the killers of Ian Campbell.

Marshall Schulman was not deemed nasty because of his out-of-court manner nor by his appearance. On the contrary, he was youngish, tanned, with a touch of gray in the sideburns. His nose had just enough of a curve to make him forcefully handsome. His voice was good. He was enthusiastic. He could charm a jury. But when Schulman was on the attack, and that was just about all the time, he went for the throat. His voice could sneer though his lip didn't. He could confuse, worry and punish a witness without relent until that witness said what the prosecutor wanted him to say. He was not above inserting snide asides and derisive gestures which defense counsel would scream was cumulatively prejudicial. But he was deft in trial work and knew just how far he could go with a given judge and a given jury on a given day. He, like Pierce Brooks, knew that Gregory Powell was as good as dead, and his strategy was directed toward getting a death verdict for Jimmy Lee Smith.

Marshall Schulman had been assigned the sensational murder case while it was still in the early investigation stage. The district attorney asked Schulman to contact Pierce Brooks, and if necessary, to direct the police investigation himself to assure an impervious court case. Schulman met with Brooks briefly, saw what kind of investigation the detective was putting together and returned to his office saying, 'That detective doesn't need me or anybody else.'

Marshall Schulman had a few decisions to make. One of them was whether or not to file additional charges such as kidnapping for purposes of robbery. Ultimately, he decided that he wanted nothing to complicate his tactical thrust. He would file one count of first degree murder on each defendant and that was all. The jury would then not be tempted to choose among various lesser offenses. There could be no later wavering should some juror be loath to sentence men to death.

His other decisions were incidental, such as whether or not to subpoena the Campbell widow for dramatic effect, ostensibly to identify the picture of her husband taken in life. Schulman's wife decided that question: 'Counsel, you will *never* be accused by anyone of being overly sensitive, on that you can rest assured.' Schulman decided that Karl Hettinger could identify the pictures, and Adah Campbell was spared the subpoena.

But despite Marshall Schulman's self-admitted insensitivity he was nevertheless attuned to problems which might arise with his witnesses, which might disrupt testimony. And after his first interview with Karl Hettinger, the insensitive prosecutor became troubled by something.

'Karl, I think you did a hell of a fine job that night. You should be commended.'

There was no response from Karl Hettinger.

'I don't think many men could've handled themselves so well. If it weren't for you keeping your wits, those killers would be free.'

Karl did not respond.

'Did you see in the paper a couple of weeks ago about the two policemen getting kidnapped and taken to a graveyard where they were released? That's what anybody would've thought was going to happen.'

Karl did not respond.

'From the first moment, you were right, Karl. From the moment you had to give up your gun right through to the end. No one could expect you to do otherwise, or hold you in any way responsible.'

Karl Hettinger still did not respond, and the insensitive prosecutor began to wonder about something which none of Karl Hettinger's colleagues and superiors had even noticed.

Seldom had a preliminary hearing aroused such interest. It was held March 19. The defendants were still wearing their leather jackets. They had not yet learned to adjust to their new lives as cop killers, notorious on one side of the law, celebrated on another. They had not as yet settled into their bewildering new lives in the 'high power' tank of the Los Angeles County Jail. They were still tense and drawn.

The young defendants were getting more deference than either would ever again receive in his life. No one wanted the slightest hint of ill treatment or prejudice to cloud the subsequent court record and interfere with swift retributive justice for the two men. There had seldom been such public opinion in any Los Angeles murder case. Hardly a day passed without letters to the editor, or editorial comment on television. The public could not fathom the ultimate cruelty: *We told you we were going to let you go but* . . .

Deputy Public Defender John Moore was an excellent foil for Marshall Schulman. He was no less aggressive a trial lawyer, but he was less apparent. He was thin, bookish, mild in voice and demeanor. A slashing attacker like Schulman could often look callous against a defender like Moore, but both men were experienced careful trial lawyers.

There was another public defender, Kathryn McDonald, a middle-aged energetic spinster, assisting Moore with the defendant Gregory Powell. But Greg was frustrating his attorneys by adamantly refusing any suggestion of an insanity plea.

'I'm having a hell of a hassle with the public defender's office, Mr Brooks,' said Greg.

'Oh?'

'They're pushing me and want me to plead insanity, and Mr Brooks, I'm not insane, never was. And they're coming up with all this malarkey about my brain operation and all this other jazz and I don't know enough about the law to know whether I can fight them or not. They're gonna drag this goddamn case out for two or three years. If it was possible to plead guilty, I *would* plead guilty and to hell with all these lawyers. I know there's a law that allows a man to represent himself.'

'Well, let me give you a piece of advice if you'll accept it,' said Brooks. 'There's an old saying in the courts that only a fool represents himself. Even great and famous judges say that if they were in trouble they'd have an attorney represent them. You should be represented by an attorney that understands the law.'

John Moore was incensed to learn that his client was still seeing the detective. And Greg was to confront Moore, saying, 'I want you to know that Sergeant Brooks has my permission to see me anytime he wants to.'

Moore replied in disgust to Brooks, 'You don't have to ask my permission if that's the way he wants it.'

Moore found his client to be intelligent, headstrong, egocentric, and extremely unappealing from the standpoint of jury impression. It was even impossible to direct the young man how to sit less straight and rigid at the counsel table, and how *not* to look at the jury with his intimidating fearful blue-eyed stare.

Perhaps the most difficult job belonged to court-appointed Ray Smith, an aging, white-haired defense attorney from the old school, given to homespun ways and homilies, who became thoroughly despised by his client Jimmy Smith almost from the first. He perhaps never believed that Jimmy Smith might *not* be lying when he protested his innocence, when he adamantly denied firing the four shots in the officer's chest. Ray Smith saw his job as that of saving Jimmy Smith from the gas chamber, of somehow salvaging a life sentence from the overpowering people's case, and accepting a life verdict as total victory.

The only witnesses to testify at the preliminary hearing were the autopsy surgeon, Dr Kade, and Karl Hettinger, who looked different to the defendants, thinner and younger without the glasses he had lost that night and not replaced. Marshall Schulman would be told a hundred times in later years that he could have put on an impregnable case in one week with just these two witnesses. But that, he would bitterly answer, was hindsight.

The hearing was held before Judge Edmund Cooke. The defendants were held to answer on the charge of first-degree murder and bound over for trial. It was an uneventful hearing marked only by the frightening testimony of the surviving officer.

At five minutes before ten in the morning, after the witness had recited the events of March 9th, his voice breaking at the end, Marshall Schulman approached the witness, who was sitting hunched over in the witness box, his hands clasped between his knees.

'Would you identify the party in this picture?'

'That was my partner, Officer Ian Campbell.'

'You saw him in life, is that correct?'

'Yes.'

'And you saw him in death?'

'Yes.'

Judge Cooke looked down at the eyes of the young policeman and said, 'We'll take a recess. I think the officer has had enough.'

There was, from the witness's point of view, only one question asked of him that day which was to return to him that night as he lay next to his sleeping pregnant wife, himself unable to sleep. The one innocuous question asked by the elderly lawyer for Jimmy Smith: *You were not restrained in any way in the back of the car, were you?*

Why did he ask that? thought Karl. What did he mean by that? What could I have done back there? Weren't those guns in the front restraint enough? I knew someone would say it, that I should have done something back there. Hit them with a tire wrench? Sure, in that little car, and a cocked gun in Ian's belly, and three more guns. Sure. *You were not restrained in any way . . .*

But maybe he didn't mean it that way. Maybe he was just trying to show that his client wasn't really so bad after all. No, not really so bad. They didn't handcuff you. They weren't so bad. Maybe that's all he meant. Maybe he didn't mean the other thing.

By now Karl was sure that almost all policemen were critical of his behavior that night. The way they looked at him in the Hollywood coffee room and especially in the police building cafeteria. The way so many heads turned as he entered. He was sure there were whispers. It was that memorandum that started it. *Surrender is no guarantee of safety to anyone. Surrender . . .*

All right now, hold on, he told himself. Let's be logical about it. And Karl Hettinger, for the first time, deliberately thought through the entire night of March 9th, gouging his memory to focus on each word spoken, each nuance of each word, each gesture and nuance of gesture of Powell and Smith and Ian, and of himself. His side of the bed was dripping wet when he finished. It was after one in the morning. He decided he was absolutely blameless in the death of Ian Campbell. He vowed never to deliberately think of it again. But he couldn't sleep. He got up and drank a can of beer and watched a television movie.

Greg's daughter Lisa Lei was born to Maxine on May 13, and Greg's attorneys brought him into court two weeks later when they feared his recent moods of depression would infringe upon the coming trial.

'Dr Crahan, tell me what medication the defendant is receiving now, if you know, sir,' the witness was asked.

'He's been receiving three types of antacids and some tranquilizers.'

'What is the purpose of the antacids?'

'To allay his verbal complaints.'

'I note you have checked the defendant's blood pressure. Did you check to compare that with prior blood pressures of the defendant?'

'Yes. The blood pressures range around 110 to 120. That's normal.'

'Dr Crahan, you are here, of course, to aid the court in determining the merits of Mr Powell's contentions that he hasn't been able to hold any food down for fourteen days. Did you check his weight yesterday?'

'Yes.'

'What was it?'

'One hundred and fifty pounds, the same he claimed on admission to the jail.'

'Do you have a record on any other weight?'

'Yes.'

'What was that?'

'He was weighed a week or two ago, and he weighed one hundred and forty-nine pounds at that time. He gained a pound since.'

'Thank you. That is all.'

Jimmy Smith was a celebrity in the high power tank reserved for murderers and escape risks. His picture had been on front pages for days and everyone recognized him. Jimmy was delighted with the waves and deferential glances from the other inmates. They passed magazines and cigarettes to him and clucked sympathetically when he told how he'd been wronged and then betrayed by that snitch Gregory Powell.

A few days after his arrival, Jimmy saw two young blacks initiate a futile act of violence born of a hatred and defiance Jimmy could not even begin to fathom, not even at this time of his life.

The two young men, after screaming their hatred of cops and whites, dragged their cell bunks against the wall and propped them up to make room. Then they tore the insides from their pillows and filled them with hardback books and coffee cups. They used torn blanket strips to tie the pillow sacks to their wrists. The first three deputies in that cell were met with an astonishing attack which brought reinforcements and a mattress shield. The two inmates were finally overpowered and choked by towels the deputies carried in their pockets for that purpose.

Jimmy Smith could not forget the shouts and moans of the prisoners and deputies bloodied in that fight. Jimmy saw one of the young blacks dragged away splattered with blood from a head wound, and that night Jimmy dreamed of blood. His dream was drenched in rivers of blood, and for a moment he saw a bloody arm reaching toward him. The arm was so bloody he couldn't bring himself to look at its owner. The bloody arm awoke him.

Jimmy spent the first week reading about himself and watching television accounts of the murder case. He watched the television film of the funeral and saw Ian Campbell's widow grieving. He read the many newspaper headlines. He listened to a county supervisor and even the governor say, 'Even though I'm opposed to capital punishment . . . '

The letters to the editor demanding retribution frightened him, especially the important editorial which said there is no such thing as rehabilitation for such brutish slayers. But what frightened him most was that many paroles were going to be canceled because of the notoriety of the Powell–Smith murder case. Jimmy looked closely at his fellow inmates the day *this* news was broadcast. Though it didn't directly affect anyone awaiting trial in the county jail, still, you never knew. He and Greg had caused the loss of other men's paroles. That was something to truly fear.

One memorable day, Jimmy was able to get close with an inmate who had managed to smuggle half an ounce of heroin into the tank. With what he could beg from his Nana, Jimmy Smith escaped from his tormentors. It lasted almost a week. The dreams of the body on the ground came no more to Jimmy Smith. Once he finally and irrevocably decided he just could not fix in his mind a picture of Powell standing over the body firing down into it, he stopped dreaming. It was the attempt to visualize it which brought about the dreams in the first place, he reasoned. That's the only explanation possible. After all, he told his lawyer, he had done nothing to feel guilty about. He was sick of hearing people talk about conscience, this thing that white people dreamed up. Jimmy

Smith now slept well, ate well, and never dreamed troublesome dreams. He was telling the unvarnished truth when he said that the thing called guilt or conscience did not exist. For him, it did not.

Karl Hettinger had been dividing his working hours between Hollywood Station and homicide division. When the trial date got nearer, he would be assigned temporarily to homicide to be available until the completion of the trial. After the trial he would be given another assignment.

'Driving for the chief of police?' said Helen when he told her.

'Yeah, if I want it.'

'If you *want* it? Why wouldn't you want it? How many officers get such a chance?'

'That's just it, Helen. *Why* am I getting it? Because I'm notorious, that's why. Everyone knows about me now.'

'What's that got to do with it?'

'I don't know. Maybe they don't think I'm fit to be a street policeman because of what I . . . what some of them think I was respons . . . Well, maybe this is sort of a nice way of putting me away.'

'Karl, it's a day job with weekends off. And you wear a suit and just go to work and come home like any other businessman. Why read all this other stuff into it?'

'I just think what other policemen might say, people who'd put their names in for that job and see me just move in.'

Helen Hettinger, now almost seven months pregnant, wondered about this man she had married, realized how little she knew him. After all, they had been wed only a few months when the murder happened. She had just been getting to know him then. Now he was so different. He never laughed or joked like he used to. He was losing weight and staying up almost every night, long after she went to bed. He was avoiding their friends.

She wished he'd talk to her about the whole thing, the coming court trial and about how he felt at work. But he just wouldn't tell her anything. When she'd ask, he'd just shrug or smile and say nothing was wrong. He wouldn't even get mad when she tried to nag him into talking. If he'd only get mad once in a while it would be good for both of them. He never would. He'd just quietly and stubbornly resist.

It was impossible for Helen or anyone else to make Karl reveal his innermost thoughts. You just didn't burden others with your problems. It wasn't the family way.

Judge Mark Brandler had a sensitive face, pale hair, a long thin nose, slightly hooked, eyes which crinkled and smiled beneath slightly drooping lids. Jimmy Smith looked at the face and had hope.

The judge had come to America as a small child during World War I, a refugee from Belgium. He had worked in the office of celebrity lawyer Jerry Geisler and had been a deputy district attorney for many years. He was the last appointment of Governor Earl Warren before the governor himself became the world's most famous jurist. Mark Brandler was proud that in his sixteen years as a deputy district attorney he had never been beaten in a jury trial. He was prouder that in his years on the Superior Court bench he had never been reversed by a higher court.

'Is there any reason why you are personally requesting that the public defender be relieved and that you be substituted as attorney representing yourself?' Judge Brandler asked in another pretrial motion.

'I feel I know myself, and I feel I am more familiar and I just want to handle it myself,' Gregory Powell said.

'You haven't had any training in connection with rules of evidence, have you?'

'I've read several books. I feel that I am adequately capable of representing myself. I wish that I could have, to help me, the use of the law library.'

'How old are you, Mr Powell?'

'Twenty-nine years old.'

'What education have you had?'

'I started at Cadillac Michigan High School but I left in the ninth grade. I finished my high school in a period of eighteen months while I was in Leavenworth, Kansas, and received a certificate of equivalency from Topeka Kansas Educational Board. I took various college courses.'

'What college courses did you take?'

'I studied logic under Dr Burke. And I had first year and second year of college algebra, and I studied some acoustics and several subjects related to music.'

Jimmy Smith's pretrial appearances were more concerned with jail conditions than with the impending battle for his life.

'Well, what's happened is they put me in a tank, your Honor,' said Jimmy. 'In a tank where I know, beyond a fact, where they keep fellows that are what they call snitches, and I know for a fact, from my own benefit, that the food is tampered with because the other inmates hate the guys in there and they can't get to them, and they have me locked in the end cell by myself with no walkin privileges. I won't eat the food down there. I don't know what's in it. I've heard all types of rumors. That's all I wanted to say.'

Things were not going well in county jail for Gregory Powell. On June 26 a sheriff's sergeant was contacted by an inmate facing an armed robbery charge who wanted to betray Gregory Powell in exchange for a letter to the judge advising of his cooperation.

The inmate related that Greg had in some way come into possession of two hundred aspirin tablets and was planning to swallow them and be sent to the General Hospital. The informer had been urged by Greg to slash his own wrists severely enough to be taken to the hospital. Once there, Maxine was to arrange for two guns to be smuggled to the informer through a certain hospital employee.

And another jail inmate told a deputy that Greg was taking a large quantity of aspirin, 'trying to kill his fool self or to put on a good act so that he might go to the hospital.'

They were not the only two to inform on Gregory Powell and he was not taken to the hospital.

A week later another inmate in cellblock 10-A-2 had a secret conversation with a jail lieutenant informing him that according to Greg, Douglas Powell, Greg's younger brother, was supposed to plant some guns in the courtroom and that Gregory Powell would crash out, 'Dillinger style' in Greg's words. The inmate volunteered to remain close to Greg and inform on further escape preparations up to and including the break. The inmate only expected any help that could be given when his own probation violation was brought before Judge Brandler later in the month. The inmate's further services were declined, but courtroom security was tightened.

Finally that same week another inmate informed the jail captain that Greg claimed Douglas Powell was to cut the chairs in the courtroom and insert guns in the cushions, and that he might use a smoke bomb diversion during the inevitable shootout which would follow.

Another informer detailed several of Greg's alternate plans. One involved striking a juror which he thought would lead to an automatic mistrial, a later lunacy hearing, and finally a transfer to Atascadero State Mental Hospital from where an escape would not be difficult.

The escape from the courtroom was to be a last resort in that it seemed suicidal. Greg much preferred a scheme of having his brother secrete a gun in the law library. A gun could easily be hidden there, and he would only have to shoot one or two people to escape. But even if he was given the right to defend himself, the judge might not permit him to go to the law library.

Still another inmate, one very close indeed to Gregory Powell, was to inform on him and assure jail officers that he would tell them of the slightest escape move of the accused murderer. Gregory Powell always suspected this inmate of informing. He never suspected any of the others, or even dreamed that half the inmates were anxious to use him to better their own lots. He would never have believed it. He had always wanted people to love him and believed they did.

In July, in the middle of a hot smoggy summer, in a gray foreboding ancient chunk of concrete known as the Hall of Justice, the jury selection was ready to begin in Department 104 of the Los Angeles Superior Court.

The courtroom was large and old, impossible to keep looking less than grim or even clean.

The defendants wore suits and ties and now Jimmy Smith had his astigmatism corrected by horn-rimmed glasses.

'For the information of the jurors,' said Judge Brandler to the panel, 'in California we have what is called a bifurcated trial in homicide cases. The jury makes a determination in the first or main trial of the issue of innocence or guilt of a defendant. In the event that the jury, after hearing all of the evidence and the court's instructions, returns a verdict of guilty of murder in the first degree, then, and in that instance, the defendant has a *second* trial, at which time the jury makes a determination based upon the evidence as to what the penalty should be. So the matter of penalty as such is not for the consideration of the jury on the first and main issue as to innocence or guilt.'

'I am challenging the panel,' said John Moore, 'on the ground that all persons entering this courtroom are apparently being searched.'

'The fact of the matter is,' said Marshall Schulman in rebuttal, 'these two defendants have probably been treated better than any other prisoner in the jail. Mr Moore has had access to the prison file, and he certainly knows that Powell could be a security risk, and Jimmy Smith too.'

Schulman immediately and perhaps intentionally irritated Gregory Powell's other public defender, Kathryn McDonald, who was attempting to obtain a severed trial for her client.

'I would like to call attention to a very much earlier case which I don't believe has been overruled, although it has been explained,' said Miss McDonald. 'That is *People versus Stewart*, which I am sure Mr Schulman is familiar with, which goes way back, to 1857.'

'I'm not that old,' Schulman said dryly.

'Nor am I, Mr Schulman,' she replied.

'I'm not familiar with it,' Schulman said.

Then began several days of jury *voir dire*, begun by Judge Brandler:

'Does the mere mention of the fact that the person alleged to have been the victim in this homicide is a police officer bring back to any of your minds or memories the fact that you may have read anything about it in the newspapers? If so, would those of you who recall reading anything at all about this case in the newspapers, or hearing it on radio or television, will you please raise your hands?'

'That's the one in San Bernardino?' asked juror number one.

'It's hard to say,' said juror number two.

'You made some inquiry?' asked the judge.

'Was this the one in San Bernardino?' asked juror number one.

'No,' said the judge, 'I don't believe this was in San Bernardino.'

'Then I don't know anything about it.'

'Mr Hall, you had raised your hand.'

'It's kind of hard to say, but it seems to raise a thought in my mind that I did hear about it, but it's vague, real vague, so I'm raising my hand.'

'Mr Johnston, may I ask, what is your business or occupation?' asked John Moore.

'Truck driver.'

'Do you recall having heard something of this case on the radio, news, or seeing it on television?'

'I very seldom watch the news on television.'

'You say you vaguely recall reading about the case?'

'I glanced through the papers. That's the way I read a paper, I just glance through it.'

'I will challenge each juror in the box under Section 1073,' said Moore, 'for actual bias. The existence of the state of mind on the part of the jury in reference to the case will prevent them from acting with entire impartiality and without prejudice for the substantial rights of either party.'

'This court has no jurisdiction in the first instance,' said Ray Smith, Jimmy's court-appointed lawyer. 'I don't think you have any authority here except to dismiss this action since the alleged crime occurred in Kern County. I don't think any result from this jury is going to be worth the paper it's written on.'

'You have indicated to us that you read something in the newspapers concerning this matter, is that correct?' another juror was asked, when the motions were denied.

'When it first came out, yes.'

'Did you read it on more than one day?'

'No.'

'Do you recall whether you saw pictures in connection with it at that time?'

'No, I didn't.'

'You say you did not hear about it on radio?'

'No.'

'Nor on television?'

'No.'

*

Two days later Gregory Powell was trying a trick he had been advised of in the high power tank by inmate jailhouse lawyers.

'The defendant informed me, your Honor,' said Moore, 'that just before we came out of your chambers where we had been discussing some informal matters, the bailiff, the deputy sheriff standing in the courtroom, twisted his arm and tore from his hand a cigarette which he had, this all being done in the presence of the jury.'

'I told him to extinguish the cigarette,' the bailiff said, 'because court was going to convene, and he refused. And I said, "If you don't put it out, I'll have to take it from you," and he said, "If you're going to take it from me, go ahead." I reached for the cigarette. I didn't touch his hand at all. And the cigarette fell into the trash can and was extinguished.'

'For the record, I will ask for a mistrial,' said Moore, 'and I will challenge the panel of jurors in the courtroom on the basis of what they may have seen or did see. It may have the effect of prejudicing them against this defendant and prevent him from having a fair and impartial trial.'

'I would like to join in that motion,' Ray Smith said.

On July 11, before leaving his apartment for the homicide squad room where he was now waiting each day for the jury selection to end, Karl looked fearfully at Helen's enormous stomach.

'I wish I could stay home with you. I wish they'd give me a few days off.'

'I'll be all right, Karl. You just stay close to a phone.'

'No pains?'

'I'm okay,' Helen smiled, and she indeed looked fine that morning. She'd gotten up early and her light brown hair was combed. She was wearing lipstick.

'I should be worrying about *you* instead of this trial.'

'Don't worry about anything, Karl. I'll be all right. If I need you I'll phone.'

'If anything should happen . . . I mean before . . . '

'Yes. Yes. I'll call the doctor. Or an ambulance. Or a cop.'

And so he left for the police building, somewhat reassured. He rode his motorcycle to better get through the traffic.

Later that day something *did* happen. Pain. Sudden, devastating, unbelievable. And Helen Hettinger called the police department and left a message. Then she found herself struck down, half on the floor and half on the bed. Her hazel eyes were round with fear. It wasn't supposed to happen like this, not like this!

Helen fought for the bed and pulled herself over on her back. Another call was impossible. This was as far as she would get. The thrashing, pitiless thing within her was demanding to be free.

The young woman breathed deeply and bit her knuckles and concentrated on not screaming in panic, on saving the life of the thing she was sure would kill her.

Helen tried to prepare, and she raised up and looked down but the tears and sweat were blinding. She was wiping her eyes when the *real* pain struck.

Karl and Pierce Brooks were having coffee in the police cafeteria that morning. Karl was sitting silently as usual. Brooks was worrying that he looked thinner and more tense with each passing week.

'Have some more coffee, Karl.'

'No thanks.'

'How about a doughnut?'

'I'm not very hungry.'

'Gets to be a drag sitting around waiting to be called over to court.'

'Yes it does.'

'Guess you'll be glad to get back to regular duty.'

'I guess so.'

'Pretty good break about you driving for the chief. You'll make sergeant first time you're eligible.'

Karl smiled and glanced around the cafeteria and through the windows out into the burning smog-filled sky. Yes, quite a break, he thought. But why did I get it?

'Telephone for Officer Hettinger,' said the cashier's voice over the microphone, and Karl left Brooks, only to return a moment later, his face gone white.

'My wife. Helen . . . '

'The baby?' said Brooks.

But Karl was gone, running to the elevator, in a few moments speeding down the freeway, talking aloud to himself: 'I can't . . .

Oh, my God, I can't even be there when my wife . . . what good
am . . . Oh, my God.'

Twenty minutes later he was stumbling over the steps, bursting
through the door and falling breathlessly into the bedroom.

And there was Helen. She was smiling at him, beads of
perspiration on her forehead and lip. She was pale but she was
smiling so that he wouldn't panic. The bed was soaked. They were
covered only by a sheet, the two of them. Helen and the red naked
baby she herself had delivered, which lay on Helen's stomach still
joined to her mother by the uncut lifeline she no longer needed.

'Say hello to your daughter,' said Helen. 'Her name is Laura.'

'I've gotta go to the bathroom,' said Karl.

On the fifteenth day of July the jury was picked.

The jury panel members were certainly not atypical, thought
Pierce Brooks. It was a fair representation of one's peers perhaps,
if one could possibly have lived in the Los Angeles area without
reading a newspaper, or seeing television, or listening to a radio. If
one had nothing better to do than endure a trial which would
certainly last two months at least, enjoyed being sequestered, had
little or nothing at home to be sequestered from. If one was not a
professional man, nor high-salaried, nor prominent, it was quite
possible to get a jury of one's peers.

Yes, thought Brooks, it was an ordinary jury and as such would
be capricious, unpredictable, naïve, totally ignorant of law and
justice and violence and violent men, crime and the criminal. The
responses to what the jurors would hear in this case would be
conditioned not by life, not even by books or newspapers – *I just
glance through them* – but by movies. And that, thought the
detective, was the most insidious enemy of justice itself as far as a
jury is concerned. *It has to be true! I can believe it! Because I saw
in this movie one time . . .*

Karl Hettinger was to wait another week before he was to
testify. There were other witnesses, many of them: pawnbrokers
who sold the killers the guns, liquor store and market clerks who
faced those cocked and loaded guns.

Despite the air conditioning, the big courtroom was hot. The
paint was peeling near the ceiling, and people had carved their
names into the wooden seats. Karl sat and absently read the
names and hardly heard the beginning of the bickering.

'Even Gregory Powell can buy a gun in Nevada?' said Schulman, concluding the direct examination of a pawnbroker.

'I object, as argumentative, sarcastic, and facetious,' said Ray Smith.

'We request that the district attorney be cited for misconduct and we move for a mistrial at this time,' said Moore.

'I withdraw the question,' said Schulman.

Karl heard many motions that week:

'I will ask for a continuance until tomorrow morning,' said Moore. 'Mr Powell has apparently received some information which has affected his mental state at this time.'

'This man is on trial for killing a policeman,' said Schulman, 'and I think we should proceed unless there is some strong justification for the continuance beyond a "Dear John" letter or something of the kind.'

'We are getting sick and tired of Mr Schulman coming over to our side of the floor and pointing out the defendant all the time,' said Moore. 'We are sick and tired of Mr Schulman holding papers in his hand and approaching witnesses with papers.'

'Mr Moore goes on into a long dissertation about what he is sick and tired of which bothers me not at all. I don't care one way or the other what he is sick and tired of,' said Schulman, hitching up his pants in what the defense maintained was a belligerent gesture.

And still more motions as Gregory Powell proceeded with his escape plans.

'In what way are you being prejudiced by the fact that you cannot just interview witnesses in the county jail without the presence of your attorney?' asked Judge Brandler suspiciously.

Finally, in late July, Karl Hettinger was called to the stand. The jurors and indeed the entire courtroom were absolutely quiet when he described the killing, and the escape, and the hunt, interrupted only by specific questions by Marshall Schulman. Toward the end Schulman let the witness narrate, and only the husky voice of the witness and the whir of the air conditioner and the persistent buzzing fly interrupted the breathless silence. The witness faltered often, and every juror, every spectator, and especially every man at the counsel table leaned forward, not to miss a word. The direct testimony was making two jurors begin to weep which in turn was making the prosecutor fear a mistrial.

'When I got back to the shooting scene I saw that . . . I saw that Ian was lying in a ditch face down. I didn't go over to him. The sheriff and the ambulance drivers went over to him. And I saw . . . I saw what appeared to be drag marks on the ground . . . quite a bit of blood leading . . . this appeared to be leading from a spot where I had last seen him . . . had last seen Ian fall on the ground, over toward the ditch. The ambulance drivers put Ian in the ambulance, and I got in the ambulance also, and we drove into Bakersfield, and I entered the hospital there . . . I don't know where Ian went.'

'Your Honor, I am going to go into another long phase. I wonder if this might be a convenient time for a recess,' said Marshall Schulman. 'The witness seems distraught.'

Karl Hettinger's testimony was also disturbing to defendant Jimmy Smith, terrified him in fact, and while the jury was still in recess he leaped to his feet to address the court.

'Look, your Honor, I have stood up several times in this courtroom. I don't know how to talk to explain myself properly. I mean, I'm not an idiot, or whatever my attorney claims I am. I've been tryin to ask this man to . . . I don't want him for my attorney anymore. He comes back from the bench here now and tells me that the district attorney is gonna prove . . . that he has got some kind of a theory, or ballistics, or somethin, that I fired shots into the man's body. And I asked him again, "Do you believe what I am tellin you? Do you believe that I did commit the crime or not?" He don't answer me. In other words, he don't believe me. I might as well defend myself as to have him sit here for me, and it don't do any good and I'm goin to the gas chamber anyway. I might as well do it myself as to have him do it.

'He told me, "I don't know, Jimmy, how're we gonna prove it? It looks bad. There's nothin I can do." And this. And that. What can I do with a man like this defendin me for my life?'

'If I understand,' said Judge Brandler, 'this is a motion again to have Mr Ray Smith relieved as your attorney of record, is that right?'

'Motion? What good does it do me? What good does it do me to ask? I'm already in Death Row now. So just forget about it. Go ahead and take me up, because that's what all of you are gonna do. I understand perfectly what I'm sayin, your Honor. It's just too obvious . . . I don't know what this is. I don't understand it . . . .' Jimmy sat and his lip quivered as he held a sob in his throat. His forehead wrinkled and he resumed his hangdog pose.

'That very statement of yours that you don't understand it, that you don't know what this is, is one of the reasons why the court . . . '

'It's a conspiracy!' shouted Jimmy suddenly.

'The court heretofore and again now refuses to grant your motion that you represent yourself in view of the serious nature of the offenses.'

'I don't want him to ever say anything to me as long as I am in this courtroom anymore. I don't want him to have anythin to say to me. No time,' said Jimmy Smith, and folded his arms and turned away from his elderly attorney, who shook his white head and shrugged helplessly.

During the cross-examination of Karl Hettinger, Attorney Ray Smith asked a rather innocuous series of questions which would be pondered by the witness that night.

'When this case is over and you go back to the Hollywood Station, what are your duties?'

'Upon the completion of this trial, sir, I am going to be permanently assigned downtown to the police building.'

'In what capacity?'

'I will be driving for the chief.'

'You will be what?'

'Driving for the chief.'

'You mean Bill Parker?'

'Yes sir.'

'You will not be out in the field any longer?'

'No sir.'

Late that night, when as usual he sat before his television drinking beer long after Helen was asleep, Karl glanced in at the baby. She was not red now, but creamy, and stunningly beautiful, he thought. He touched her cheek and then crept back toward the living room to resume his vigil before the lighted screen and the droning voices which lulled him. He seldom knew what the movies were about.

Even civilians can see through it all, thought Karl. The lawyers could see through it. Why would *I* be picked for that? They're just putting me there to get me off the street, to do it in a nice way. I'm too well known now just to stick me in some ordinary desk job. They think that's where I should be after what I . . . after what *they* think I . . .

He tried to stop thinking. He sat and stared and drank beer and hoped the dream would not come tonight. The dream which would never awaken him until it was over, but which seldom failed to awaken Helen when she heard him panting and felt his sweating body and saw him there in the dark running through his dream. Flat on his back, his legs pumping, the sobs tearing forth every few seconds.

When she awakened him he would tremble and dry himself and go into the living room to the television.

The first few times Helen followed him in.

'Karl, please tell me.'

'Helen, I told you I don't remember what I was dreaming. I don't even know if I *was* dreaming.'

'Is something . . . '

'Nothing's bothering me.'

'Well how can anybody help you? Karl, we've been married almost a year and I don't even *know* you.'

'I'm just a little tense from the trial. I'll be all right.'

'Tell me about the dream.'

'I told you I don't remember.'

'We don't talk, Karl. People have to *talk*.'

'Nothing, Helen, there's nothing . . . '

And she would turn, angry and frustrated, and stalk back to her bed and lie there seething. Then she would feel the soaking sheets where his body had been and the anger would dissolve. Helen Hettinger would feel the fear and doubt creeping through the darkness to envelop her.

She had always known herself to be a strong girl. But Helen's kind of strength – the strength which had seen her through the birth of her daughter – was no match for the baffling unspeakable thing which was stealing her man away, possessing him. She couldn't see it or touch it. She became frightfully aware of her inadequacy. 'I'm not smart enough,' she would say. 'I just don't understand and he won't tell me, *can't* tell me.' She would not be able to sleep for an hour, but she would be asleep long before her husband came to bed.

The next day during cross-examination the witness looked only a little more haggard, perhaps darker around the eyes. By now he had begun compulsively digging his nails into his palms. His wife and his sister Miriam noticed. Other than that and the weight loss which showed in his cheeks, he looked pretty much the same.

'Well,' said Attorney Ray Smith, 'will you please give any explanation you choose to make to this jury as to why on March 12 in front of your police officers there you positively identified Jimmy Lee Smith as having shot bullets into Officer Campbell, and why in this courtroom you have said you could not identify him?'

'Yes sir,' said the witness. 'As I stated before, I made this in the form of an accusatory statement. I was not under oath then. I believe this is an accepted police procedure.'

'What are the physical facts or the movement that leads you to believe that it was Jimmy Lee Smith who fired those four shots?'

'As I looked back, as I am looking back, the form to the left of the form that is firing down into the body appeared to be moving. And it appeared to be moving from the spot that I last saw the defendant Powell standing in. Due to these circumstances, I am assuming that the figure that is over the body is that of Smith, and it is my assumption that Smith has moved forward also from the spot that he was standing in, and is over the body firing into it.'

'That is an assumption on your part, you didn't see it, did you?'

'No sir, I did not see it.'

And when the jury was in recess the old attorney felt obliged to address the court with his incessant problem.

'Your Honor, I feel it would probably be appropriate to have some little remark in the record because in the event of an automatic appeal, the Supreme Court, of course, will be reading the entire record. Since the last outburst of Jimmy Smith, he has not talked with me, he refuses to talk with me, he slides his chair over some four or five or six feet to get away from me. Now it is not that he's hurting my feelings by doing that, but I feel that the Supreme Court should know that the condition exists.'

'Well, all the court can say is,' replied the judge, 'that you are very vigorously representing the defendant, and this may be just a part of the defendant Smith's stratagem to attempt to create some possible error or confusion in the record.'

The client of Ray Smith arose wearily. His hair was cut so short now he was almost bald, and that coupled with the new glasses gave a mournful look to his soft quiet-spoken way.

'Your Honor, for the past two days I have been fastin and prayin and I haven't had anything to eat. My condition is not real weak, I am not physically sick, but I have been walkin rather slow back and forth between the courtrooms. And this afternoon, I was walkin slow and the two officers told Powell and the other two officers to go ahead, and they grabbed me by my arms and they jerked me and they forcibly ran me down the hallway and manhandled me and twisted me and turned me and shoved me into the bookin room!

'I was only usin a slow motion to conserve my energy, and I would go on to say the reason I am fastin and prayin was on the advice of my mother, I call her my mother, I mean my auntie, Mrs Iona Edwards.'

'You are not suggesting that you are fasting because you are not receiving proper food from the county jail?'

'No, I am not, your Honor. They have offered me food every meal every day as much as I want. I am doin this because of the condition of my folks. My aunt is seventy years old and she is a diabetic and I am doin this so that she won't get no sicker. She cannot do what she is tryin to do, and I feel if I pray in this matter that God will help us. You know? You know what I mean?'

Jimmy Smith often mentioned prayer and fasting and God during the course of his many motions and complaints he was permitted out of the presence of the jury. Privately, however, he would review his looted life and conclude that if there is a God, He must be a burglar.

The next hours involved a lengthy hearing to decide whether Jimmy Smith was in fact manhandled. The court after taking sworn testimony was satisfied he was not. Jimmy then made another motion to fire his attorney.

'I asked him, your Honor, I said, "Would you take my uncle with you to locate witnesses?" It is a predominantly Negro neighborhood, your Honor. I said, "Will you take my uncle down with you, Mr Smith?" Now he said no. He said, "I'm gonna take my wife with me and I would rather not have him along with me and my wife." Which I understood *right* away.'

'I can't force you to eat, Mr Smith,' said the judge. 'And I can't force you to talk to your attorney. But I would suggest to you that, as Mr Ray Smith just said, it is certainly to your best interests and for the best interests as far as your defense is concerned, that you *do* cooperate with your counsel. All right. Let's bring the jury down.'

'Take me to the nearest lamppost!' shouted Jimmy, eyes like torches. 'And put the hood over my head and execute me, because this is what is happenin here in this courtroom now! This is all that is happenin.'

Jimmy sounded so brave and eloquent, he was surprised and rather pleased with himself.

After reading a local editorial entitled 'License to Kill' concerning easy paroles, Jimmy Smith decided to put his declaration in writing. The judge received a note:

Seeing that you insist on giving me a Judas to represent me, why don't you first don your white sheet and pillowcases and hang me to the nearest lamppost which I am sure would be more in keeping with the local temperament.

Another note from Jimmy Smith, in his always neat and formal hand, complained of 'vermin infested cells built for II men which contain VI or VII.'

Karl Hettinger was excused from further testimony subject to being recalled. But he found that escape from the case would not be that easy.

The crank letters had started to arrive when one Los Angeles newspaper printed his home address. Most of them were incoherent and general and hardly bothered him at all, even the vicious ones. There were others, however, that *did* trouble him, those which threatened his family and those which were very specific, written by a man he had all but forgotten – a burly young homosexual whom he had once fought as a vice officer, and who had fallen into a gully near Ferndale Number Nine, breaking his collarbone.

At first the homosexual's letters merely reiterated earlier threats and complained that his arrest was unlawful, and that Hettinger and Calderwood had cost him a great deal of money and great pain. Then the letters from him began to get very specific:

Dear Karl,

I understand you were taken for a ride to some farm country near Bakersfield. They raise lots of chickens there. I understand you ran away and left your partner. That's where *you* should end up. On a *chicken* farm.

One day between court appearances, Helen saw him standing on the porch at the time the postman usually arrived.

'What're you waiting for, Karl?'

'Oh nothing, Moms, nothing.'

'Expecting a letter or something?'

'No, not particularly.'

But when those letters came, he would try to get them first. He gave the threatening ones to Pierce Brooks, and he double locked all doors at night, and kept the blinds always drawn in his baby's room. The ones which *accused* him he read again and again before he gave them up.

For Karl Hettinger, the summer was unbearably long. It was almost as bad waiting in dismal courthouse corridors or at the police building as it was to be on the stand. It was estimated that the trial would last all summer.

'Can you approximate about how soon death would occur as a result of that type of a wound?' Schulman asked the pathologist who did the autopsy on Ian Campbell.

'Well, that's difficult to say, however I believe that the hemorrhage through this area would've been sufficiently severe that the mouth and the throat would fill with blood. This blood would get down into the air passages of the lung and death would probably result in fifteen minutes or thirty minutes, something like that.'

'Doctor, I believe you related that number four was a fatal wound?'

'Yes, it was a fatal wound inasmuch as it was a through and through wound of the heart. It was just about as promptly fatal as any wound could be expected to be. I'm sure that unconsciousness would result quite promptly in a matter of seconds and death would result in a matter of no more than a few minutes, perhaps five minutes at the most.'

The prosecutor had in his hand photographs of the bullet-torn, blood-drenched body of Ian Campbell. The defense requested to approach the bench, where Ray Smith said: 'Your Honor, the district attorney showed Mr Moore and myself a photograph that I believe he is going to ask to be identified. I want to object on behalf of Jimmy Lee Smith on the ground that the photograph itself would be inflammatory, and it would certainly be an error to show it to the jury.

'While Dr Kade was testifying as to the course of one of the bullets, I noticed juror number one shuddered and looked at the man next to her. I don't blame her for that. It seems to me that this photograph is merely cumulative and can serve no useful purpose except to upset members of the jury.'

'I object,' said Moore, 'to the introduction of the two photographs at this time based upon an illegal search and seizure caused by the illegal stopping of the Ford vehicle in Hollywood by Officer Hettinger and Officer Campbell. I also move at this time to strike all of Officer Hettinger's testimony. That is, any testimony after he stopped the automobile, on the basis that he did so without probable cause.'

'The objections and each of them are overruled.'

Back at counsel table when a Kern County detective was testifying as to Gregory Powell's initial statements, John Moore said, 'I object on the grounds it is a violation of this defendant's constitutional right to have the presence of an attorney or the advice of an attorney at any time.'

'Oh, that is a frivolous objection,' said Schulman, but the prosecutor may never have been more wrong in his professional life.

'I will cite the district attorney for misconduct,' snapped Moore.

'And I will cite *you*,' said Schulman.

'And move for a mistrial at this time,' said Moore.

'Ladies and gentlemen of the jury,' said the judge patiently, 'the jurors will disregard the specific statement of Mr Schulman that the objection is a frivolous one. The citations for misconduct are denied.'

Among the long procession of prosecution witnesses was a Los Angeles robbery detective who had interviewed Jimmy Smith as to the armed robberies. He related Jimmy's apparently evasive response to his questions.

'Defendant Smith said to me: "Well, I can't help it, I just can't think."'

'I said: "Well, what is the problem?"'

'He said: "I have nightmares. I can't sleep. I just keep seeing that officer's coat jumping."'

'And at this time I had to admonish Mr Smith because I told him again that we were only interested in robberies. That we could not overlap into the homicide investigation and we did not want to hear about it.'

And then the detective read the statements of Gregory Powell as to the robberies, including a confession to at least one Las Vagas robbery the detective still believed Greg's brother was actually responsible for:

'Well sure, with the red hood on, I'll bet you there isn't a person down there that would tell you within twenty pounds how much I weighed,' Greg had told him. 'I know they described me in the paper as at least six feet tall and weighing at least a hundred and ninety pounds. But I did that job too.'

When it was Pierce Brooks's turn to testify, Greg's attorney once again repeated the old objections.

'We will object, your Honor, on the constitutional grounds, violation of due process, and not free and voluntary, not having been advised of his right to counsel, nor having the advice of counsel.'

'Well, if the court please,' said Schulman in exasperation, 'I don't know where the constitution says anyplace that a police officer ...'

'Pardon me. I made my objection,' said Moore.

'The objections and each of them are overruled,' said the judge.

The jury would hear the taped conversations of Brooks and Gregory Powell:

'Greg, anytime you want to ask me a question, you go right ahead and ask me and I want to let you know now that I will never lie to you,' said the taped voice of Brooks. 'I will either answer your question truthfully or I will tell you that I cannot answer the question, that it would interfere with the investigation.'

Moore once again interrupted the proceedings.

'I will make my objection on the constitutional grounds, violation of due process, without the benefit of consulting with an attorney, the advice of an attorney, or the presence of an attorney, and not free and voluntary.'

'Not free and voluntary?' asked the judge.

'Yes.'

'All of the objections, with the exception of the objection "not being free and voluntary," are overruled. Do you desire to take the witness on *voir dire* examination as to any of the objections?'

'Not at this time, your Honor.'

'All right. All of the objections are overruled. I say, knowing the thinking of our appellate courts, that we have to be very very careful, particularly in a homicide case with a death penalty, because it's a joint trial for the sake of convenience.'

On the morning of August 6, with the air conditioning broken, the judge entered a suffocating courtroom. All three lawyers were looking unhappy but still had their white shirts buttoned at the throat and their ties adjusted. The air conditioning was repaired before a decision was made to let them remove their coats.

'Pursuant to the provisions of Section 1089 and 1123 of the Penal Code,' the judge said that morning, 'it appearing to the court that the alternate juror, Mrs Jeans, is ill and unable to perform her duties as an alternate juror, and has, in addition, requested the court to be excused, she will be excused and discharged as an alternate juror.'

John Moore then said, 'Defendant Powell will move a mistrial on the basis that there was not a sufficient showing for her excuse, your Honor.'

'All right. The motion is denied.'

When the next detective witness took the stand, Schulman said, 'I am now going to ask you to relate the conversation that you had with the defendant Gregory Powell after the defendant Jimmy Smith left the room.'

'I will object on the grounds previously stated,' said Moore.

'If the court please,' said Schulman, beside himself now, 'I would like to find out just where the constitution says . . . '

'We don't want to speak in front of the jurors,' said the judge.

'Withdraw it,' said Schulman.

The jury then heard Gregory Powell's boast to the robbery detective: 'Anytime you can find one where it's got Schenley's on it, it's me!'

The hot summer days bled one into the other for Karl Hettinger, until August 9, when after a defense motion was granted, he found himself standing once more in a place he thought he would never see again. Now, though, the summer sun was in the cloudless sky over Bakersfield. It was hot and still and the lawyers were in shirtsleeves, as was the witness, who stood in the dust of the road near enough to the chair where the jurors sat by the onion field. He looked at the stakes in the ground which had been placed to represent where four men stood exactly five months ago when the sky was black and bitter cold and the wind was howling.

Karl looked across the field he had run through and he thought of the glasses he had lost and he wondered if they might be just on the other side of the wire fence where the tumbleweed was packed solid. But he did not go to the fence and look. Karl, like Ian Campbell, was frugal. He did not take the loss of a thirty-dollar pair of glasses lightly, and, in fact, he had still not replaced them. But he would not go into that field, nor near that fence. Even now in the light of day with so many others here. He stood where he must, looked where he must, answered what he must.

Then he found himself testifying. Telling it again. Pointing to the exact places where it all happened.

'Was there any talking at that point?' asked Schulman.

'Yes, that's when he made the statement about the Little Lindbergh Law,' said Karl, his soft husky voice disappearing in the open air.

'I don't know if the jurors can hear you. What took place then?'

'When Powell moved to that position of the stake, he had the gun pointed at about a forty-five degree angle and this is when he said, "We told you we were going to let you go, but have you ever heard of the Little Lindbergh Law?"'

'What did Campbell say?'

'He said yes.'

'What did Powell then do?'

'Powell shot him.'

'How did he do that?'

'He raised the pistol to shoulder height and he fired it.'

It was perhaps the sixth time Karl had testified to this and when it was the defense counsel's turn John Moore said, 'Would you walk down to the spot where you looked back, sir, and look over your left shoulder in the way you did?'

'Do you want him to *run* down the road, Mr Moore?' asked Schulman sardonically.

'If there is an objection, I would like to hear it,' Moore snapped back. Then to Karl, 'Just before you got out of the car, Officer, you were told to freeze, were you not?'

'No. I may have been, but I don't recall it.'

'Well, you didn't go over your notes today?'

'I would like the record to reflect, your Honor,' interrupted Schulman, 'that I have gone over the record fairly thoroughly and I don't recall this witness at any time testifying that Powell told him to freeze or anything of that nature.'

'At this time I cite the district attorney for misconduct for stating what he thinks the evidence shows.'

'I object to those remarks also,' said Ray Smith.

'This is in regard to a question by Mr Moore in a sarcastic manner that "You haven't looked at your notes today,"' Schulman retorted.

'I will cite the district attorney a second time for misconduct for saying I asked the question in a sarcastic manner!' said Moore.

'All right. The objections and each of them are sustained,' said the judge, sitting in a folding chair at a folding table. 'The jury is admonished to disregard the colloquy between all counsel.'

'When you ran twenty feet, you turned and looked back over your left shoulder, didn't you?' asked Attorney Smith on cross-examination.

'Yes.'

'You were going full force?'

'I slowed some.'

'Now is it your testimony that Officer Campbell fell backward or forward when he received the first shot from Powell?'

'He went down and he appeared to fall more backward. He went more backward than he went straight down.'

'And would you say that when he fell down, Officer Hettinger, that he was lying six feet two inches long on the ground? Or was he more or less fallen in a heap?'

'I don't know, sir.'

'And before he hit the ground you were *gone*, isn't that right?'

'I don't know.'

'I have no further questions.'

The judge was then to grant another defense motion: to see the onion field at night, the way it might have been on another night. So at 9:30 P.M. Karl Hettinger stood by the road with Pierce Brooks, beside the lighted county bus. He looked across the darkness toward Wheeler Ridge and then east and north toward the mountains far far on the horizon. The wind was not howling and it was not cold this night, still he shivered, while the sweat trickled down his chest and back and ribs.

He thought about his new daughter and thought about Helen and he thought about putting brake shoes on his car. He thought about anything but what was creeping into his mind. He beat it back and thought about the stomach cramps which had been plaguing him of late. But a thought stole into his mind, so he turned from the fields and looked into the crowded bus. The judge and jurors were sitting quietly. The defendants were in separate darkened cars, handcuffed, under heavy guard. Suddenly Karl began hurting badly, a twisting, pounding hurt in the pit of the stomach. He feared a diarrhea attack. They waited several minutes in the quiet, lonely dark after the judge ordered the lights turned out on the bus.

Finally, the judge said: 'Let the record show that it is now about six minutes after ten. The jurors have remained in the bus and the lights have been out, and we are going to take our adjournment at this time until 9:30 A.M. on Monday, when we will reconvene in Department 104 of the Superior Court.'

When the trial resumed in Los Angeles after the sojourn to the onion field the defense tried new motions.

'Your Honor,' said John Moore that morning, 'the evidence here would show that this witness, Sergeant Brooks, without the permission of the public defender has interviewed this defendant about this case, and has so destroyed this defendant's confidence in the public defender at least for a period of time, that he would not cooperate with the public defender. As long as there is an attorney on the record, I say, in my belief, the police officer should not be able to talk to him.'

'Mr Moore, if the law required that I notify you, I would do so,' Brooks would reply. 'And if you or anybody from your office had ever called and asked me not to see Gregory Powell on any occasion I would not have done so. I did not force myself on him. He asked me to see him. Even in your presence he asked me and you did not object.'

Attorney Ray Smith once more offered his old motion.

'I want the record to show that the defendant Smith is asking your Honor for an advised verdict of not guilty for the reason and on the basis that the evidence thus far introduced in this case fails as a matter of law to show that the offense alleged in this information occurred within the jurisdictional territory of Los Angeles County.'

Pierce Brooks often sat at the counsel table and glanced absently at the photo exhibits, at the open eyes of Ian Campbell lying on his back on the stretcher, his chest soaked with blood and the trickles of blood running from his upper lip down into his ears. Mostly the detective looked at the opened eyes of the dead policeman. The eyes hadn't yet clouded when the photos were taken. Brooks thought Ian looked just a little sleepy and serious, perhaps even sorrowful.

The detective stared at the picture and thought of the thousands of dead bodies he had seen, some with eyes open, some closed, some with one open and one closed. He had never seen the killer's face in a dead man's eye as the old stories had promised. But he knew whose image would show sharpest if such a thing were possible. Then he turned and glanced at the two killers: Gregory Powell, sitting tense and straight, head swiveling on that long neck, looking toward the jury every few moments to catch their reaction to testimony. Brooks glared at Jimmy Smith, sitting hangdog as usual, lip pouting, forehead permanently wrinkled despite his youth, as though he lived in pain. Brooks despised Jimmy Smith, and wondered how the cowering bully felt at the last when he had a cop helpless at his feet. Yes, Powell was a dead man. The battle was over. He had yielded to the detective. Brooks felt nothing toward him now. But Jimmy Smith, that was another matter.

The prosecution used multicolored arrows and markers to describe the death scene and escape route of Karl Hettinger. The exact place of Ian's murder was to be known as the place marked by the red arrow. The *red arrow* was repeated so often it became a litany that Pierce Brooks would never forget. And when it came time to use markers on the blackboard to describe the defendants' and officers' positions, Pierce Brooks smiled to himself and was tempted to use a yellow marker for the coward Jimmy Smith, but had too much reverence for the court and law to be anything but respectfully serious.

Finally on Monday, August 19, defendant Jimmy Smith took the stand and was cross-examined by Marshall Schulman.

'Was there any conversation between you and Gregory Powell, whispered or otherwise, that let you know that he was going to shoot one of the officers or both of them?'

The answer, Pierce Brooks was to say, was pure Jimmy Smith in its evasiveness.

'In my mind, no sir. I'm positive that I had no thought in my mind. Right now. That I can remember.'

'And you never worked the farms around that area of the Maricopa cutoff?' asked Schulman.

'I'm almost sure I haven't. I swamped some spuds once down there and I also picked cotton goin the other way. But I never remember workin close to the mountains. Maybe I did, but I don't remember it.'

Then Marshall Schulman read a disjointed transcription taken from a detective who was present the night of Jimmy's capture. It was taken by a stenographer and it was read by Schulman verbatim in the form of a question. The statement was difficult to follow, at times incoherent, and Pierce Brooks looked at Jimmy Smith and imagined the absolute fear that was on him that night when he huddled there handcuffed, a blanket over his naked shoulders, his feet bloody and painful, while he was interrogated, not for his usual five-dollar shoplift, but for the *murder* of a cop. He could easily imagine Jimmy babbling incoherently, and he could understand how a man like Jimmy Smith could have survived his wretched life by *never* giving anything *but* an indirect evasive reply to anything anyone ever asked of him.

Brooks could *understand*, but that was all. He despised the lying coward too much for a quantum leap into pity. Jimmy had blurted things to the Bakersfield detectives:

'When I hit the county jail, I'm gonna make them give me . . . give me . . . I know that I . . . A psychiatrist thing, you know, and

I bet you he can tell, you know, that I, you know . . . that I, you know . . . that I'm not mental, that I couldn't do it, you know, do *that*. I *hope* I didn't do it. I might do it, you know, in a pinch, or maybe if I was shoved into it, or something, but I mean, as far as just outright, you know, just kill a man, you know. Was there anything else you wanna know?'

The jury listened in rapt attention as the words on the transcription tumbled brokenly:

'I don't know whether . . . maybe not, maybe Powell will tell you. Maybe . . . man, he's sure to think, you know, about that. You know what? It's just, you know, outright cold-blooded, you know? This takes a lot of . . . I don't know what this is. He's gotta be a nut, man, you know, just to, for no reason . . . If you are gonna attack me, or just to cold-blooded kill . . . How could *I* do that? Could *you* do that? I mean, you maybe could. I mean, there's no motive other than maybe to escape or somethin. Well, uh, this don't . . . I don't know. Maybe it was doin time, or somethin that made me, could build myself up to somethin, or . . . or, if I thought I could escape the Death Row, or somethin, you know? But, how could a guy in this cold-blooded . . . and this was a simple matter, takin the guys back in the hills and dumpin em. This is . . . this just can't be done, you know?'

'What *were* you doing,' asked Schulman finally, 'when Gregory made the statement: "Jimmy, what the hell are you doing? The guy is getting away"?'

'As far as I know, I was just standin there like an idiot because I was in shock.'

'You were shooting into a man's body, weren't you?'

'I never saw his body. I never saw . . . I never looked . . . '

'Mr Smith, describe the twitching motion that you saw on that body on the ground.'

'I either saw an arm or a shoulder . . . I don't know which it was. I couldn't even determine the color . . . but it seemed to me . . . I even had dreams about it, that it was . . . periodically, just a few times it was done like . . . or it was movin. I don't know. I can't even describe it.'

'You say you have dreams about this coat jumping?'

'When I first arrived in here at night I was dreamin about it. I couldn't understand it, why I kept thinkin about this.'

'Did you see the coat jumping?'

'I don't know. Yes, I guess I did see it. I thought I saw it, anyway. I'm pretty sure I did.'

'You never mentioned the four bullets in the body until March 13, when you corrected everything. That's when Sergeant Brooks told you there were four bullets in the body.'

'Yes sir. That made me remember. Him tellin me.'

'Remember?'

'Yes it did. Then I remembered hearin those other shots!'

Now Marshall Schulman could look at the sunken cheeks and skull-like face and flat blue eyes of Gregory Powell and whisper to Pierce Brooks: 'Old death's head ain't got a chance, and neither does his partner now.'

At last, Brooks could look at Jimmy Smith and no longer be troubled by the thought of him hunting Karl Hettinger, quartering his victim with the little car and the lights, hunting him down like a wolf.

The defenders of Gregory Powell knew that if Jimmy Smith's fate was uncertain, their client's position was perilous indeed. And their client realized it and permitted the groundwork for a defense of diminished capacity if not outright insanity.

A neurosurgeon from Vacaville was called who had performed a craniotomy on Greg on September 14, 1961, during his last prison term.

'This patient gave a history upon coming into the institution of having had a head injury in childhood,' testified the doctor. 'It is the routine at the California Medical Facility to subject such a patient to neurological screening. In other words, to determine whether there is, in fact, any injury or other disease of the nervous system in such a person. This involves performance of X-rays of the skull, brain wave tests, or electroencephalograms, and neurologic examination. In this particular case, an area of calcification was seen in the right fronto-temporal region.'

'Roughly the right temple?' asked Moore.

'Yes sir. As I say, this area of calcification was revealed on X-ray, and its exact nature could not be determined in spite of a number of different tests, so after obtaining opinions from other neurosurgeons and X-ray specialists, it was concluded that the only way we could be certain that this calcification was not caused by a brain tumor was by exploratory surgery.'

'Now, Doctor, what does calcification mean?'

'It is a deposition of calcium within the tissues.'

'Where does it come from, if you know?'

'Well, the reasons for this are not fully known. It may occur without any known injury or disease. There may be areas of very dense calcification in the brain without apparent adverse effects. A possibility, and the thing that led to surgery here, is that certain types of brain tumors may form calcium within their substance.'

'Now it is my understanding that atrophy, the shrinkage or wasting of the brain, can come from a trauma?'

'Yes sir. It will progress up to a point and then will stop and become permanent and will not change beyond that. With overwhelming head injuries sufficient to produce prolonged coma it may be progressive.'

'All right now, what effect would such an injury have upon the behavior or the way a person behaves or acts?'

'I base what I am about to say on a study of previous cases with problems of this sort and usually those with this problem to a more severe degree. A person with cerebral atrophy, this is a mild atrophy in this case, but a person with real cerebral atrophy tends to have episodes of, let's say, intermittent explosive behavior, unpredictable usually. This may be precipitated or accentuated by alcohol intake and in effect it is usually a bizarre behavior.'

'And could the repeated intake of alcohol continue the process of atrophy or wasting away?'

'A very heavy intake of alcohol for a prolonged period would cause atrophy itself without head injury, so it becomes a bit more difficult to say specifically, since we have two known agents under discussion that can produce atrophy.'

But when Marshall Schulman took the doctor on cross-examination it was evident that the prosecutor had done his homework.

'As a matter of fact,' said Schulman, 'there are probably individuals in high prominent positions with atrophy of the brain, are there not?'

'Yes there are,' said the doctor. 'I know of a congressman, but I won't give his name.'

'Okay, what about General Eisenhower as a result of a stroke?'

'This would be unavoidable on the basis that I know of no other stroke that hasn't had atrophy, so I must assume the President had.'

'And you studied the results of those electroencephalogram tests, did you not?'

'Yes sir, I did.'

'And did they show normal?'

'Yes sir, they did.'

Then Schulman related a 'hypothetical' statement of murder which included the Little Lindbergh statement and posed a question to refute the explosive behavior theory.

'Well, the things in the hypothetical question,' said the doctor, 'which would seem incomparable with the explosive behavior concept, would be the space of time involved for one thing.'

'Let me put it this way, Doctor. The space of time involved at the time of the taking of the officers to the time of the shooting was approximately two hours. What would you now say?'

'Well as I previously testified, if explosive behavior was to occur, usually it's going to occur right away in my experience.'

'You have indicated that alcohol may affect Mr Powell's condition, is that right?'

'Yes sir.'

'On the other hand it may not, isn't that also right?'

'That's true.'

'And if I related to you for the purpose of this question that just prior to this trial an EEG occurred, sleep-induced, an alcohol-induced EEG . . . you are familiar with both these tests?'

'Yes.'

'And the alcohol-induced EEG showed no change at all from the sleep-induced EEG, and the sleep-induced EEG showed certain spiking consistent with the fact that an operation or a craniotomy was performed, does that indicate to you that the intake of alcohol would not necessarily affect Mr Powell?'

'I would have to draw that conclusion, that it might not necessarily affect him.'

'A person can still have his intelligence, think and plan of his own free will, and do a criminal act and still have some atrophy, mild or otherwise, without the atrophy having anything to do with the criminal act, isn't that right?'

'I think that is true.'

'It doesn't stop him from becoming a free-willed individual?'

'No sir, not a mild atrophy.'

'Thank you. That's all.'

The defenders of Gregory Powell then called another medical expert. This one was examined by Public Defender Kathryn McDonald.

'What is your profession, Doctor?' asked Miss McDonald.

'I am an otolaryngologist.'

'What?' said Schulman.

'In simple terms, an ear, nose, and throat specialist.'

Then the doctor testified: 'I do not think that this mouth wound would likely be a fatal wound. It has not injured a vital structure. The structures that it has injured are not likely to bleed vigorously. I just feel that one could sustain this wound and live. My opinion is that it would not be fatal unless it would . . . you would have to consider other possibilities or infection or complications.'

'In your opinion, Doctor, would such a wound be fatal as a result only of hemorrhage?'

'The structures that are injured here would not bleed enough. I have seen many tongue wounds. They bleed vigorously for a short time.'

Marshall Schulman was to ask: 'Do you know any of the parties involved?'

'No. I considered it a civic duty if I could testify in this case at my convenience.'

'You testified that there would be no vessels severed by this bullet wound?'

'There would be only capillaries.'

'No veins at all?'

'No major veins.'

'And no arteries at all?'

'I don't believe it's necessary . . . '

'Please answer my question, Doctor.'

'All right. I will state again there would be no major vessel severed.'

'If a bullet goes through a man's tongue, he is not going to bleed very much?'

'He will bleed vigorously for a few minutes.'

'Now the palate is the roof of the mouth. This bullet tract went through the roof of the mouth. Would that not bleed also?'

'It would bleed, but we cut the roof of the mouth in surgery without concern, in midline.'

'I understand that,' said Schulman, 'but assuming a man is out in an onion field and there's nobody around to sew him up or stop the bleeding . . . '

'I don't know whether you are trying to bait me into becoming provoked, but this is repetitious.'

'What would you do if a man came into your office with a wound tract that has been described to you in the hypothetical question, what would you do for him? After taking X-rays and determining the extent of the injuries?'

'The probability is, I would be prone to leave the bullet right there.'

'You would?'

'Yes.'

'It would not be bleeding?'

'It wouldn't.'

'Well, did you know, Doctor, at the scene where this man fell down there was a blood spot found, and did you know when he was transferred to the morgue in Bakersfield sometime later he was still bleeding? You never saw this body, did you?'

'No.'

'You don't know anything about this body?'

'No, other than the autopsy report.'

'You have never seen this wound tract?'

'No.'

'You have never been a pathologist?'

'No.'

'You are an eye, ear, nose, and throat specialist?'

'No, I am an ear, nose, and throat specialist.'

'A person is rendered unconscious. I'm assuming he is bleeding and he is on his back, lying on the ground in a dirt field out at Bakersfield shot in the mouth. He would suffocate, wouldn't he? That would kill him, wouldn't it?'

'This is a possibility, if he were unconscious. He might suffocate.'

'In your entire experience, have you *ever* seen a wound tract identical to the wound that has been described to you?'

'No.'

'I am making a motion to strike all of this witness's testimony,' said the prosecutor, 'on the ground he is not a qualified expert, and he's incompetent to relate an opinion as to cause of death concerning bullet wounds. He is not a qualified pathologist.'

Still more defense theories were introduced. Then Gregory Powell took the stand and decided now it was he, not Jimmy Smith, who fired the .32 automatic the one time, leaving the empty cartridge on the ground by the body. The defendants were more than desperate now, and each was forced to discredit the other as much as possible, to show the other to be a complete liar and hope by this to insert a slight doubt. Hoping by this to save one's own life.

'As I was alone looking at the dead officer,' Greg testified, 'I raised the gun to my temple and I started to pull the trigger. Maybe I was deliberately looking for an excuse not to shoot myself, but anyway, the thought flashed across my mind that Jimmy is going back after Max because he knows she's got money. And I jerked my hand a little bit. The gun went off, I fired across in front of my face.'

Schulman hitched up his pants and was hard-pressed to control the sneer when he took Greg on cross-examination.

'Is it your testimony that you fired the .32 yourself when you were putting pressure on your finger in an attempt to commit suicide, but a thought flashed through your mind that Jimmy was going to Los Angeles to get Maxine or the money, and you pulled the gun away and the gun just went off right across your face?'

'It's the truth.'

'Really?'

'Pardon me,' said Moore. 'Your Honor, I will cite the district attorney for his little comment about "really." It has been going on continually throughout this trial.'

'Incidentally,' said Schulman, 'so there will be no mistake here, there has been a little baby brought into this courtroom throughout this trial. Held out here and fed. Is that your baby?'

Pierce Brooks had to grin when Greg gave exactly the same answer as he had when Brooks showed him the Las Vegas Colt, one of the murder weapons.

'That's my baby,' said Gregory Powell.

'That is the child you have a tremendous amount of concern for, I presume?'

'Yes sir.'

'That is the child that has been brought here into this courtroom throughout the course of this trial?'

'I don't quite appreciate that statement, Mr Schulman,' said Greg. 'The baby is receiving the very best of care.'

'Really?'

'Pardon me,' said Moore. 'Once again the little remark of "really" crept out. I will once again cite the district attorney.'

'The objection is sustained. The jury is admonished to disregard the use of the word "really" by Mr Schulman. The citation of misconduct is denied.'

'You told the robbery detective, did you not,' said the prosecutor, 'that during the time you were committing the robberies you were drinking and having a ball?'

'No sir, I wasn't having a ball.'

'Were you having a ball when you shot the officer down in that onion field?'

'No sir.'

'The baby was born in the hospital in Camp Pendleton, wasn't it?'

'That's right, sir.'

'The charges for that birth were something like twenty-five dollars and were sent to her husband who is a serviceman overseas, isn't that right? As a matter of fact, three lovely children were taken away from Maxine, the girl you were so concerned with, by court order. Isn't *that* right?'

'I object to this as being outside the issues,' said Moore.

'He's claimed he's concerned about children,' said Schulman.

'Right here we have to stop while I cite this gentleman, so called, for misconduct,' said Moore. 'He stands in this court making his snide comments through eight weeks of trial and I ask for a mistrial at this time.'

'Did Officer Hettinger tell you about *his* family?' asked Schulman when permitted to proceed.

'I don't recall him doing so.'

'The officers didn't give you any trouble?'

'No sir, nobody would under those circumstances.'

'I imagine *not*, Mr Powell.'

'Pardon me, Mr Schulman is making snide statements again,' said Moore.

'You were snapping out the commands, weren't you?' continued Schulman.

'No sir, I was not snapping out commands,' said Greg.

'Oh yes, you were thinking about getting home to Maxine? When you came around and shot Officer Campbell right in the face were you thinking about Maxine then?'

'I have already testified as to what . . . '

'Were you thinking about Maxine then, yes or no?'

'I wasn't thinking about anything.'

'Were you thinking about your unborn child then?'

'I wasn't thinking about anything, sir.'

'Were you thinking about Officer Campbell's wife and family?'

'I have already testified as to what . . . '

'You testified earlier how they finally allowed you to make a phone call. Did you allow Officer Campbell to make a phone call to his wife so she would know he wouldn't be coming home that night?'

'I don't . . . no, sir.'

'I object!' said Moore. 'You're getting prejudicial misconduct into the record. I will cite the district attorney for misconduct.'

'The citation is denied. The objection will be sustained to the question on the ground it is argumentative,' said the judge.

'When you took the officer's body and flipped him into the ditch, were you thinking about his wife and family?'

'I have already stated what my state of mind was.'

'When you were looking for Officer Hettinger to kill him, were you thinking about *his* wife? Were you?'

'I don't think I was consciously thinking about killing or doing anything.'

'You felt it needed to be done, isn't that right?'

'Yes sir.'

'That you *needed* to kill Officer Hettinger?'

'You used the word "kill" in such a way . . . but I imagine this is the truth.'

'I don't know any other way to use it.'

And then Gregory Powell, not Marshall Schulman, showed his coolness under fire. 'Is there a question pending, sir?' asked Greg calmly.

'So you remember telling Sergeant Brooks that the Lindbergh

Law was mentioned but you don't recall who said it at the time just before the shooting?'

'I think I heard that there had been talk about the Lindbergh Law, sir, but I don't recall exactly what it was.'

'Well, what kind of stance were you in when the gun went off accidentally?'

'I was moving sir. I was just stepping around the car.'

'That accidental shot was real accurate, wasn't it?'

'That's a question? Well, it wasn't accurate, sir, because it wasn't aimed at anybody.'

'Did you say anything to Jimmy or to Hettinger at the time? "Oh my God, it was an accident. I didn't mean it." Anything like that?'

'No sir.'

'Why not?'

'I didn't get a chance. I was stunned and then Officer Hettinger screamed and started running.'

'Did you think that when you accidentally shot a man that you have to go ahead and kill somebody else to cover it up?'

'Well, I wasn't thinking, sir, but being an ex-convict and knowing the kind of trouble I was in, and one officer shot, yes sir, I felt it was necessary to try to get the other officer.'

'What was Jimmy Smith doing when he was at the circle indicated on the chart?'

'Well, he was firing a weapon, sir.'

'Where was Officer Campbell's body?'

'I don't know. I told you that about half a dozen times, sir.'

'You aren't losing your temper now, are you?'

'No sir, I am not.'

Pierce Brooks sat at the counsel table, watched the witness and agreed that he was indeed not losing his temper.

'What is your feeling toward Jimmy Smith today?' asked Schulman.

'I sort of pity him.'

'And what do you think Jimmy Smith's feeling toward you is?'

'I imagine hostility, sir.'

And *that* was indeed the truth. During the next recess after the defense rested, Jimmy Smith kicked his chair halfway across the room, but offered no resistance when uniformed deputies ran forward to restrain him.

'Your Honor,' said Ray Smith, 'he wants to call your attention to my incompetence throughout this trial.'

'If there are any improprieties, you can call them to my attention at the proper time,' the judge said when Jimmy was once again returned to his chair.

'I don't want this man to continue!' Jimmy shouted. 'This man is only sentencin me to my death, and I haven't did anything.'

'The case has been concluded,' said the judge. 'The district attorney has started his argument to the jury. I will hear everything you have to say, but that is after the arguments are concluded.'

'After I'm in the gas chamber, then you'll listen to me! When it's too late! After, not before!'

On August 23 Marshall Schulman addressed the jury:

'It is just a plain, outright, cold-blooded killing. The physical evidence shows you that Officer Hettinger was outright cold-bloodedly killed at the scene of the . . . '

'Not Hettinger,' said Ray Smith. 'You used the wrong word.'

'Officer Campbell. Thank you. Officer Campbell,' said Schulman. 'I think you know who I mean.'

But Karl Hettinger was to wonder if Schulman's slip of the tongue was not prophetic.

When it was Ray Smith's turn to argue he did it in the way he knew best: intimate, personal, homespun. The old lawyer smiled often, and looked each juror in the eye, addressing the juror as a close friend, and sought to place all blame on his client's partner.

'Gregory Powell is, beyond doubt, the most colossal liar you will ever see or hear in a lifetime, and I mean that sincerely. You will have to live a hundred years to see a bigger liar than Gregory Powell.'

John Moore talked to the jury of many things, not the least of which was his adversary.

'As the district attorney said to you: "The individual who fired the shots into the chest of Officer Campbell as he lay on the ground in Kern County is without doubt a frightening example of a human being."

'Mr Schulman has a trait that is valuable to prosecutors if your only concern is that of conviction. He has the ability by tone of voice or by manner to make whomever or whatever he desires sound dirty or obscene. He can say the words "wife," "mother," or "baby," and if he so desires, they sound dirty and obscene.

'He repeated the threats to kill that Powell made to a number of employees during the robberies. His tone of voice in reading lines of transcript made your skin crawl.

'If you had a verdict of kidnapping, and death were to result during that, you would be tempted to find them guilty of murder in the *second* degree. Mr Schulman would not like that. Even if it be true, he would not like that.'

On August 30 an incident occurred involving juror number four which both the defense and prosecution thought exceedingly strange.

'Did you receive certain information from Deputy Sheriff Margaret Decker concerning comments made by one of our jurors?' the judge asked the clerk that day out of the jury's presence.

'Yes, I did, your Honor,' said the clerk.

'Miss Decker, as part of your duties you have been admitting females into the courtroom?' asked the judge.

'Yes, your Honor.'

'Do you know Mrs Bobbick, juror number four?'

'Yes.'

'Who started the conversation?'

'I started the conversation by saying to Mrs Bobbick that I had hoped that the judge would instruct the jury on Tuesday rather than Friday because I had planned a three-day fishing trip in Mexico. Mrs Bobbick stated that her husband also had the three-day weekend and she too hoped that it would not go out. By this time we were out of the elevator, and I started to go to my office. Mrs Bobbick paused and put her hand up to her mouth and stated: "I have smelled a rat through this whole thing," and that lawyer Ray Smith had either "uncovered," or "brought to light," or "hit the nail on the head." Now, it was one of those three expressions.'

At 10:00 A.M. Mrs Bobbick was ushered into the judge's chambers.

'Mrs Bobbick, I want you to feel completely at ease,' said the judge to the frail, middle-aged housewife with the darting eyes. 'Miss Decker, who is a deputy sheriff, testified under oath . . . '

'I would like to be under oath, too, before I say anything.'

'Yes, I am going to put you under oath. Anyway, the two of you were in the elevator. There were some other people in the elevator. And the two of you just passed the time of day?'

'She followed me, I think, because she never did that before. But this particular morning, this Miss Decker followed me out and she said to me, well, she said, "I think it will be tomorrow. You'll have to bring your bags." And then she said, "Oh," she said, "I don't know how it will go," she said, "the evidence is so overwhelming." That's what she said, and I'm saying this under oath. And I thought it was a little bit out of order for a deputy to say that to me, and I don't know why she followed me out. She never did this before. It seems that there're other people that kind of follow you out from the courtroom, but I never noticed. I couldn't understand why she did that. It was impressed on my mind because she did that.

'And she was telling me that the evidence was overwhelming. That's what shocked me. And I say, "Well, I don't know," because it was just an impulsive reply. I mean, it was like a psychological bombshell that she should say that to me. And I said, "It seemed to me that I smelled something for a long time," and that is what I said. I don't know if I mentioned Attorney Ray Smith's name.

'After all, I'm sitting in the jury box and I'm under oath, and why she did that to me, I don't know. She simply got a psychological impulsive statement.'

'Well, Mrs Bobbick, the court certainly will make inquiry into this alleged misconduct on the part of Miss Decker.'

'Yes sir, I would like to say one more thing. I want to say what happened in the jury room.'

'I don't want to discuss what happened in the jury room.'

'I would like to say one more thing. I feel that because this has happened now and what she did, I'm probably disqualified as a juror, because that is brainwashing. I mean, that is saying something to a juror that shouldn't have been said. I purposely avoid having lunch with people and doing anything improper, and she comes up to me and says that. I feel that because of what is going on now, perhaps I'll be antagonized. I feel that I should be relieved of jury duty because this is important as far as my conscience is concerned. It's a psychological brainwashing thing that is like a snowballing that has been going on and perhaps maybe I'm . . . '

'Mrs Bobbick, if I may, we certainly will make an investigation into the alleged statement of Miss Decker,' said the judge.

'In other words, all these things have an effect on the human brain,' said Mrs Bobbick. 'I'm just like anyone else. My brain is like a sponge. If something happens to it, it has a reaction. I'm not dead or anything!'

'No, Mrs Bobbick,' said the judge, glancing toward the court reporter who was taking the testimony.

'Perhaps it was an impulse on her part. Perhaps a psychological thing. I'm a living human being and I'm influenced like other people are, but I try to go off by myself, to keep away from people. I have lunches myself. I don't know what it does to me.'

'Mrs Bobbick, we have no intention whatsoever of disqualifying you as a juror.'

'I mean, now, see, this is another incident, and it has a psychological reaction on me. I can't help it. I am one of those human beings. I am very level-minded and I am very conscientious.'

'We know that you are because you listened very attentively to the evidence and you took copious notes throughout the trial.'

'I know.' Mrs Bobbick smiled. 'And I would like to ask another question. If at the end of your instructions I still feel as if I'm disqualified, could I get up and say I am disqualified?'

'No.'

'If I don't get up, in other words, I will feel that everything is leveled off.'

'There should be no reservations in your mind. All we are interested in is having a fair and impartial jury.'

'Do you feel I'm leveled off? I feel I am. But I'll ask your opinion.'

'As long as you feel that way about it. Your state of mind is the important thing.'

'I am leveled off. It probably was just one of those things.'

'Thank you very much,' said the judge.

At a later in-chambers hearing the judge sat before the woman deputy.

'How long have you been a deputy sheriff, Miss Decker?'

'Sixteen and a half years, your Honor.'

'And knowing you and your exemplary record, the court won't question your conduct in connection with this matter. However, since Mrs Bobbick is one of our regular jurors the court would suggest that you confine your activities to assisting in meeting the needs of the jurors.'

'Yes, your Honor,' the deputy replied.

Judge Brandler sighed and ran his fingers through his fine pale hair, and his sleepy lids dropped slightly and he wondered if this trial would ever end with the record intact, with not only the defendants, but now the jurors themselves behaving in such a way to threaten mistrial.

The judge had little reason to feel reassured. The next day the court was given a handwritten note:

> Juror No. 4 has requested disqualification as a juror in this matter. Juror No. 4 has made this request because of what she, the juror, says is unnecessary dissension, thus possibly causing her to become ill and thereby not be able to fully concentrate and render a proper and just decision.

The judge then instructed Dr Crahan, the court psychiatrist, to examine Mrs Bobbick, but was advised by the doctor that the juror refused. The doctor however did have the opportunity of observing her and ventured an opinion that she appeared to be a schizoid personality with attendant suspicions of hostility, but at that time he found nothing to warrant the court excusing her as a juror by reason of conditions of health. However, Dr Crahan did warn the judge that if the court was advised that the juror became hysterical or showed any other signs of breaking down, to let Dr Crahan know so that the doctor could again attempt to examine her to ascertain whether there was any change in the condition.

The judge decided to proceed with the juror and hope she could endure. However, on September 3 he received another note:

> I, Mrs Bobbick, Juror No. 4, would like to make a request of the court to be relieved of further duty because of extreme hostility which appears to border on coercion and force.

*

This time the doctor did examine the juror and the judge wrote:

> The court was advised by Dr Crahan of the following facts based upon his physical observations of Juror No. 4: She appeared to be hysterical, her eyes were popping, she was gesticulating with her hands, she was rambling in her conversation, she couldn't sit still, and she appeared nervous and tense. The doctor advised the court that continuing on as a juror might well seriously impair her health.

If Marshall Schulman was distressed by the news, the defense was elated and saw a chance for a mistrial at least.

Pierce Brooks shrugged wearily and privately expressed confidence that Judge Brandler could work things out. In his years as a detective he had seen a thousand jurors who were not merely crazy but were incredibly stupid to boot. It was all a chancy thing, after all, this whole jury business. That's what the detective muttered privately, but publicly he controlled his disgust. He revered the courts and the law and was the kind of man who had always served without question.

The defense wasted no time in calling the doctor to challenge his recommendation to excuse the juror.

'What is your field of specialty?' asked John Moore.

'Internal Medicine and psychiatric diagnosis. I have practiced medicine for some thirty-five years, and have examined between five and ten thousand people for the Superior Court of Los Angeles County as medical director of county jail.'

'Have you ever examined a juror who was holding out as a minority in a jury room?'

'Objected to as assuming facts not in evidence,' said Schulman.

'Objection sustained.'

'What were the statements she made about the other jurors?' continued Moore.

'That they were abusing her, cursing her, and they were suspicious of her. She had made the statement that she had cried for the first time in forty years. And she told the judge in chambers that she saw a rattlesnake up near her home and she called the police and the police came and the juror said, "Do you know, Judge, the police shot that poor rattlesnake?"'

'This is a capital case, your Honor,' said Ray Smith. 'I for one would like to have Mrs Bobbick come down here and tell us on the record what is going on that has caused her to become so hysterical that she had to write and ask your Honor for permission to get off the jury.'

'All right. The request of counsel is denied,' said the judge. 'The court will order the discharge of juror number four and direct that the name of an alternate shall be drawn to take her place.'

'I ask for a mistrial!' said Moore, but he was denied.

On the fourth of September, Gregory Ulas Powell and Jimmy Lee Smith were finally found guilty of first-degree murder, and Gregory Powell wasted no time in making a motion to represent himself during the penalty trial.

'Due to complete disagreement with both my lawyers, I intend to discharge them immediately,' Greg informed the court, and it was astonishing how much he himself *sounded* like an attorney after listening and learning for two months: 'I don't believe I ever stated at any time I felt I was incapable of defending myself, as the district attorney maintains. I think I stated, if the record will reflect, that I realized I didn't have enough time to prepare my case. I think Mr Schulman has made assumptions as to what I know and what I do not know, and I feel that this is none of his business. The court is not aware of some of the colloquy and the different issues that have gone on between counsel and defendant. Therefore what Mr Moore has done in his own behalf and what he has done at *my* urgent insistence is something the court cannot possibly know. The argument that Mr Moore presented to the jury, *I* wrote the night before. Many of the things which were in the argument I had written out myself, your Honor. I could have made the same argument and done as well.'

Pierce Brooks looked at Greg and listened to his newly acquired legal jargon and rather surprising eloquence and said to Marshall Schulman, 'What an ego! Yes sir, that's my boy!'

'What *is* going on here?' demanded Schulman. 'This is just another method of presenting some questions on appeal, and Mr Moore is just sitting there and letting it go by!'

'All right, I know this much,' said Greg. 'Mr Moore cannot possibly obtain my cooperation if I will not give it. And I have stated that I will not. I have made my position quite clear.'

'Your Honor,' said Jimmy Smith, now entering the debate, and the heavy-lidded blue eyes of Judge Brandler moved back and forth toward the many speakers at the counsel table. 'Mr Powell has been treated with a silver spoon in his mouth ever since we were arrested. I haven't did anything! I am innocent and I know I can . . . that I could prove it. I could prove it to this jury beyond a reasonable doubt. Goin to Bakersfield was *my* idea, not my attorney's idea and the record will so reflect that I am the first one that mentioned it. Then he comes back and asked the court, would the court grant the motion to go to Bakersfield as a part of the defense. We went to Bakersfield. There was no defense put on for my part. My attorney ran around drunkedly all over the ground. And I attempted to call him back and ask him whether they could put Mr Hettinger on the stand and ask certain questions that would prove beyond a doubt that I was innocent. He would not ask him. It was a physical impossibility of Hettinger runnin down the road and lookin over his shoulder. And another thing, I asked him to give me a hypnosis test. Any kind of way, a psychiatrist, anything. He hasn't made any attempt to do that. He refused. I did not commit the crime. This whole trial has been put on for *this* man Powell. He is a self-confessed murderer. May I just say one more thing, your Honor?'

'Yes,' sighed the judge.

'I can prove beyond a reasonable doubt that I, Jimmy Lee Smith, is more qualified to defend myself in this courtroom than Gregory Ulas Powell. And also the district attorney has papers in his possession that will prove these facts to this court. I think.'

'Prove what?'

'Prove I am qualified and capable of defendin myself, maybe not verbally, but when given opportunity to take my time and to write out things that I wish to say, and to have use of the law library which I have asked the court on occasions, and I was refused. Sergeant Brooks even knows I am not guilty. They all know it. That is why all these lies were produced in this court. Just to get *me*. I want another attorney. Please.'

'I am not interested in the many falsehoods that Jimmy Smith has just related,' said Schulman, worrying now about an issue being raised on appeal concerning Smith's accusation of a drunken attorney. 'I think the court should make some observations on the alleged drunken stupor Mr Ray Smith was supposed to have been in at Bakersfield.'

'I believe the word he used was "drunkedly,"' said the judge.

'Not only there but here in the courtroom too, your Honor!' shouted Jimmy.

'The court has no information that during the course of this trial Mr Smith has partaken of any intoxicating liquor.'

'You have almost got me almost in the gas chamber!' Jimmy cried. 'What am I gonna do? Your Honor, please have mercy on me!'

'The motion that the court is ruling on now is just the motion that Mr Ray Smith be relieved, and that the court at public expense appoint another attorney to represent you. That motion the court does deny.'

'Well, your Honor, you will have to relieve him in some kind of way,' Jimmy said. 'He *must* be relieved, your Honor. If you don't, I don't know what I'm gonna do, but I must do somethin. I can just sit here and don't do anything just as well as to have him. The man is not doin a thing.'

'All right . . . ' the judge said, but Jimmy interrupted him once again.

'Would you consider, your Honor, if I brought in a witness to state that after talkin to Mr Smith they smelled liquor on his breath?'

'All I can suggest to you is, the mere fact, if it is a fact, that an attorney may have a drink during noon hour is certainly no basis for any criticism of that attorney. Was there anything further?'

'Yes, your Honor,' said Greg, on another subject. 'There are no mutual meeting grounds for Mr Moore and myself, and I think that I quite clearly stated this morning my requests and my wishes, and they are still the same. I still insist, and desire – *demand* – that I be allowed to defend myself for the duration of this trial, and I am prepared to go forward immediately. There is no mutual meeting of the minds. Mr Moore has indicated that he wishes to withdraw, and I concur, and have already asked that he be discharged.'

'As a lawyer, I know,' said Attorney Ray Smith, 'that the facts of this case justify a first-degree murder conviction. I have been striving for one thing, and that is to keep Jimmy from the death penalty and that has been the theory of my defense, if you can call that a defense. For that reason, I will *not* request to withdraw.'

'I'm tired. I can't stand it anymore,' moaned Jimmy Smith. There were others in the courtroom who felt the same.

On September 6, during the penalty trial, Gregory Powell, representing himself, called his mother to the stand.

'Take the stand and be seated. State your name, please,' said the clerk, and Greg was looking more fit, well dressed, and relaxed than at any time in the case thus far, prompting Schulman to remark that being a lawyer agrees with him.

'I would ask your relationship to anyone in this case,' said Greg.

'I am your mother,' replied the witness.

'What is your occupation?'

'Housewife and college student.'

'Will you give your age, please.'

'Fifty years old.'

'Mrs Powell, could you tell us in general what your health has been since the year 1941?'

'I'm sorry, but I am going to have to object to that as immaterial to the issues,' said Schulman.

'I expect to prove by this testimony that during the formative years of my life the responsibilities of raising children were placed upon me,' said Greg. 'She has had, your Honor, I don't know the dates, but she has had ten major operations. She has had Bright's disease. Included in these ten major operations were three thyroids during this period of years.'

'I am going to object to this, if the court please,' said Schulman. 'Why do I have to be placed at a disadvantage just because he chooses to release his counsel and defend himself?'

'I don't feel you are placed at a disadvantage,' Greg retorted.

'It's not up to you to feel one way or the other,' shot Schulman, 'and if Mr Powell has anything to say he should address it to the court. If he wants to play lawyer I think he should act like one!'

And then Greg, proving he had in fact learned to act like at least one lawyer, replied: 'I would cite the district attorney for that remark, your Honor!'

After the judge restored order Greg continued, looking yet more confident.

'Mrs Powell, would you give a brief history of my personal health until school age, please?'

'At the age of three and a half you had scarlet fever and you were in the hospital with a mastoid operation. When you were four and a half you fell out of a car and had a serious head injury.'

'If the court *please* . . . ' said Schulman, then gave up and slumped in his seat.

'Do you remember, Mrs Powell, your financial status at the time?'

'We were very poor.'

'Did you have an occasion to think about the winter months that were spent there?'

'You want me to tell them I broke up a chair to have a fire one day to keep us warm?'

Schulman was now unable to control his eyes, which were rolling back every few seconds, giving him a stricken look.

'Would you say that despite all this, it was a happy home?'

'Yes, I certainly would. You don't have to have money to be happy.' Ethel Powell smiled.

'Going back to the time that I started the first grade, what was your general health at the time?'

'Terrible.'

'Could you tell us specifically?'

'Yes, the very first thing – I know it doesn't matter because it happened before you were born – but I lost all my teeth through an accident, and that left an infection in my mouth, and the infection spread through my body, so my health was in terrible shape.'

'How serious was your Bright's disease, Mrs Powell?' asked Greg.

'I was given three months to live.'

'While I was attempting to take care of the three younger children, did you receive any disciplinary reports about me from school?'

'I never received one. I only had a visitation from the head of the school as to why you wanted taken off the honor roll as a patrol boy.'

'Just a minute, I'm sorry,' said Schulman, 'but I have to object to that hearsay testimony.'

'This isn't hearsay,' said Ethel Powell. 'This was said to me direct. I'm repeating what I heard.'

'Mrs Powell,' said the judge, 'may I suggest that Mr Schulman's objection is directed to the court.'

'Was I engaged in a lot of sports and so forth?' asked Greg.

'Yes. Oh gosh, you got a lot of ribbons too. Blue ribbons. Yes, we had a wonderful track meet in which you did very well.'

'Did you go to the track meet?'

'I sure did. My husband and my father and mother did too. Gee, I know your father was happy. You took part in two things. You could've had top honors in both of them if you'd taken his advice. *Which you didn't do.* One was a throwing thing.'

'Softball. I took second.'

'I thought you got a blue ribbon.'

'I did in the relay.'

Schulman was now holding his head, tapping his tablet with a pencil.

Then Ethel Powell began a narrative of Greg's unhappiness when his grandfather died, and of falling into bad company and running away with Archie. And of her trip to Florida to bring him home, and of a later trip to Colorado when he was arrested there. And of her many pilgrimages to visit him.

'I went first to Cheyenne, Wyoming, and later to Englewood, Colorado, to a reformatory. Another time to Washington, D.C., to another kind of reformatory.'

'Did you come back to visit me a second time there?'

'Yes, for three weeks. I also saw you in Kansas, in Leavenworth. I had just gotten out of the hospital. I had an emergency appendicitis. You told me you had attempted suicide more than once.'

'Just a minute, I am going to object . . . '

'Was I in bed receiving blood transfusions?'

'Yes you were.'

'After that did you visit me in Atlanta, Georgia?'

'I was refused permission. Then I received a letter from you at the medical center in Springfield, Missouri.'

'Later when I moved to Los Angeles, Mrs Powell, did you see very much of me?'

'No, I didn't.'

'Why?'

'You were in the company of a colored man named Pinky. We thought it was like a homosexual.'

'After I came out of Vacaville in May 1962, did you come to know Maxine?'

'Yes.'

'Mrs Powell, you told me that you were president of a Toastmistress Club?'

'Yes I was.'

Schulman started to his feet, thought better of it, and sat back down.

'Mrs Powell, what is your health at present?'

'I am under the care of a doctor for a heart condition. I just received permission to go back to college this fall. I've been out of college for a year.'

Greg smiled then and looked at her. He was proud of his mother. 'Thank you. That is all,' he said.

'I have no questions,' said Schulman.

'You may step down,' said the judge.

Next, Greg called an administrator of the California Adult Authority to show that, if not sentenced to death, he would be safely locked in prison the rest of his life. The witness was however cross-examined intensively by Marshall Schulman, who once again had done his homework.

'Is it fair to state,' said Schulman, 'that sixty-three percent of those individuals sentenced to prison for life on first-degree murder were released after serving somewhere between seven to nine years?'

'Your Honor, I would object,' said Greg.

'Do you have any table of individuals who have murdered a police officer and been committed on a death sentence?' asked Schulman.

'No sir, we do not. We don't have enough such cases to compile a table. We had one such case recently that we investigated.'

'What were the results on that investigation?'

'He was granted a further commutation to life *with* possibility of parole. He has subsequently been paroled.'

Marshall Schulman smiled slightly. He had no further questions of consequence.

Then a parade of family witnesses took the stand, a service-station owner in Oceanside for whom Greg once worked, his off-again on-again brother-in-law who at that time was between marriages to Greg's sister, finally Robert Powell, his father.

'Mr Powell what is your occupation?' asked Greg.

'I'm a schoolteacher,' said the tired-looking man with the long muscular neck and the handsome friendly face.

'How long have you been a schoolteacher?'

'About seventeen years.'

'How long did it take you to get your degree, sir?'

'About fifteen and a half years.'

Then Robert Powell told of playing music in one-night stands during the Depression, and of his wife's poor health.

'Mr Powell, during the time you knew me as a child did you spend a lot of time with me?'

Robert Powell replied, 'I . . . I spent available time with you . . . that is, what I'm trying to say is . . . it was limited. My time.'

'Would you say I was a sensitive child?'

'Very much so.'

'Neurotic, to put it mildly?'

'Yes.'

'Mr Powell, starting with my arrival home from Vacaville last year, did I seem changed from the child you knew in those early years?'

The tired, grieving man smiled a little and said, 'Oh yes. You used to feel real cut down because you couldn't play music as well as your dad. But when you came back from Vacaville, why, your brother Doug and you and I had little jam sessions at the house. We sang trios. We worked out some real fine material and there was nothing but harmony in both senses of the word, when you first came home.' And Robert Powell dropped his eyes and swallowed.

'Dad, I hate to ask these questions, but I've got to. You have four children. Two boys, two girls, and I'm the oldest?'

'Yes.'

'Starting with Sharon, has she been divorced?'

'Yes.'

'Going to Lei Lani. Has she been divorced?'

'Yes.'

'Has your youngest son, Douglas, been in trouble? Was he discharged from the army because of trouble?'

'Yes.'

'Why have four children all had trouble adjusting to society? Do you know the answer, Mr Powell?'

'No,' said Robert Powell, and the muscles of the long neck rolled as he gulped again.

'Mr Powell, do you personally think I can write well?'

'Yes I do,' said the father. 'The idea content is excellent.'

'Mr Powell, I asked your wife and I will ask you also, why do you want me to live?'

'All right now, I am going to have to object to that as immaterial,' said Schulman, shaking his head. 'It appears difficult enough on this witness . . . '

'Mr Powell, do you believe two wrongs make a right?' asked Greg quickly.

'Objected to . . . '

'No,' answered Robert Powell.

'That is all, your Honor,' said Greg.

When the lawyer of Jimmy Smith called Jimmy's Nana as a witness to his nonviolent personality, Marshall Schulman set out to prove that Jimmy Smith did have a violent moment in his life, as a teenager, when he pulled a knife on a truant officer. Jimmy's Nana took the stand.

'Mrs Edwards, do you remember that on March 7th, 1947, Jimmy was picked up for truant at a theater at 539 South Broadway?'

'I don't know anything about dates, Mr Schulman,' said the crippled old woman. 'I don't remember very well. I don't know nothin about dates.'

During the old woman's testimony, there was a disturbance at counsel table. Jimmy Smith was babbling incoherently to no one in particular. 'It ain't right. My Nana never had nothin! My Nana never so much as said a swear word, or took a drink, or went to a dance in her whole life!'

'Did you hear that Jimmy pulled a knife?' Schulman asked.

'Only when the truant officer kicked him and pulled a gun on him,' the old woman said softly. 'And the gentleman I worked for in Beverly Hills said this boy had just as much right to pull a knife as the truant officer had to pull a gun, because they are not supposed to have guns, he said. That's what he told me, Mr Schulman.'

'Thank you. That's all,' said Schulman.

Now the tears were spilling from Jimmy Smith's eyes. 'You leave her alone! You leave my Nana alone!' he sobbed.

Jimmy Smith could never discuss his great-aunt or think of her as other than *my* Nana. It was always *my* Nana, the only thing which was ever truly his.

'Mr Smith, you will have to speak in a quiet tone of voice,' said the judge.

'It's not fair! It's not fair!' Jimmy shouted.

'The jury is admonished to disregard any statements of Mr Smith,' said the judge. 'He is represented by counsel, who will speak for him.'

'Nothing further,' said Ray Smith.

'Defendant Smith rests?'

'Yes, your Honor,' said Ray Smith.

Before the final arguments to the jury there was one last witness to testify for the defense – Gregory Powell.

While Greg walked to the stand, Pierce Brooks looked around the room at some of the familiar faces, two of which he had come to pity: a crippled black woman and a sad-eyed music teacher from Oceanside. 'Jimmy Smith's aunt was a poor old confused lady,' Brooks said later. 'I don't know what she ever did to deserve that miserable coward she raised. And I felt sorry for Gregory Powell's father. He was a nice guy. His only problem was he'd never booted his wife in the butt, years before, when he should have.'

Greg's testimony was emotional from the start:

'Well, number one, I guess it's pretty obvious I'm a sentimental slob. When I go to the movies I cry when there's a touching scene. I've been asked many times whether I feel remorse about this. Yes, I feel remorse for Ian Campbell's death. I feel more remorse, however, for his wife and kids. I'll go to prison, if this is the sentence, and I'll stay there and work and try to make up for Ian Campbell's death.

'I still don't feel that I should be on the streets. I honestly can't say and don't know whether I can cope with the stresses and strains that are out there. I feel in my heart that even though I know I am guilty of a terrible thing, that I should be allowed to live, because I feel that there is good that I can do.

'There've been many men in prison who have done good, that have made contributions to society. Whether I am capable of doing it or not, I don't know yet. All I can say is that I will be trying.' And then Greg began to weep.

'Are you composed? Compose yourself,' said Schulman, whose turn it was to cross-examine.

Greg nodded, wiping his nose and eyes with a handkerchief.

Pierce Brooks burned to read a document he had in his hands, a document which had been given to a sheriff's deputy in the jail by an inmate named Segal who had been an aspiring writer and actor prior to his arrest for selling marijuana. The document was a one-act drama authored by inmate Segal in collaboration with Gregory Powell, and Brooks had been holding it for five months.

Inmate Segal thus became the first person to write Greg's version of the kidnapping. It was not the version told in court by Karl Hettinger, nor by Jimmy Smith – nor by Gregory Powell for that matter. It was what Pierce Brooks would call a punk's fantasy, which to Brooks revealed the real Gregory Powell.

Fate intervened though, for Segal escaped after his transfer to Chino Prison, and Pierce Brooks was unable to get the drama introduced during the trial without Segal's testimony. But Brooks kept it and referred to it often, especially the portions in Greg's own handwriting wherein he would try to correct Segal's atrocious grammar and spelling, and insert the word 'cop' in place of Segal's 'officer' to make the tale yet tougher.

Brooks isolated certain passages as his favorites, hoping that somehow there could be a theory of law discovered wherein it could be introduced:

POWELL: 'Take the punk's gun, Jimmy.' Hettinger hesitated, his hand on his gun.
POWELL: Laughed. 'Go ahead, punk,' he said.
CAMPBELL: 'For God's sake, he's got us cold.'
POWELL: 'You know what, man – you should have been a dead man back there acting so funny.' Hett started sniveling.
POWELL: 'Before we drop these punks off, do you think we ought to take care of business – How much bread do you 2 have?'
CAMPBELL: 'I have about ten.'
HETTINGER: 'I have about the same.'
POWELL: 'Boy, you guys are hurtin, we've got five times that amount in *pocket* change.' To S. 'We'll knock over a super Market before we leave the valley.' 'I'll make a deal w/the 2 of you. We'll pull off onto one of these dark canyons – I'll put a couple of slugs in both of your guns. Then stick it in your

holsters. I'll shoot it out with you one at at time. If one of you gets me Jimmy will throw your guns in the bushes & split. I need some action and it shore would be a quick way to solve this mess. I don't think the punks have enough nerve to try it. What about it?'

CAMPBELL: 'No thanks, man – I don't want any part of it. Just turn me loose and I'll be happy. They don't pay me enough to try to be a hero.'

POWELL: 'What about you, Buddy – you were making noise like a hero back there when we picked you up. Are you game?'

HETTINGER: 'No, if I shot you you might still manage to shoot me.'

POWELL: 'In other words you're chickenshit—'

HETTINGER: 'Yeah, I guess I am at that, I never thought I'd get in a position like this—'

POWELL: Laughing. 'Well it sure would have saved us a lot of trouble. If either one of you would have gone for it you would have been dead anyway. It may sound like bragging but you punks don't know what an expert shot is – cause if there's any rating above expert I'm about 20 grades above it.'

Brooks read on and on, a dreary little drama riddled with 'punks' and 'Powell laughing' until the moment of arrival in the onion field.

POWELL: 'OK H – climb on out.' He did – P climbed out on the passenger side. As I walked across to the back of the car I could hear Smith speaking.

SMITH: 'Are you kidding – Have you ever heard of the Little Lindbergh Law—'

CAMPBELL: 'Yes.'

POWELL: I raised my gun to cover C as he looked at me – coming around the back of the car & shot him – ~~waitin for S to shoot H~~ H screamed.

SMITH: 'He isn't dead,' & he started firing *instead* into C as he lay on the ground. I ran across behind S & the only thing I could think of was (I've got to get that other cop) (It was like a scream in my mind) I was alone w/1 dead cop laying dead in the road and another hiding close by who I had no way of flushing from

cover – I walked back to the other ~~officer~~ cop laying on the
ground—

And now Pierce Brooks looked at Gregory Powell, crying on the
stand for the children of Ian Campbell, and Brooks read the last
line of the drama where Segal and Greg described his capture:

I was sick for I knew I would never see Max again. Cause I
never dreamed that the cops would get me to the station alive –
Had I been granted 2 wishes at that moment I think they would
have been 1 – to have been face to face w/J with a gun in my
hand 2 – or more important to know what I knew now & just
be meeting my wife for the first time – But wishes were useless
for I knew I had been born to loose—

Scum, thought Pierce Brooks, watching Greg wiping his nose, his
shoulders shaking, as Marshall Schulman stood by the counsel
table staring at the young man. It was not the killing which now
angered and embarrassed the detective. It was the killer's
unwillingness to pay for it with his life, and most of all the
unmanly sniveling from him who had edited and co-authored the
one-act drama. Scum, thought the detective. Scum. Scum.

'Who are you crying for, Mr Powell? Yourself?' asked Schulman finally.

'No sir.'

'When did you decide to start crying for Ian Campbell and his
wife and children?'

'Mr Schulman . . . '

'*After* you were found guilty?'

'I think the first regret . . . that I felt was the night I looked
down at the ground and realized the terrible thing that had
happened, and I attempted to commit suicide.'

'Is that why you asked Chief Fote for a break? You didn't do
anything, Jimmy Smith did everything. Is that when you felt so
bad?'

'Mr Schulman, at that time I was thinking of the living.'

'Yourself?'

'Max and the unborn baby.'

'Why weren't you crying when Chief Fote talked to you?'

'I was.'

'He's not here,' said Schulman sardonically, glancing at the counsel table. 'Mr Brooks *is* here. Why weren't you crying when you talked to Mr Brooks in all those conversations?'

'I wanted to die then.'

'You weren't crying then, were you?'

'No, because I was attempting to die.'

'When did you first feel sorry for Ian Campbell and his family?'

'Well that is a compound question so I will have to answer it in two parts,' said Gregory Powell.

Pierce Brooks dropped his eyes, and smiled and thought, I knew you wouldn't let me down Greg, my boy. You never fail me.

'The outside world was so tough to you,' said Schulman, 'that after Officer Campbell was shot by you, you tried to kill Officer Hettinger because you didn't want to go back to the institution?'

'Your Honor, I would object to this,' said Greg.

'Why, does the truth hurt?'

'Just a moment,' said the judge.

'I will ask you also,' said Greg, 'to cite the district attorney for misconduct, and I would ask for a mistrial at this time!'

'Did you feel remorse when you were emptying your gun at Officer Hettinger as he was running away?' Schulman continued.

'I would object to that as being outside the lines of cross examination,' said Greg.

'The objection is overruled,' said the judge.

'All right then, apparently I did, sir. I missed him.'

'What makes you think you're such a good shot? Because you shot at cans?'

'Mr Schulman, I have hunted all my life. I belonged to the Junior Sportsman's Club when I was a kid. I don't think you could miss at that range, unless you had some subconscious desire to.'

Pierce Brooks shook his head. There it was. What he'd been waiting for. The Hollywood syndrome. Of course, anyone could hit a running man in the black of night with his heart pounding and a dead man at his feet. In movies it was always easy, wasn't it? Pierce Brooks knew he himself would miss the silhouette target this month at the police academy range. In daylight, stationary, at seven yards.

'I guess you were happy about missing him then. You were so happy you tried to hunt him down like a dog.'

'Your Honor,' said Greg, 'to protect the record, would you

caution Mr Schulman to please stay within proper bounds. I cannot interpose an objection every time because I don't know all of them.'

'Do you want a lawyer? The court will give you one just for the asking. For free too,' snapped Schulman.

'The jury is admonished to disregard the statement of Mr Schulman,' said the judge.

'I guess I should cite him for misconduct at this time,' said Greg.

Attorney Ray Smith's desperate final argument to the jury attacked the credibility of Gregory Powell. There was nothing left to the old lawyer:

' . . . And Powell asks his father, in substance, "What's wrong with *you*, Dad? Why have I been like I have been? What's wrong with *you*?" Boy oh boy! I thought he had sunk to a new low. "What's wrong with *you*, Mother? Why am I like this? Why do I smoke marijuana? Why am I a homosexual? Why did I rob people? Why do I take girls out and get them pregnant without benefit of clergy? Why? Why? Why?"'

And Greg's mother, who had been given permission to assist her son during final argument, sat at the counsel table taking copious notes.

During his final arguments to the jury, Marshall Schulman said: '"Well, we told you we were going to let you guys go, now you know we lied." That is in effect what Powell was saying to the officers. He was going to torture them a little bit before they go. He raised the gun and shot Officer Campbell right in the face.

'Why does Powell sit here and cry? Because he's hoping you don't have as much guts as he has. You can't blame the murder of a human being on his mother and father. He has his own choice. He has his own free will. He knows what he's doing. Maybe I should've called Ian Campbell's mother and his wife, I don't know. I feel sorry for Mr Powell's parents, but who's putting them through it? You've got the answer right over there – Mr Remorseful, Gregory Powell.

'What are you going to do, send them back to school? Send Jimmy Smith to the prison boxing team, or the baseball team? Let them visit all their friends? Let them be big shots? Let them tell their fellow prisoners: "What have you done? Well, I committed a robbery. Well, I got myself a cop!"'

'Do you want to let them be big men at San Quentin and then start working on the Adult Authority and eventually get out? Are you going to allow them the chance of escape?

'The only thing that's going to help Jimmy Smith is to make Powell look worse than Jimmy. It's nothing but a question of vicious and more vicious, horrible and more horrible. They had a trial out there in the onion field. They found the officers guilty of being witnesses who could identify them.

'Do you understand Jimmy Smith? He's easy to understand sitting there with his Folsom slouch. He's a cold-blooded, no-good killer. The only difference between them is that Powell shot the officer when he was standing, and Jimmy like a hyena finished him off.

'Again, words fail me when it comes to describing the viciousness of this killing. I can only visualize it. I can visualize Officer Campbell out in that onion field. I can visualize his desires, his wants, his needs. I can see it and I can believe it. And I can see Hettinger too, and the fear and terror that he was in, and may still be in, with that experience that he lived through.

'Ladies and gentlemen, there is really no proper punishment but the imposition of the death sentence on these two men. I will ask you to be strong enough to return such a verdict as to each one of these defendants. Thank you.'

On September 12 the jury returned with a verdict of death as to both defendants. One of the attorneys fired by Greg, Public Defender Kathryn McDonald, was in the courtroom, and was startled by a theatrically ominous thunderclap which followed the reading of the death verdict.

Gregory Powell was not startled by thunderclaps nor even by death verdicts. He was already at work on his motions for the new trial, and his various escape schemes which were being frustrated by jail security. He called a jail captain to the stand to complain of his being taken to the law library chained to a wheelchair, and of his runners, family members, being searched.

'Your Honor,' said Jimmy Smith in a motion of his own, 'I don't know how to explain this fully. I mean, I guess you can understand. I am about to die. I think Mr Smith has did a wonderful job. I think he did the best he could. But it is obvious to me, whether it is to this court or not, that he didn't do his best. Because I am innocent and I have been found guilty and sentenced to death. And I would like the opportunity . . . I would rather have nothin to do with Mr Smith anymore so far as this case is concerned. And I demand this, your Honor.'

Judge Brandler partially granted Greg's motion:

'The defendant Gregory Ulas Powell will be permitted to make use of the law library twice a week for three-hour sessions. We are going to allow you to make use of any and all law books that you desire in your cell.'

'I wish to state that I will be filing an affidavit of prejudice against the court for the restrictions that have imposed upon me,' said Greg angrily.

At last, in late September, Jimmy Smith was finally given his wish. An attorney wanted very much to handle his motion for a new trial. The attorney, Irving Kanarek, would one day defend an even more famous murderer, Charles Manson. Being such a celebrated killer, Manson could choose almost anyone and selected Kanarek after asking specific questions about various kinds of lawyers. Manson was to smile that day, a smug, self-righteous, fanatic's smile.

'Get me this Kanarek,' Manson was to say that day.

Marshall Schulman knew little of Irving Kanarek when he appeared in court as Jimmy Smith's attorney to make the automatic motion for a new trial.

Schulman hitched up his pants and thought of what he had heard about the attorney. Kanarek was short and burly, with heavy cheeks and heavy lips. He looked a bit sad in the eyes and smiled seldom. He seemed tired from carrying a mountain of law books and a bulging briefcase.

Marshall Schulman would come to liken him to a bulldog. Not just in appearance, but in technique. Jimmy Smith would call him a club fighter. 'Nothin smooth about him,' Jimmy would say, 'but he wades right in there.'

That first day he waded right in there.

'You want two months' continuance until December?' the judge asked incredulously.

'The transcript is five thousand, seven hundred pages, your Honor,' said Kanarek.

'This court has the matter of the proper and expeditious administration of justice to take into consideration,' said Judge Brandler.

Then Kanarek framed the kind of exceedingly polite, bewildering sentence which would eventually send Marshall Schulman home with a belly full of acid at the end of a day.

'I would like at this time, also, your Honor, to suggest to the court that it would be a violation of the due-process clause of the fourteenth amendment to the United States Constitution and the several cases involving right to counsel which have been decided by the United States Supreme Court, in which that court has held that the right to private counsel, or the right to counsel of a criminal defendant, especially in a capital case, is such that in order to protect a criminal defendant who is in such a status, that the courts make sure that the defendant has all of his rights protected.'

Judge Brandler denied the motion and set the motion for a new trial for October 31. Gregory Powell asked for Public Defender Kathryn McDonald to come back and his motion was granted.

But on October 23, a notice of Motion for Discovery was filed. Irving Kanarek and Kathryn McDonald wanted all documents concerning the matter of the excused juror, Mrs Bobbick.

'Let me inform the court,' said Schulman angrily, 'that it is my information that Mr Kanarek has subpoenaed Mrs Bobbick to appear on the thirty-first of October on behalf of the defendants. He has also subpoenaed a number of jurors who participated in this trial!'

And on that day a psychiatrist defense witness was called to testify to his opinions regarding an interview someone conducted with Mrs Bobbick, the defense hoping to show that Mrs Bobbick was *not* emotionally disturbed and was excused without proper cause. Since the doctor had not examined the juror personally, his opinion was not allowed by the judge.

Kanarek once again battled through his syntax to frame admissible questions:

'Now what we are asking is, and based upon this record, we are asking, in view of all the circumstances of what happened and the lack of opportunity for independent medical examination of Mrs Bobbick, and again we have a capital type of case which the United States Supreme Court has always given ... '

'Mr Kanarek,' said the judge, 'you repeated it so many times, the court is well aware of the fact it is a capital case.'

'I'm sorry, your Honor,' Kanarek apologized. 'But in any event, all that we wish your Honor to consider is what is possible, physically possible. Otherwise, there is no opportunity, your Honor, at the motion for new trial, or on appeal, although I hope that your Honor grants this motion, and I think that when we have finished, I hope that your Honor does . . . I think it should be granted. It forecloses all opportunity . . . as your Honor is well aware, you can't augment the record on appeal.'

'Next question, Mr Kanarek,' said the judge. 'Objection sustained to the pending question.'

'Doctor, having read the reporter's transcript, do you have an opinion as to whether Dr Crahan had adequate basis to determine the mental condition of Mrs Bobbick?'

'I am going to object to that as immaterial to this particular hearing!' said Schulman.

'Objection sustained,' said Judge Brandler.

'Doctor, you read the reporter's transcript . . . '

'That has been asked and answered several times! Objection sustained,' said the judge before Schulman objected.

'Very well,' said Kanarek finally. 'Call Mrs Bobbick.'

Schulman watched the frail, trembling woman take the stand, her eyes flitting over all of them. He knew he wouldn't have to cross-examine her.

'Mrs Bobbick, you were the juror number four that . . . '

'I am worried about my husband,' Mrs Bobbick sobbed.

'Would you rather testify on another occasion?' asked Kanarek.

'Well, no, my husband had surgery today. I would rather testify now.'

'May the record reflect the apparent physical condition of the witness,' said Schulman.

'Your Honor, Mr Schulman is not a doctor,' said Kanarek. 'There is no apparent physical condition of the witness as far as I can see.'

'Let the record reflect that the witness is now crying. She seems to be emotionally disturbed and upset,' said the judge.

'May I respectfully object to that and suggest that some doctor testify whether she is emotionally disturbed,' said Kanarek.

'I *am* emotionally disturbed,' said Mrs Bobbick.

'Mrs Bobbick, would you tell the court what other members of the jury called you during the course of deliberations,' Kanarek resumed.

'Objected to as immaterial,' said Schulman.

'Objection sustained,' said the judge.

'Your Honor, the offer of proof is that this was not a volitional request on her part to leave that jury room. She was actually submitted to a series of criminal violations in that jury room by the other jurors, and the offer of proof is that she would testify as follows: "I have been called God Damn. I have been called a stupid jerk. They shouted at me. They jumped up and down like a bunch of monkeys. So I'm schizophrenic because I don't do what they do. The sex jokes that was going on during the trial up in the jury room . . . when you go upstairs, you know . . . small room, no . . . nothing to do, no place to go." That is what the witness would testify to here,' said Kanarek. 'Her exact words, your Honor. I can only make the offer of proof by making it. Mrs Bobbick, I will show you this document. You wrote this in the jury room?'

'My memory isn't very good because my husband has had major surgery and I haven't slept for many nights, and I don't remember the details,' said the witness tearfully.

'The voices on the tape recording will convince your Honor that there was nothing but amicability between herself and those two representatives of the District Attorney's Office who took this statement,' said Kanarek.

'During the time that you were deliberating, Mrs Bobbick, did you come from the jury room down here and talk to someone?'

'Yes.'

'Do you recall when you were in the jury room you were interested in the word "deliberate"?'

'Yes.'

'And would you tell the court what transpired when you came down?'

'Nothing unusual, because the English language is a hobby of mine and I look up words all the time at home. I have fifteen dictionaries at home, and it has no significance as far as I am concerned. It may have, to a perfect stranger. That, I don't know. It's nothing unusual. I look up words all the time. I have dictionaries of all languages, and it doesn't . . . I look up words all the time.'

'Was one of the words you looked up "deliberate"?'

'Well, it's possible.'

'Did the jury foreman say "God damn you" because you wouldn't go along with what he wanted you to do in the jury room?'

'I don't remember. I haven't slept many nights during surgery and I don't remember.'

'Mrs Bobbick, did you not tell Dr Crahan that the jurors were threatening you, abusing you, cursing you, and suspicious of you?'

'I don't remember. It's impossible. I had to even get my husband to give me these dates from his medical examinations and what happened. I couldn't possibly remember that. I just don't know.'

'Were you very disturbed at the time you made the statement to the doctor?' asked the judge.

'Very much, because I knew on August 26th that my husband faced surgery because of a double inguinal strangulated hernia!'

'Did you mention this to the doctor?' asked Kanarek.

'No, because I didn't realize it at the time. I mean, it was only a cursory examination.'

'Mrs Bobbick, what was it that changed your state of mind when you wrote that note?'

'I don't know what you're talking about. Because I have maybe two thousand books in my home, and I went to the dictionary all the time, and you're talking about a word, and I don't know what prompts me to look a word up at home except that perhaps I want to improve my mind, perhaps. But I don't know why I run to dictionaries.'

'I'm referring to the time when you were in the judge's chambers,' said Kanarek. 'I'm asking you to tell the court why you wrote the note.'

'Why I wanted to be excused from the jury?'

'Yes.'

'Well, because I've read about strangulated hernias, and every day that you delay could be a matter of life and death, and that was in my mind. That is the basic anxiety.'

'What? The strangulated hernia?'

'Well, that's about the only thing. I mean, naturally, that's something that we should worry about, any intelligent person.'

'Isn't it a fact, Mrs Bobbick,' said Kanarek, 'you are afraid of harassment by the district attorney and the authorities if you should give testimony from the witness stand that would in any way hurt the position of the prosecution of this case?'

'I'm afraid of nothing. I don't have fear. As a matter of fact, I feared the people such as yourself. You have come into my home under most unusual circumstances. I couldn't resist you . . . ' Then the witness began sobbing in her hands.

'Your Honor, may the record reflect that the witness turns the tears off and on at will,' said Kanarek.

'Oh no!' said Schulman, and held out his hands to the judge.

'You may suggest that,' said the judge angrily. 'But the court doesn't observe that at all. And the court is going to suggest that you take into consideration common decency as far as this witness is concerned.'

'Now, Mrs Bobbick,' said Kanarek, unruffled, 'then is it so, that you do not like me? Is that correct?'

'I don't know what you're talking about!' said the witness fearfully.

'Do you like me?'

'That is immaterial!' said the judge.

'But your Honor, I suggest that it's relevant on the bias and prejudice of this witness that it is proper impeachment to show that she is biased against the defendant because she is biased against the attorney for the defendant.'

'Let's proceed to some relevant question! The objection is sustained,' said the judge.

'Your Honor,' said Kanarek, 'this witness was not emotionally upset and disturbed. This witness was a holdout, eleven to one. She had a position. That is reflected in the district attorney's own transcript. She had a position that she maintained.'

'I'm tired of this, your Honor!' said Schulman. 'This record is replete with falsehoods and I don't like it. This attorney just seems to be able to relate anything in his mind that he wants to believe exists, whether it ever did or didn't exist.'

'Mr Kanarek's wild assertions, although they are reflected in the record, are not supported by any evidence. So proceed to the next question,' said the judge.

Irving Kanarek then read Mrs Bobbick's statement to district attorney's investigators to prove that Mrs Bobbick had taken a not-guilty position: '"Well they forced me,"' read Kanarek verbatim. '"I finally just gave up. I collapsed. My nerves just gave up. I just collapsed. I couldn't take it anymore. That's what's wrong with something like that. They should be in a glass cage where they cannot . . . where they cannot, uh, uh, terrorize . . . and terrorize is the word . . . uh, a person because when they all shout, if you had twelve people here and you were the odd twelfth one, eleven shouting men, and you were you, you're a woman, you just couldn't take it. This is terrible, this is like you're being bombarded in your ears and your nerves and you're not . . . " And then there are asterisks which according to the legend on the first page would indicate that to the transcriber it is unintelligible, your Honor.'

'I would like to be excused, I have to go,' said Mrs Bobbick, but the judge smiled kindly and motioned her to remain seated.

Kanarek read further into the transcribed material: '"In other words I was treated like I was some kind of a nut. At first it was my request, you know. That's how it got started about the jury instructions."

"Hadn't you read them before you went in?"

"Oh, my, yes, at gun-fire speed. I'm not a lawyer, you know. After all, sure, at gun-fire speed, machine-gun fire, really. Well, who can retain that? Sure, I retain a lot, but I can't retain . . . I don't have a photogenic mind."'

'Your Honor, I must get my husband to the doctor,' cried Mrs Bobbick, interrupting the reading of Kanarek.

'The sole question the court has to determine is whether or not the court properly or improperly excused her as a juror,' said the judge, 'by reason of the two requests that were made and the testimony that was submitted to the court by way of Dr Crahan.'

The next transcribed statement of the woman read by Kanarek was on a broader topic: '"Well, when they pick the jurors they don't pick it from the experts in law. They try to pick average people. Taxpayers, responsible people. That's all I consider myself. Now, I may be wrong but I think maybe I'm going to write the legislature and ask them to please change so that the jurors, because there are twelve, they can really use the majority rule. I think perhaps that would be the best solution. I really do think so. I've been thinking about this for a long time, ever since they all scooped down on me, and I seem to be the oddball, see? It seems to me that on ten to two or nine to three, like it is in civil or something . . . That's what I think."'

'And that,' said Schulman later, 'from the lips of that poor woman was the sanest thing I heard all day from anyone.'

'I am exhausted,' said Mrs Bobbick when Kanarek stopped reading. 'I don't know what I'm doing. I should invoke the fifth amendment or something for my own protection.'

Other of her fellow jurors were called that day by Kanarek.

'She was jumping up and down,' said one witness, 'throwing her arms around, screaming at us, using rather unpleasant language. She didn't have very complimentary things to say about all of us and just was in a very unpleasant physical and emotional state, I would say.'

'All right,' said Kanarek. 'She was at odds with the other eleven on the jury, was she not?'

'We couldn't tell! She never would tell us,' said the witness.

'Didn't she request at one time the instructions, or that the formal charges be read?'

'Yes she did. She had forgotten what the charges were.'

After lunch that day, Mrs Bobbick was recalled to the stand. 'I invoke the fifth amendment. I cannot go on . . . my husband . . . in order to protect my health.'

'You now don't recall anything that occurred in the jury room?' asked Miss McDonald for Gregory Powell.

'I have a migraine headache, and I don't remember. I can't recall. I can't go into that.'

'Your Honor, may I object that the conclusion of this witness be stricken?' said Kanarek.

'You may object,' said the judge. 'Overruled!'

'She said "migraine."'

'She can testify whether she has a headache or not,' said the judge, who looked as though he were getting a headache.

'She said "migraine."'

'Objection sustained as to the migraine,' said the judge, sighing in surrender. Then he added: 'I was wondering what the attitude of defense counsel would be if, with the background such as was presented to us here through the testimony of Dr Crahan and Mrs Bobbick's appearance here in court, what would be the defense counsel's thinking if the juror had remained as a trial juror in this case and had participated in a verdict of murder in the first degree? That is something for counsel to think about. I am not asking for any reply to it. Proceed to the next question.'

Irving Kanarek recalled the defense psychiatrist and asked a question in what Schulman would call his inimitable style:

'Directing your attention to the fact that in the August 30th, 1963, proceeding there is a conversation between the court and Mrs Bobbick, and the fact that in Dr Crahan's testimony he makes the statement that the judge told him about some incident in connection with a rattlesnake, and the fact that the only time that the court has talked to Mrs Bobbick is in the August 30th, 1963, transcript, at which time there is no mention whatsoever of a rattlesnake, and the fact that Dr Crahan used this information which he says he obtained from the judge, who had no contact with Mrs Bobbick except in the August 30th, 1963, hearing, do you have an opinion as to the adequacy of such a statement by Dr Crahan in connection with his analysis?'

'Objected to as immaterial,' said Schulman, staring in disbelief.

'What a classic question that is!' said the judge. 'Objection sustained.'

Then the jury foreman testified. 'She wouldn't cast a ballot,' he said. 'She would have a nervous tantrum and demand that we all remain silent and give her time to gather her thoughts and we should keep quiet for two or three hours at a time. We couldn't talk, we couldn't deliberate because it would offend her.'

'Was there profanity used in the jury room?' asked Kanarek.

'I would say normal profanity.'

'What is that?'

'"Hell," "damn," "goddamn."'

'"Stupid jerk"?'

'That's just an assertion . . . '

'Did you use the language "stupid jerk"?' asked Kanarek.

'It's possible I may have used it.'

'Who did you direct that language at?'

'I don't think I directed it at any individual any more than anything else.'

'Did you call *yourself* a stupid jerk?'

'I very likely did for ever getting into this thing in the beginning,' said the witness.

The juror was then asked to define Mrs Bobbick's alleged hysteria.

'Well it's awfully hard to describe. When a person jumps up and screams and starts tearing their hair out and telling you she can't stand it and grabbing her head and nobody had even said anything to her. I would say she is hysterical,' said the jury foreman.

The Bobbick affair used up four volumes of transcript. Marshall Schulman finally said to the court, 'I don't think anybody can stop Mr Kanarek from saying anything that comes to his mind. I am going to ask your Honor to find this counsel in contempt. He repeatedly disobeys the direct orders of the court.'

'Mr Kanarek is not doing this intentionally,' said Judge Brandler.

'I think he *is* doing it intentionally!' said Schulman.

'I'm satisfied he is not doing anything intentionally,' said the judge wearily. 'It is just that he persists, and apparently he just *cannot* understand the court's ruling!'

'What are you going to do, your Honor?' asked Schulman. 'Here you have this *bulldog* going forward in spite of the objections thrown in his face. He understands the ruling of the court and he just proceeds in spite of them. I don't know where he will be stopped!'

'Mr Kanarek,' said the judge patiently, 'you asked the identical question to which the court just sustained an objection except maybe the words are in a little different order or a little different inflection.'

Irving Kanarek, almost hidden behind boxes of law books he had placed on the counsel table in front of him, just blinked and paused for a brief moment then shook himself into action and proceeded with a new citation from a book in hand.

'Here, for instance, your Honor, is a case . . . '

'I don't want any further discussion about them, Mr Kanarek! I can read, and I will analyze them!'

'I know that, your Honor, but if you would bear with me for a moment, please. For instance, there's a case, *People versus Chesser*, 29 Cal 2d 815.'

'Mr Kanarek, you need some help there again with your volumes,' said the exasperated judge. 'What I suggest you do, why don't you put them in some order so they won't be one on top of another? Then you will be able to *find* your cases.'

'I invite your Honor's attention to . . . I refer your Honor to the case of *Moore versus Michigan*, which is a United States Supreme Court case. Would your Honor care to take this citation?'

'Yes,' said the judge sadly, as Schulman fiddled with his necktie and hitched up his pants.

'Did I mention the Modesto case?'

'Yes, Mr Kanarek. Yes.'

During the days of new trial argument, Irving Kanarek was permitted to make motions which the prosecutor would rage about when back at his office at day's end.

'Your Honor, these five witnesses would testify,' said Kanarek, 'that during this motion for a new trial they were taken to an anteroom and the female deputy required these female witnesses to lift their dresses, lift their slips, if any, until the most intimate items of apparel, their underpants, were visible to the female deputy, and whatever other intimate items of apparel were present, and that she conducted this search of them before they were allowed to enter the courtroom in connection with the allegedly public proceeding!'

After seven days Judge Brandler finally ended it. Before imposing sentence, the judge read a written plea from defendant Smith in his own hand.

> I have never had a profound fear of death, but to end one's life in such a grim, ghastly, and unjustified way frightens me. I am hoping that there is some humane element within your spirit that enables you to understand and have compassion on a dying man whose life you can save. I hope and pray that you are an instrument of God's everlasting compassion and save my life for me. Do I deserve to die?
>
> With humility and humanliness,
> Jimmy Smith

On November 13, 1963, Judge Brandler answered him:

'Sadistically, Gregory Ulas Powell led Officer Campbell to believe that his life would be spared up to the very moment that he took deliberate aim and cold-bloodedly executed Officer Campbell by firing one shot into the officer's mouth. While Officer Campbell, unconscious but writhing on the ground, was in the throes of death, Jimmy Lee Smith, standing over the body of Officer Campbell, with revolting, inhuman savagery, pumped four additional fatal shots into the prostrate officer.

'Then these two defendants continued their barbarous attack by stalking the fleeing Officer Hettinger, who miraculously escaped from being likewise cruelly executed by these defendants.

'The evidence in this case overwhelmingly justified the death verdict by the jury. It is the order of the court that you shall suffer the death penalty and that said penalty be inflicted within the walls of the State Penitentiary of San Quentin, California, in the manner and means prescribed by law at a time to be fixed by this court in the warrant of execution.'

TWELVE

They were driven to San Quentin in a black and white station wagon with three armed deputies in the front. A car with two armed deputies drove ahead and another car followed.

When they arrived, they learned that the prison authorities were installing new cells upstairs of North Block where Condemned Row was located. For many years the row contained only thirty-three cells, but now that was not nearly enough. The back of the row had another line of cells called 'the shelf,' used for punishment for cons from the mainline. The new arrivals were being housed in the Adjustment Center, which was at that time the mainline segregation unit.

Jimmy and Greg arrived on Tuesday, and Jimmy spent the night going over legal documents. He was determined to become a better jailhouse lawyer than Gregory Powell. On Wednesday a riot broke out in the segregation unit.

'Jumpin fuckin Jesus,' shouted Jimmy Smith that day to Gregory Powell. 'I can't get away from maniacs like you no matter where I go!'

The cons tore up the cells and created havoc for two days. Jimmy Smith and Gregory Powell did not take part, and on the second day when it was quelled and the tear gas was still lingering in the corridors stinging their eyes, Jimmy yelled, 'Hey, Powell, that gas ain't nothin compared to what you're gonna feel when they strap you in that chair upstairs!'

The two ex-partners were enormously popular in the segregation unit. After all, they had earned headlines with the murder of Ian Campbell. Jimmy found himself basking in the limelight. What he hated was that Powell was getting half the attention. So Jimmy began to show other cons his copy of Gregory Powell's first statement to the Kern County Sheriff's Office.

'Think he's a big man?' Jimmy would say indignantly to any con who cared to read it. 'Look at this and see what a snivelin rat he really is. Look how he rolled over me, and I was his partner! Look how he snitched me off!'

Jimmy discovered that cons on the third floor took care of the less fortunate ones downstairs in the strip cells. Those in Jimmy's unit had three meals a day and earphones to listen to the radio while the ones below had half rations with only a mattress and blanket for comfort. The building was new and modern and had many small windows set into the bars for light, and on his first night Jimmy watched a con tie a bar of soap to a rope made from torn sheets, and toss it some twelve feet up to a window, breaking the pane. The con then sat on the toilet with his pillow beneath him, completely covering the opening. When he raised up it created a suction which drew all the water down. The water gone, the con then could put his head in the toilet and call down to his friend below through the toilet telephone to tell him a package was enroute.

The con below would break one of the small windows in the same fashion and would eventually receive a package of food or cigarettes tied to a string from above. He would have to entangle that dangling prize with his own string tied to a comb flung out the window. It was a tedious maddening process and sometimes took a hundred tosses, but determination usually prevailed.

Greg was having difficulty adjusting to prison life. A guard made the following entry in his record:

January 21, 1964: Powell threw food at Jimmy Smith and Smith subsequently stated he would like to beat Powell half to death.

January 22, 1964: Powell found guilty of refusing to shave in violation of Prison Rule D 1202 concerning obeying orders. Sentenced to isolation pending transfer to Death Row.

January 28, 1964: Powell does not get along very well with other inmates, therefore does not go for exercise. Will throw a tantrum if he does not get his way.

Death Row! It was impossible. Of course, Jimmy had, like all small-time criminals, enhanced his ego in a thousand back alley bull sessions by saying, 'I'll probably end up there someday.' But like all small-time criminals he didn't mean it. Nothing could have been farther from his mind.

Now after his time in the San Quentin Adjustment Center, he was being taken across the yard blinking in the harsh sunlight, toward building number eight, which housed the row itself.

In Jimmy's earlier term in San Quentin he had heard the bulls shout certain words in reference to other condemned men, like Caryl Chessman. The bull was leading him through the throngs of curious cons, and Jimmy felt a surge of excitement and dread. He was the *big* man and the other cons dropped their eyes when he looked at them. At first it was incredibly thrilling, but then the third time the bull had to say it, Jimmy's throat tightened and the blood rushed to his face and he began to tremble.

'Get back!' the guard commanded. 'Dead man comin through!'

Jimmy Smith had spent the nights in the Adjustment Center thinking of death, and he had vowed that for the first time in his life he would act with courage when it came. Who knows, it might at last be the most pleasurable experience of his miserable life. Who could say it wasn't? One thing for certain, Powell would accept it like a hero, of that Jimmy had no doubt. Powell would give his life in a minute to make himself a big man in the eyes of the other cons. Jimmy wondered how Powell had liked the walk across the yard to the row. But he knew the answer to that. It would be the high point of Powell's life.

One thing Jimmy vowed, if they killed him first, if he went before Powell, he would go like a man. He would burn in hell before letting that bastard show him up.

When the new arrivals were brought into the row itself Jimmy had abandoned his fantasy of beating Powell to death. Jimmy decided the best revenge would be to watch him endure the misery that was sure to come.

Four guards escorted Jimmy and a chubby con named Willie to the row that day.

'Next stop, the penthouse,' said one guard as the elevator took them up. Then Jimmy found himself facing a huge steel door with a thick glass window about eight inches square. One of the escorts pressed a button and unlocked the door from the inside. Jimmy was to learn that the inside guard had the key taken away from him at night, locking him in with the cons.

Next came the stripping, and the inevitable: 'Okay, bend over and spread em out.' Then came the issue of blue dungarees. No one ever left or returned to the row with the same clothing on his back. At last, Jimmy got his first glimpse of the cells and the bars and steel mesh surrounding them. It was quiet and he heard only the occasional click as one of the condemned men changed TV channels with his remote button.

Then Jimmy was marched to a cage at the end of the corridor leading to the cells unit. Once inside, two guards entered and the sergeant locked them inside calling to a gun guard to admit them. The gun guard had a walkway that ran the length of the cells and was protected by bars and steel mesh. The gun guard pulled a release bar and Jimmy walked into what he thought would be his home until he died.

None of the condemned men spoke as he marched down the corridor, and at first Jimmy thought the cells were empty. He squinted into each of them and finally made out the cigarettes glowing and forms lying on the bunks looking up at the TV's perched on platforms near the top of the bars.

Cell number nine looked like the dozens of others Jimmy had inhabited in his life. It contained one bunk bolted to the wall, a small table, a heavy wooden stool. As he was making his bunk he heard a familiar voice calling him. It was a huge condemned man he called the Bear, who had been to college and played Canadian football and was something of an artist. Jimmy had known him in the county jail high power tank. That first day in the high power tank when Jimmy urinated in the toilet, the Bear had said to him, 'Now after you're through pissing, you take some toilet paper and wipe off that stool real careful, and if you ever miss and hit the floor you wipe that up too.'

And the Bear had said to the inmate delivering the tray of chow to high power: 'I know what kind of sex acts you punks do. Now you go wash those filthy hands and bring a new tray.'

'Screw you,' said the man with the food tray and it was the last thing he said for the five minutes it took to revive him.

'That's all I fuckin need,' Jimmy mumbled to his neighbor on the other side, 'the Bear livin next door to me. I'll probably accidentally drop a match near him when he's house cleanin and he'll break my back. My life is just one big junkyard full of misery and bad luck,' moaned Jimmy Smith.

Cell number nine of Death Row. This was the real thing. This is where every event of his life had inevitably led him, he thought. He was sullen and teary when breakfast came, but then he perked up.

'Hey, this is okay,' he said to his neighbor. 'Like, good enough for some fancy café on the outside.'

'Enjoy it while you can, brother,' said the voice next to him, with a loose slobbery giggle.

The first day on the row was routine. The second found two of the residents locked in a fierce fistfight as soon as the doors were opened for exercise. A black man, a robber and murderer known as Taco, was battling a white youngster named Junior, a cop killer who had been slightly crippled while trying to run a police roadblock. Junior was clearly the winner in this fight, and while he was pounding the black man, the gunrail guard, after several warnings, fired a shot. The shock literally blew Jimmy back into his cell onto the floor.

Ten minutes later when he recovered from the fear, Jimmy was told the first shot is a blank, it's the second one you've got to worry about.

'Jesus,' Jimmy whispered. 'It ain't safe. It ain't safe nowhere in this miserable world.'

'Don't worry about it,' said a middle-aged white man who murdered wives. 'Nobody's going to kill you before the state gets its chance.'

Time passed even for condemned men. For some the months passed much too quickly. One unforgettable winter day Jimmy Smith found Gregory Powell on the floor at his feet.

The fistfight erupted so fast that Jimmy could not even remember what caused it when it was over. They were in the corridor. Someone made a snide remark, but that was common enough between them. Then Greg had thrown something in his face. What was it? Jimmy thought, now that it was over, and he was lying on his bunk nursing his bruised knuckles. Paper! That was it, wasn't it? A ball of crumpled paper!

Jimmy had plowed into him so fast they were both surprised. Powell went down. Oh yeah! Jimmy thought in exultation. Powell went down to his knees after one punch, like the whinin snivelin punk he was! It was easy, so damn easy Jimmy couldn't believe it. Big man. Big tough man with a gun. Now Jimmy vowed to punish

Powell. Maybe once a week. Maybe twice. Just kick his ass, just a little, when the gunrail wasn't around. That was all that saved Powell this time, the gunrail cocking that gun. The metal sound was like a lightning bolt to Jimmy Smith. Powell was on his knees trying to hold Jimmy's arms. Oh *yeah*!

That night, most of the men on the row could hear Jimmy Smith screaming triumphantly into the toilet telephone, the voice echoing through the corridors.

'Powell's a lyin braggin punk!' Jimmy screamed. 'Powell says he was a boxer at Vacaville! I heard what he was! He was a punk in the gym! The other guys'd bend him over a workout bar and brown him! He was a gymnasium punk!'

'Shut up, Smith,' the Bear growled. 'I'm trying to watch television.'

'And that ain't all!' Jimmy shouted to everyone and no one. 'He's worse than that! Yeah! He's a . . . a incestuous bastard! That's what he is!'

'Smith, if you don't shut up I'm gonna twist your head off tomorrow,' said the Bear, and Jimmy Smith was finally silenced that night.

The following entry was made in the record of the prisoner.

February 15, 1965: Found guilty of fist fighting with Powell in violation of Prison Rule D 4515 concerning fighting. Sentenced to 3 days cell status.

His ecstasy was halted the next night when a friend whispered, 'Jimmy, that ain't too cool what you done.'

'What's that?'

'Throwing blows with your partner.'

'My *partner*? That punk?'

'He's the only one in the world can save your life, baby.'

'What?'

'Jimmy, looky here. You guys got a chance for a new trial. Man, most of us got a chance what with the Escobedo case, and now the Dorado case. I'd bet a million bucks you two guys'll get a new trial.'

'So what? It's gonna end the same.'

'For him, sure. He ain't got a chance. But you, Jimmy, it's different for you.'

'Different? I got a hot beef too.'

'You just gotta get separate trials. Have your lawyer make him out a fucking devil to your jury.'

'That ain't hard to do,' said Jimmy.

'Subpoena him to *your* trial and have your lawyer ask him if you really shot the cop or is he lying on you.'

'I suppose he's gonna say yeah, and ruin any little chance he might have, huh?'

'He don't *have* to, Jimmy,' the voice whispered. 'He just gotta *look* tough and scare the fuck outta the jury. And take the fifth.'

'The fifth?'

'The fifth fucking amendment! It can't be used against him at his trial. It won't hurt him none, and it'll probably save your ass. Other partners have pulled it off. You know how fucking goofy juries are. It'll work, I tell you.'

'Jesus.'

'But it ain't gonna work if you go around using him for a punching bag. You gotta play up to him, baby.'

'Jesus,' Jimmy breathed, and he did not sleep that night. Not for a moment.

Three days later, it was a haggard, nervous Jimmy Smith who knew what he had to do and who offered his old partner a cigarette during exercise. And he smiled at his old partner and made a self-effacing joke. And at the end, just before returning to their cells, put his hand on his partner's arm, something he had never done, and looked in his eyes, and said friendly things. And the next few days, he sat with his partner and flattered him and touched him often. He took his partner away from the others and talked privately with him, whispered many things, letting his lips touch the ear of his partner.

Several weeks later, the following entry was made in the prison records:

April 20, 1965: Jimmy L. Smith pleaded guilty to possessing pitcher of home brew in violation of Prison Rule D 1205 concerning contraband. Also found guilty of committing oral copulation on Gregory U. Powell in violation of Prison Rule D 1206 concerning immorality. Sentenced to 10 days isolation.

*

The changes in both men were reflected in prison disciplinary records. Greg's behavior improved. In fact until an escape attempt in 1967, there would be no more minor troublesome prison violations on the record of Gregory Powell. Though the plans for escape never ceased, the other irritations seemed to have vanished. Greg had sexual contact with many inmates in San Quentin, both in the Adjustment Center and on the row itself. But this was different. It was the final utter submission of the recalcitrant member of his 'family.'

Jimmy's behavior deteriorated, became sex-oriented, erratic.

September 14, 1965: Smith attempted to engage other prisoner in homosexual practices, necessitating moving the other prisoner to a new cell.

September 18, 1965: Subject accused other prisoner of snitching and was attempting to have others join him in a mass rape of this prisoner.

October 14, 1965: Smith talked about suicide for the last two days. Is not only a compulsive prevaricator, but is quite unable to accept responsibility for creating his own difficulties.

Jimmy Smith had at last surrendered, yielded utterly. The last overture was made to Gregory Powell. He was literally on his knees – abject, humbled, degraded. He didn't want to die.

THIRTEEN

Once he was driving for the chief of police, spending in-between hours answering telephones, cutting newspaper clippings which might interest Chief Parker, doing the perfunctory public relations tasks required of the chief's driver, he thought for sure the dreams would go away. They did not. They started to come almost every night.

Karl Hettinger was not a man of great imagination. His dreams were more literal than symbolic. They had a beginning, a middle and an end. They started at the intersection of Carlos and Gower in Hollywood and continued with him caught screaming on barbed wire, ripping free only to run in slow motion through an onion field, finally hunched over in the front of an ambulance looking back at Ian on the stretcher. When the dreams first started, he would always look back with great hope unable to see the bloody holes torn in Ian's chest. He would see only the blood streaming from his mouth down his cheeks into his ears, filling the ears, and spilling out onto the crisp white stretcher sheets. As he got accustomed to the dream he never looked back with hope at his partner. Though he couldn't see the bubbling holes, he knew Ian was dead. There was never any hope in the later dreams.

The intensity of the dreams did not abate. Helen, after the first few months, was becoming accustomed to the thrashing and sweating and whimpering in the night.

'Why don't you go to a doctor about these dreams, Karl?' she would plead.

'It's nothing, Helen. What can a doctor do? I just had a shocking experience and I'll get over it. It's not so bad now.'

'That's not true, Karl. They're coming more often now.'

'No, they're not. I should know, shouldn't I?'

And Karl would set his jaw and press his lips and Helen knew it was over. He wouldn't argue, he just stubbornly resisted, saying it would work out.

It was a blessing to work for the chief of police, though he hated being indoors so much. But at least it wasn't strenuous. His body was unaccustomed to functioning with half a night's sleep. Before Ian Campbell was killed he had slept long and deeply. Fatigue often set in early in the afternoons these days.

And he liked, or perhaps loved, the chief himself. William H. Parker was unlike any man Karl had known. He was eloquent, outspoken, perhaps the best educated and best read of any chief in Los Angeles history. The chief was married but childless, totally committed to his duties. He obviously liked his driver, would take him into his confidence, telling him things that even his closest colleagues were never told. The chief was said to be a good judge of men and seemed to sense that his serious and silent young driver would never betray a confidence. He was right.

The chief also seemed to sense that perhaps Karl Hettinger felt patronized for being there. So often the chief's conversations subtly veered in that direction, and he would say things to reassure his driver and tell him what a splendid bodyguard and companion he was. The chief would become angry when he overheard insensitive policemen questioning Karl about the Campbell murder, or the recent trial. He saw that it still caused the young officer some anxiety to talk about it.

The chief's kindness was rewarded by zealous loyalty. His new driver felt a compulsion about protecting the chief and even though Parker would tell his staff to remain at their desks when he took one of his frequent walks to City Hall, there would be a figure behind him, following unnoticed at a discreet distance – a slender figure in a suit which was too big, a young man with close-cut strawberry-blond hair, and blue eyes which were darkening and sinking.

Helen Hettinger deeply regretted she had not married Karl sooner than she had. She had not known him well enough before the killing to gauge how much the event had changed him. There were some changes, however, which were very obvious.

'Karl, you just ran through another red light!' she would say. 'I did?'

'Karl, what's happening to you? That's the second time today you did that.'

'Are you sure the light was red?'

'It was red, Karl. You used to be the most cautious driver in town. What's happened to you, lately?'

'Are you sure it was red?'

Karl Hettinger was given an annual physical examination. It was the same examination as always. The doctor asked him if he had any medical problems, the patient answered that he had not. The patient gave his blood and urine specimens to the lab technician, had a chest X-ray, an eye examination, was measured and weighed and released. Nothing unusual was noted or reported except that one nurse took his folder from the examining room and saw something which caught her eye. The patient had lost twenty pounds.

'Think I'll ask this officer for his diet,' she said.

'How's that?' the other nurse asked.

'This officer's lost twenty pounds.' Then she began comparing the new physical with the last one. 'That's funny. He's an inch shorter. He's barely five feet nine now. What the hell kind of diet is that?'

'Let's see,' said the other nurse going through his folder.

'His vision. It went from 20/20 and 20/30 to 20/40 and 20/40. What's going on here?'

'Look, honey,' said the older nurse, nodding toward the far examination room. 'When *you know who* examines them you can't tell *what* he'll write down.'

'Karl H.,' read the first nurse on the label and opened the file to find the patient's last name. 'I wonder if the *H* is for Houdini?'

'Why do you say that?'

'He's lost twenty pounds. He's going blind. He's shrunk an inch! This guy's pulling a disappearing act.'

On his next physical the patient's sight returned to 20/20 and 20/30 just as it had been before. However, the weight did not return. Nor did the inch of stature. No one took official notice of the metamorphosis, and the patient would be the last one to ever mention it.

On the thirtieth of August, 1964, just one day after the thirtieth birthday of Gregory Powell, just nine days after what would have been the thirty-third birthday of Ian Campbell, a son was born to Karl and Helen Hettinger. They called him Kurt, and Karl began to dream of taking his son on camping trips and teaching him to fish and play baseball, and spending hours talking to him. He wished for his son, without knowing it, all the things which had been absent in his own boyhood.

'How about some Mexican food, Karl?' his wife said when she recovered from the childbirth and was anxious to get out of the house.

'Oh, I'd rather not.'

'Well, how about Italian food?'

'Oh, I can get some and bring it home, I guess.'

'You used to love Mexican and Italian food when I first married you.'

'I still do, Moms.'

'But you used to really eat. Now you just eat enough to live.'

'Let's not start that again, please, Helen.'

'I hate to be a nagging wife, but I think there's something very wrong with you.'

'There's nothing wrong. I'm just getting a little tired of working in the chief's office and listening to all these questions about the murder. All these policemen that work in these office jobs love to hear about all the exciting police work they're not in on. But they won't go out in the street and do it. It might spoil their chances to butter up to the brass and get promoted.'

'Well whadda you know? You actually got a little mad for a minute. That does my heart good. Why don't you get mad at *me* sometime? Why don't you swear at me?'

'Why should I get mad at you, Moms?' said Karl, smiling into the hazel eyes of his young wife, who at twenty-two seemed to him more mature and infinitely stronger than he. These days he doubted his strength.

'We never talk. Really talk about things.'

'What things?'

'You know. About things that bother you. The things you think about. About the dreams, maybe.'

'The dreams aren't coming so often anymore.' Karl sighed. 'I told you that.'

'I sleep with you. Don't tell me.'

'I'm thinking about going to the Detective Bureau. Chief Brown himself asked me to transfer into his bureau. I'll bet when I get out of this chauffeur job and start doing police work again I'll be a new man.'

'Why don't you stay in the chief's office? You're almost thirty years old. You've had enough cops and robbers. Stay inside.'

'There's nothing to worry about, Moms,' said Karl. 'Tell you what. Let's plan a camping trip now that you're on your feet again. We haven't been for a while.'

Karl transferred to the Detective Bureau, but the dreams didn't vanish as he had predicted. The little nagging pains got worse, especially the one at the base of his skull. No one ever knew how bad it sometimes got.

'How about telling me the truth, Karl?' Helen said one night after dinner. 'You're not too crazy about the detective work, are you?'

'I like it okay. Only . . . '

'Yes? Tell me, Karl.'

'Well, sometimes the older ones say things. Like . . . '

'Yes?'

'Like they resent me because I drove for the chief and got brought into the bureau because of it. And maybe like they think . . . '

'Go on.'

'Like they think I shoulda done more . . . you know . . . like about the . . . about when Ian was killed.'

'That's ridiculous!' said Helen angrily.

'Well, they send people . . . they send everybody to the academy for in-service training, you know, and they teach it there. They teach what policemen should do. They teach that I never shoulda given up my gun and they tell things you should do that I didn't do.'

'Go on.'

'Well, that's it. I don't know. I think I'm gonna like the pickpocket detail. My partner's one of the slickest old guys in the business. He knows every pickpocket in L.A. I think I'm gonna like that job. You know what we need? A camping trip. Some fishing.'

There were hushed conversations around the fires during those fishing trips. They usually went with the Cannells and the Jameses, and sometimes the Howards would join them. It was a closely knit police group. The husbands had worked patrol together and were all outdoorsmen. Jim Cannell was the implicit leader. He was not only a fisherman, but a hunter, a hiker, and a camper, who worked harder at it all than the others. He talked incessantly with few pauses in the stream of words. One of them said if you would transcribe Jim Cannell's talk you'd never find a period. He usually dominated the conversations and the others winked and let him. Karl enjoyed listening to his friend.

This night, however, Jim Cannell was not entertaining in his usual booming voice. The talk was low and quiet, and Karl and Helen were walking out in the darkness by Lake Isabella and did not hear it.

'My idea of a fun day *used* to be to go fishing with Karl Hettinger,' Cannell was whispering to Stew James and his wife, Donna. 'I mean it's unreal.'

'What's unreal?' asked Cannell's wife, Jo.

'The change in him. Look at him. How stooped he walks.'

'He used to walk superstraight.' Stew James nodded. James had blond thinning hair and was known as the worrier of the group. Cannell always said he would get bald from worrying.

'I haven't seen him laugh in a long long time,' said Dick Howard, the third man at the campfire, the youngest.

'He used to keep me going all day with one-liners,' said Cannell to the women. 'He had that wit that sparkled at least one time during every rollcall and kept the rest of us awake. . . . Maybe I'll invite him duck hunting. Maybe he needs to get out more.'

'He isn't much for shooting,' said James.

'That's another thing that's different about him,' said Jo Cannell. 'Helen said he can't shoot anymore. He used to be a dead shot and now he can't hit . . . '

'A moose in the ass,' said Cannell, draining his beer, and the others grinned because Jo Cannell was a Jehovah's Witness and disapproved of her husband's profanity and beer drinking.

'Helen said she and Karl were shooting at tin cans out by the lake and he couldn't shoot a lick,' Stew James agreed.

'He's having trouble qualifying at the pistol range every month,' said Cannell.

'And have you noticed the way he's always rotating his head like his neck hurts?' said Donna James, a perky brunette who sat with a blanket wrapped around her this crisp night when a damp wind blew in from the lake.

'So maybe his neck hurts,' said Dick Howard, who worried and noticed less than the others.

'You didn't know him well enough to see the difference,' said Cannell. 'You didn't know him B.C.'

'B.C.?'

'Before Campbell. You just wouldn't believe it. It's *unreal*,' said Cannell.

'He's growing into a little old man before my eyes,' said James, the worried eyes turning down.

'He looks like doomsdays itself,' said Cannell. 'There isn't enough left of him to know it's Karl Hettinger.'

'Well what's the sense mulling over all this?' said one of the women. 'Nobody ever asks what's troubling him.'

'He's not that kind of guy,' said Cannell. 'He keeps things in and doesn't like prying.'

'That's what I've always thought,' said Stew James. 'My role is to be his friend and not bug him. If he wants to talk he will.'

'Some people *can't* talk,' said Jo Cannell.

'Well anyway, I think I'll see if he wants to go duck hunting,' said her husband. 'It doesn't matter a goddamn bit if he can shoot or not. We don't hunt anyway, we just drink.'

'Well I think *it* still bothers him,' said Donna James.

They all knew what *it* was and several heads turned involuntarily toward the lake, but his silhouette had vanished in the darkness. It was the unstated law that none of them ever talked about *it* in front of Karl Hettinger. He had never described his night of terror to any of them, had never told any of them how it had affected him, not so much as a word. They mistakenly assumed that he must at least discuss it with Helen. With somebody.

Now the voices had dropped to a whisper and the words were muffled by coffee cups or beer cans.

'If it's bothering him, he'll just have to talk about it.'

'It must just be the shock of seeing Ian killed,' said James. 'It must be that. Ian was such a . . . gentle guy. Do you know he loved classical music?'

'So do I,' said Jim Cannell, 'but I ain't so gentle.'

'I'd drink to that, if I drank,' said Jo Cannell dryly.

'Maybe Ian and Karl should never have been working together,'

Dick Howard agreed. Then he grinned at Cannell. 'They were too nice. Should've been somebody like you with them, Jim.'

'I wonder if maybe Karl could feel . . . oh, responsible in some way,' James mused.

'Why should he?' asked Cannell. 'It was Campbell's fault, not Karl's.'

And there it was. The blame being laid. All policemen, even the closest friends of Karl Hettinger, had to lay blame for the catastrophe. They were policemen. The most dynamic of men. No man-caused calamity happens by chance. Only acts of God are unpreventable. It was a certain as sunset.

The next day, fishing, Cannell had his friend very much in mind and he began consciously to test him.

'Why don't you tune up that wreck of yours, Karl? It sounds awful.'

'Are you kidding?' said Karl. 'If I tried fixing it, it'd never run again.'

'Come on. You used to say you were a pretty good shade-tree mechanic.'

'I can't fix anything,' said Karl.

'Where're we gonna fish today, Karl?' asked Cannell carefully.

'I don't care. Wherever you wanna fish.'

'Well, you pick the spot, Karl,' said Cannell, staring at the smaller man. '*You* pick it, Karl. We'll go wherever *you* decide.'

'Well . . . I . . . how . . . how about *you* deciding, Jim. *You* make the decisions.'

They fished mostly in silence. Jim Cannell looked often at his friend, watching him rotate his head, and massage his aching neck. Watched the fingernails periodically digging into his palms.

That night, Karl sat alone on a picnic table away from the fire. He sat as he always did, hunched forward, hands pressed between his knees, eyes down, nails digging.

After a while Helen Hettinger said, 'Where's Karl?' and turned, seeing him sitting in the darkness. 'He must be cold,' she said, and went to the truck to get a blanket.

'That's a bitchin broad, brother,' said Cannell. 'And how's that for alliteration, my dear?' he added, gulping his beer and winking at the tiny Jehovah's Witness, who was looking in resignation at the pile of empty beer cans.

Helen Hettinger by now also knew her role with her taciturn

husband. She knew he would never confide in her or in anyone on earth. She also believed she was not bright enough or sophisticated enough to know what to tell him even if he would confide. She only *sensed* what she must do.

Helen left the light and warmth of the fire to sit next to Karl on the picnic table. She said nothing. She wrapped the blanket around his shoulders. Then she got inside the blanket with him and pulled his head down to her shoulder.

'She was just a kid herself,' Jim Cannell said later. 'But she held Karl like you would a baby. Without saying a word. The rest of us after a while got up and hit the sack for the night. They just sat there like that in the dark. She rocked him a little. I'll never forget that gesture.'

Karl *did* enjoy the pickpocket detail for the short time he worked it. His partner was all that was advertised, a graying, portly police veteran who melted into crowds. Oscar O'Lear was invisible on crowded sidewalks or in bus stations or department stores, even on buses, anywhere pickpockets lurked. He was the scourge of the feather-fingered brigades which descend on the downtown streets of Los Angeles both in and out of season.

The pickpockets tended to be a clannish lot, much like hotel burglars. They avoided the company of less artistic thieves who need guns and saps and knives to make their living. Karl loved watching the way his partner could spot a pursepick clear across a department store, usually by first identifying the thief and then sensing who the victim would be even before the thief himself had made his choice.

Most of all, Karl was glad to be out of the police building onto the streets once more. He was almost sure the inside work was causing many of his problems. He had assured his wife the insomnia and dreams would stop and the appetite would return when he once got back to an outside job.

But in fact it was all getting worse. So much did he fear the dreams that he sat every night before his television until the early morning hours drinking one beer after another, taking nonprescription sleeping pills because he still refused to see a doctor, hoping each night that sleep would come easily and that he would not dream.

There were other things now to contend with, some of which

Helen could observe, some she guessed. He was hit by a maelstrom of symptoms which almost drove him to seek medical assistance except that each particular malady would vanish as mysteriously as it appeared, and another would take its place. He had bouts of diarrhea, chest pains, and vicious headaches at the base of his skull which truly frightened him.

There were other things which frightened him more. One was just being afraid. This was fear without a name, which could not be battled because he had no idea where it came from. He would be sitting there alone sometime after midnight when Helen, pregnant for the third time, was asleep. He might even be vaguely interested in something on television. It would creep up on him. It would attack without warning. A bodiless, merciless, strangling thing. He would find himself cowering in his chair, and his hands would be so wet he couldn't hold the beer can and would spill most of it on himself like a baby. Then he would sit with his hands between his knees and wait for the ocean of scalding blood to crest and break and slosh through his skull, for this is how he felt it. Then it would slowly pass.

If only he knew what made him afraid. If only he knew, he was sure he could defeat it. He did consider seeing a physician, but what could they tell him? It was his nerves, of course. He was too weak to live down the shock of the killing. That's what he suspected must be troubling him. A real man could have come out of it in a short time and resumed a normal life.

After all, he had done all he could that night. He had nothing to feel bad about. Nothing at all. It was easy for some of them to criticize him. To have their training classes and criticize him and Ian and say what *should* have been done. Yes, it was easy enough to say. Anyway, it didn't bother him at all. There was nothing else he could have done. There was nothing to feel bad about. He didn't feel bad about it. There was no reason to feel bad.

Then he was crying. It was the first time he had cried like this. Karl Hettinger sat hunched in his chair and his wet cheeks glistened silver from the light of the television, and his shoulders began heaving and great shuddering sobs ripped out. He lost control. He wept and the shame of it made the tears gush hot. There was nothing left, not a shred of self-respect. What kind of a man would cower and cry in the swirling darkness?

He no longer sat in front of the television. He had fallen to his knees. He couldn't breathe, so overwhelming was his grief. The sobs were soundless now and absolutely dry. He cried until he hardly had the strength to stagger to his bed. Though he was not a religious man he thanked God for not having let him awaken his wife. The most unbearable part was to think of another human being seeing him like this. He whimpered very quietly in his bed until his control returned.

For a year after this, Karl would be completely impotent, then sporadically impotent. One day while walking through a department store with O'Lear looking for thieves, he saw a masonry drill he needed. He started to buy it but instead just put it in his pocket. It was as baffling and inexplicable as the weeping.

FOURTEEN

On June 22, 1964, the United States Supreme Court had handed down an opinion in the case of a young man named Danny Escobedo who had murdered his brother-in-law but who had confessed only after first telling police: 'I'm sorry, but I would like to have advice from my lawyer.'

On January 29, 1965, a murder conviction had been reversed for a convict named Robert R. Dorado, who stabbed another inmate to death and confessed to that murder. The California court took Escobedo a step further and decided that the police have a duty to warn of both the right to remain silent and the right to counsel.

Finally, on June 13, 1966, the United States Supreme Court reversed the case of Ernesto Miranda, a confessed kidnapper and rapist. Miranda laid down for all of America the same guidelines as Dorada had in California. Chief Justice Earl Warren condemned the police for tricking the confession of guilt from the defendant.

Ernesto Miranda had been arrested the day Ian Campbell was buried.

By January 30, 1967, there were sixty-two men in San Quentin Death Row. They were cop killers, stranglers, slashers, rapists. Some had murdered once, some several times. Some could not be stopped from murdering, not even by prison walls. There wree thirty-four white Anglos, twenty-one blacks, five Latins, two Indians.

In the past three years, fifty-seven men had gotten new trials with twenty-six of those receiving death verdicts a second time. One, in fact, was freed after a third trial.

It was a time of great hope on the row for those sixty-two residents, even for Gregory Powell, who felt that a retrial was inevitable, though it would probably end the same as the last.

Four residents of the row, all white men, had one last agitated conversation that afternoon.

'Man, fuck it, I say, we gotta go while we got the chance!'

'Yeah, but the new trials . . . '

'Fuck the new trials!'

'Yeah, but you got a hot beef, you and Powell.'

'So stick around and bum dimes and watch TV and eat your fucking zuzus. Me, I'm trying it tonight.'

'I didn't say I wouldn't go.'

'Well, I'm going and Powell's going, right?'

'Right,' said Greg.

'Well, you guys got hot beefs.'

'Man, Reagan is the governor. Don't you understand that? He's laying to smoke people!'

'I didn't say I wouldn't go. What the fuck I got to lose?'

At 1:00 A.M. a sergeant was completing a visual check of each of the sixty-eight cells which comprised the row. The sergeant heard nothing but the heavy breathing of the sleepers. He walked quietly in the tennis shoes he had been wearing ever since the inmates complained that his leather shoes were too noisy. When he left the aisle he had an easy feeling. He returned, peeked around a corner, and saw an inmate running quietly in the other direction.

The sergeant made a call and armed officers immediately poured into the row. Additional guards were assigned to the prison towers, and building number eight was completely surrounded. The guards found the two lower bars of Gregory Powell's cell sawed through and removed. The prisoner was missing. The bars on the cells of Edmund Reeves, Charlie Pike, and Joshua Hill were partially sawed and the grooves filled with paint-tinted soap. Pajama-clad dummies stuffed with newspapers were found in the cells of Pike and Hill.

In Gregory Powell's cell the guards found a can of excrement, estimated to be three days' usage after the toilet had been torn from the wall. With the toilet removed the prisoner could reach into the utility corridor. The hole also revealed a drainpipe which on past occasions had been used to transport contraband from one floor to another.

Gregory Powell was captured above Reeves's cell at a barred transom. He had been sawing through the bars with a hacksaw blade. Several blades were found in Reeves's cell as well as two rolls of rope, each over one hundred feet long, woven patiently from bed sheets which had been cut into three strips. A test revealed that it took only three and one half minutes to saw through a cell bar with a hacksaw.

Prison authorities were unable to trace the source of the hacksaw blades. One theory was that they were leftovers from an earlier attempt by two fire bombers who killed six people at the Mecca Bar in Los Angeles. The arsonists had been commuted to life, no longer residing on the row, but would eventually testify as character witnesses for Gregory Powell. Another theory was that the hacksaw blades were smuggled inside within Gregory Powell's typewriter by a family member.

There were no escape charges filed against the four inmates. When Homicide Division, Los Angeles Police Department, asked why, they were given these reasons: (1) There was delay and confusion in obtaining the evidence used by the principals in the attempted escape. (2) When a criminal prosecution is sought against inmates, the inmates have a tendency to subpoena all their inmate friends to court which creates a security problem. (3) Filing additional criminal charges would delay the pending appeals.

All four inmates were confined to isolation cells for the maximum term allowed, twenty-nine days, with loss of television and commissary privileges for the entire time. In addition, inmate Powell was charged thirty-two dollars for the broken porcelain toilet. This was the second isolation punishment for Joshua Hill. He got twenty-nine days the third time for slashing a fellow inmate's throat.

Gregory Powell had been five minutes from the roof of building number eight. With two hundred feet of rope and a San Francisco fog going for him, he came perhaps closer than any man to escaping the row.

Jimmy Smith now lived next door to the Gamblin Man, Aaron Mitchell, who would become the most famous member of the row. They shared the same television set and seldom bickered over control of the channels, each usually checking with the other before using his remote control switch.

Jimmy and Mitchell played bridge avidly and Mitchell talked of his early life in black ghettos, of petty stealing, and of finding a white woman whom he felt obliged to steal for. He told of finally killing a policeman in a robbery and of nearly being killed himself in the shootout.

Jimmy played cards with the Gamblin Man until one day Jimmy went to the dentist. When Jimmy returned with the escort guards he was surprised to see Mitchell standing naked in his cell.

'Jimmy, get the bull to pull the bar,' said Mitchell in a strange voice, and Jimmy walked to the rear of the block.

'Just got a tooth pulled,' said Jimmy to the guard, pointing to the cotton in his jaw. 'Want back in.'

When Jimmy got to the other end of the block Aaron Mitchell walked out of his cell still naked. He looked around with a confused expression and began pacing up and down the deck for a space of forty feet.

The guard on the gunrail rose from his stool and gaped. All talk and card playing stopped. Those men who were close watched Aaron Mitchell slash his wrist and raise his left arm, letting the blood splash to the floor. Then he continued to walk, saying: 'Do you know I am gonna die just like Jesus Christ did? I will die to save you guys.'

Four guards rushed into the cellblock and carefully coaxed Aaron Mitchell into giving up the razor blade and surrendering. The doctor went into his cell and no one ever heard another word from the Gamblin Man. Not that night, not ever. The next morning they executed him.

Jimmy could smell the fumes being filtered through the water out into the air that day, April 12, 1967. Jimmy smelled the gas for days even when logic told him it was gone. Two weeks later a judge issued a blanket stay of execution to cover every man on Death Row. Jimmy's stomach at last began to unknot. The Gamblin Man was the last of them to die.

The gardener was reading the paper in his truck. After this break, he would finish up the yard and be through for the day.

He thought how strange it was to see so many Christmas advertisements already in the paper since it wasn't even December yet. They just started earlier and earlier each year, thought the gardener. It was going to be the biggest Christmas season ever, they said. They were selling everything at exceptionally low prices to celebrate the coming year and it would be a most jubilant New Year, they said. The 1960's were finished and they were starting a more hopeful decade. They would help you to celebrate the coming of 1970 by giving you rock-bottom, end-of-the-year, closeout prices, they said.

He couldn't help remembering all the things he had stolen during that one Christmas season. Then he saw the ink smeared on his fingers from the newspaper. He couldn't believe how badly his hands were sweating. They were dripping sweat and black with ink. His hands were usually wet, but not like this.

He knew there was no avoiding it. He had to think of his last crime. When they stopped him. When he was caught. When they exposed him for the thief he was.

He hadn't committed his last crime during a Christmas season. No, it was even a more insidious time of year, when California gardens are bursting with life, vigor, potency. It was in May, the gardener's favorite month.

He was caught at a supermarket. It wasn't the first time he had stolen in this place. In fact, less than a month before, he was almost caught here. He was seen on the first occasion when he stole two packages of cigars. He had come back to steal again.

It troubled him the second time he stole cigars. He thought it was doubly wrong to steal what you couldn't use. Why would someone steal what he couldn't use? He had just grabbed the cigars on an impulse. He smoked them so they wouldn't go to waste. Well, he reasoned, hadn't he often smoked a cheap cigar in those ancient college days when he played poker? It had been his dormitory trademark.

He always remembered trivia like that. What was so difficult to remember were things that happened from March 1963 when Ian was killed until May 1966 when he was finally caught for his crimes and stopped stealing, and became a gardener. Certain things would stand out when he really tried to remember, but other things just weren't there and his wife would have to remind him. Why didn't the crimes fade like that? Why did he have to remember his crimes so well?

He was almost positive the store employee was watching him shove the packages of cigars in his pocket. Then he had the cigars. Now came the critical period. The thing he always thought about and dreaded. The critical period of making his escape. There could be no doubt. He knew he was seen. He could hardly control his legs. He walked slowly, deliberately, just as always. The store employee followed him.

It was the same store employee who had seen him do it the other time when the employee wasn't sure enough to challenge him. This time he was sure and the employee followed him to the parking lot and copied down the license number.

He was frightened. He would be stopped now. They were going to arrest him. Would he run, fight, surrender? He didn't know. He knew nothing except that there were waves of blood cresting, surging, breaking in his head. He got in his car. Still he was not arrested. He paused before starting the engine. Still nothing. He knew the store employee had taken his license number. He must have taken his license number. He didn't sleep at all that night. He couldn't eat at all the next day. Finally it came. The telephone call. Report to the fifth floor of the police building. To Internal Affairs Division.

FIFTEEN

They made him wait in an outer office of Internal Affairs Division for a long time before calling him inside for the questioning. It was a technique they used on all their subjects, even though all their subjects were police officers and presumably knew about such obvious interrogation devices. They usually dragged out the Mutt and Jeff routine where one investigator sides with the subject against another ostensibly more hostile one. They used every old trick on their subjects, and the strange thing was that it worked. It worked on policemen who understood, even better than it worked on ordinary criminals who did not. The thing which made it work on policemen was the thing which made the lie detector work better – the conscience of the subject. So many ordinary criminals are sociopaths that the lie detector is utterly useless and interrogation techniques are frustrated because there isn't a sense of guilt, the most valuable tool of the interrogator.

Karl waited and knew what it was for. He tried to think of a plausible story, but the more he thought, the less likely became any excuse he could conjure up. Instead of thinking about what he would tell them, he thought about an incident which had dominated his thoughts lately. It had happened since his recent transfer to Highland Park Detectives.

They had surrounded a house in a narcotics stakeout. A suspect had suddenly come running into an apartment from the direction of the house and up a flight of stairs with Karl running behind and commanding him to stop. Karl drew his magnum as he ran into the building and shouted at the runner. Then Karl stumbled and fell on the stairway, and his hand instinctively gripped the gun but the hammer only popped back part way and the gun didn't fire. The runner was apprehended and turned out to be a frightened teenager who had seen men with guns and had nothing to do with the narcotics stakeout. Karl had to clench his fists to keep his hands steady every time he thought about what he had almost done accidentally that day.

He had bought that four-inch magnum partly because he felt it would let him fire more accurately now that he was having so much trouble with his monthly shooting qualification. He had come to a conclusion about the inexplicable bouts of fear which were attacking him with such great frequency. He reasoned that he was probably afraid of Gregory Powell's friends. Perhaps while the case was on appeal Powell would send someone to harm the vital witness or his family. He never feared Smith, only Powell. That must be what the fear is all about, he told himself. He bought a better gun to make the bad feelings go away.

At last they called him into the Internal Affairs interrogation room, and Karl Hettinger discovered he was not a man to escape irony. One of the two investigators was Sergeant Riddle, whom he had worked with in the past, and who had been the chaplain presiding at the funeral of Ian Campbell.

'I put a dollar down on the counter before I walked out,' the subject told the two investigators after only a few prefatory questions.

'That wasn't enough to pay for the cigars,' said Riddle. 'We're old partners, Karl. Do you want to tell me about it?'

'Well I just made a mistake then.'

'You didn't make a mistake, Karl. The man at the market saw you stealing cigars there last month.'

'Well if I did it, I didn't know I did it.'

And the investigators settled for that: a passive, tacit admission of guilt. He was ordered back in the afternoon.

His lunch was two cups of coffee. He thought about the interrogation during the lunch hour. Just a table and three chairs. A tape recorder on the table. Four walls which closed in on him. He thought of all the criminals he had interrogated in similar rooms, so stark and bleak and inhospitable. Strange he never thought about it before. He never thought about how frightened some of them must be.

That afternoon he sat across from Captain Colwell, the commander of Internal Affairs Division. The captain was not a man noted for a sense of humor. Now as he sat so still and somber, and stared at the subject, Karl felt himself hunching forward even more than usual. He was going to try to explain to the captain that it was a mistake. He intended to lie and say that it had never happened before. He was going to tell the biggest lie of all and say that he was not a thief. Instead he withered under that gaze. The captain looked so stern, so disapproving, so *right* to the subject.

'This is a resignation,' said the captain in a voice as businesslike as his gaze. 'Do you wish to sign it?'

The subject looked at the form. It was already filled out. It had his name on it. He signed his name and resigned from the police department.

'You won't be prosecuted for this theft,' said the captain. 'But if we find you've been involved in other thefts, I can't promise you anything. We have men checking. Do you understand?'

'I understand,' said the former police officer.

Internal Affairs Division report 66-703 was only six pages long. It sketched the circumstances of the theft and listed the witnesses' names and addresses, and the address of the store, along with the date and time of occurrence.

There was no comment either official or unofficial as to extenuating circumstances. No other thefts were discovered, nor was anything said about the inconsistency of the shoplifting with the subject's unblemished past. It was much the same as an ordinary police report which a policeman would write on an ordinary shoplifter.

Of course, the staff in the chief's office and elsewhere where Karl had worked were deeply shocked. Everyone had liked the silent young officer with the sad eyes. It made them very uncomfortable to think he was a thief.

Helen Hettinger had never seen her husband cry, did not know of the episode in the night when he cried, but she heard him just this once. He telephoned her to tell her he'd resigned.

'But I don't understand, Karl. For shoplifting? What did they say you shoplifted?'

'I'll tell you later. Anyway, it's all over.'

'All over? Just like that? You resigned?' Fear was in the pregnant girl's voice. He heard it and he burst into tears.

'Oh, Karl,' she said, and now she was weeping with him. 'Oh, please, Karl. Come home and tell me. Please.'

'I . . . I . . . don't want to come home just now, Helen. I . . . I

... just feel like going fishing. I just ... I'll take a drive and go fishing.'

A few days later he received a written notice that the police credit union was calling back its note for six thousand dollars. The credit union demanded payment within forty-eight hours. Helen's parents helped them secure the money to pay.

The Hettingers also learned that their police medical insurance would be canceled May 30, two days before the baby arrived. Fortunately, Kaiser Hospital, with whom they were insured, said they would deliver the baby for the group insurance fee just as though they were still insured.

'Care to talk about it?' asked Jim Cannell on the telephone when he first head the shocking news.

'I just blew it, Jim,' said Karl, and that was as much explanation as he offered to any of them about his disgrace.

'Okay, that's good enough for me,' said the chesty voice of Cannell on the other end. 'How're you fixed for bread until you find a job?'

'I'm okay, Jim.'

'Stew James wants to know too,' said Cannell. 'About whether you need a loan.'

'With all his kids? Working three jobs? Tell him thanks, Jim. Tell all my friends thanks but we'll get along.'

And that was all he ever said to any of his friends on the subject of his resignation.

His parents came as soon as they heard.

'You say you don't know why you took things?' his mother asked, as the three of them sat in his living room, father, mother, son.

'No I don't, Mom. I just did it and I don't understand it.'

'Well, you're out of it now,' she said. 'Maybe it's a blessing that you're out of it, son.' And she turned and looked at Karl's father who sat awkwardly on the other side of the room, his big workman's hands dangling uncomfortably from his huge wrists. He leaned forward once or twice to catch a word here and there, but he was partially deaf.

'I'll explain it all to your father later,' she said. 'The main thing is don't fret about it. It's over now and everything's going to work out.'

'Thank you,' he said, and for the first time ever she saw tears in her son's eyes.

Karl looked across at his father, and the carpenter looked helpless and confused as though he wanted to say something to his son, but didn't know what to say or how to say it.

Helen Hettinger stepped into the room for a moment, big with her eight-month pregnancy. She looked at the three of them, the saddened family trying desperately for the first time to talk intimately to one another. Helen thought of the story Karl had once told her about his wanting a baseball glove so badly, and dreaming of being a second baseman or shortstop, and how, when the gift from his father finally came, it was a catcher's mitt. He had laughed when he told her how he played second base for years with that catcher's mitt. And how when he got older and begged for his first fishing tackle, his father had mistakenly bought him ocean fishing tackle. The boy had strapped the cumbersome gear to his back and bicycled all the way to Danson Dam to fish for tiny bluegills. Helen never saw a blond boy fishing alone on a lake bank without thinking of young Karl Hettinger.

Now it was the carpenter she pitied, sitting there, with his big awkward hands. He would have been willing to drive a nail or a million of them, to saw a forest of lumber, anything to help his son. But there was nothing he was able to do. They couldn't even speak. It was the ancient inherited shame of fathers and sons.

When they finished the brief quiet talk, Karl, for the first time he could remember, kissed his mother. His father touched him with a bruised, knuckle-heavy hand, and nodded, and tried to say something. Then they left.

That summer, after their daughter, Christine, was born, Helen Hettinger insisted that her unemployed husband accompany Jim Cannell on a fishing trip to Crystal Lake in the High Sierras. It was a therapy trip for Karl, the first tolerable days since he left the police force. The two men fished and drank beer and shucked their clothes and swam in the cold water, never once talking about police work or of Karl's leaving the force.

Jim Cannell wanted to tell Karl how he felt, how they all felt about his leaving the department. That it didn't matter at all. That if he stole, there had to be an underlying reason, because he was the most honest man Cannell had ever known. He wanted to say that if Karl would just go out and see, he'd learn that none of his friends had deserted him. But Karl was such a private person Cannell didn't know how to tell him these things. And Karl never asked, was afraid to ask.

While crossing Donner Pass in the pickup they heard the news on the radio. Los Angeles Chief of Police William H. Parker had been fatally stricken with a heart attack. Cannell was stunned by the news. Parker had seemed invulnerable. Karl was deeply shocked and spent that night talking more than Cannell had heard him talk in years. He told one story after another of the chief, things he had learned while driving for him. Now it didn't matter if he broke the confidence. He admired the chief to the point of adulation. He was grief-stricken that night.

Someone later mentioned that the object of his loyalty had every opportunity to prevent his dismissal. Could have ordered a more thorough inquiry into something as strange as an outstanding policeman like Karl Hettinger stealing cigars. Could have saved his devoted subordinate.

Someone *else* would say these things. Such a thing would never have occurred to Karl Hettinger. He was a thief a hundred times over. He had betrayed the department. He had betrayed the chief.

There was only so much yard work to do at his new house. They had bought the house three months before he resigned. Now he tore into his gardening with all his pent-up energy and talent with growing things. There was only so much to do and then it was done. He had applied for some jobs. One was as a loader on a dock. The employer had checked with the police department and somehow found out the circumstances of his resignation. He was refused employment because of it.

'You shoulda seen the people working there,' he told his wife bitterly. 'They'd hire *anybody* there. Half of them were probably thieves. But they wouldn't hire *me*. Next time I won't list my former employment. They won't be able to check on me.'

But he really didn't try very hard again. He was afraid. He knew they'd find out somehow. They'd learn about him, discover that he'd been a thief. So he stayed home and Helen supported a husband and three small children on a bookkeeper's salary. He stayed at home and took care of Laurie and Kurt and the baby,

Christine, and wondered how they could possibly continue to make their house payments.

Now, when we're so desperate, he thought, ironically, I *can't* steal. The very thought of his recent thefts filled him with shame and made him physically sick to his stomach. One day when the two older children were napping, the baby began crying. He went to her and felt but she was dry so he shushed her for a moment. Still she wouldn't stop. He left her room and tried to read a newspaper and drink his beer and ignore her. She cried louder.

At least there won't be more of *them*, he thought angrily, and he thought of his impotency, the shame of it. Still the baby cried.

The longer he tried to ignore her the louder and more insistent were her wails until his headache was becoming brutal.

'Shut up,' he said, grinding his teeth unconsciously. 'Shut up. Shut up. Shut up.'

Then he found himself bursting into the baby's room, slapping the baby on the bottom. The child's terrified screams startled him.

He looked down at the baby, at the red fingermarks on her buttocks. 'Oh, God,' he said, and stumbled straight to his room, but the magnum wasn't there. He had sold it to make a house payment. Then he thought of the service revolver, and found it in the closet. He looked at the barrel, at the hideous black hole, at the gun he had surrendered to Jimmy Smith. The baby was still screaming.

He wanted to see them for the last time and he went to the room where the other children slept. Kurt was two and Laurie, her daddy's girl, was three. She used to awaken when he came home from the station at night and wouldn't sleep until he kissed her. She adored her father. He looked at them and began to cry. He wept brokenly. He wasn't even enough of a man for this. He couldn't even spare his family by ridding them of himself. The baby still screamed. He ran to her.

'Oh,' he said, seeing the pathetic frightened face on the sheet wet with tears. 'Oh.'

He picked up the infant and walked with her and rocked her. 'Oh, I'm sorry,' he said. 'I'm so sorry. I'm sorry. Oh, I'm so sorry.' Finally the baby stopped crying, but her father could not.

Pierce Brooks, now a lieutenant of detectives, sat in his home and held in his trembling hands a decision handed down by Justice

Stanley Mosk and the California Supreme Court on July 18, 1967. He had his own copy of the decision replete with margin notes, underscoring, footnotes. It was a raging detective who sat and smoked and reread late into the night:

> Defendant's principal contention is that the receipt into evidence of each of the extrajudicial statements violated the constitution rules set forth in *Escobedo v. Illinois* (1964) and *People v. Dorado* (1965). Those rules are controlling here even though the trial took place in 1963, i.e., before they were enunciated.

Brooks angrily scrawled in the margin: 'Interrogation took place more than a year before the *ex post facto* requirement of *Escobedo* and two years before *Dorado*!'

Then Brooks read further:

> . . . the record does not establish that defendants were adequately advised at any time of their right to remain silent and their right to counsel. Neither Chief Fote's admonition that Powell's statement 'could be used in court against him' nor Officer Cooper's remark that anything Powell said 'may be used as evidence against you at a later time' amounted to advice that he had an absolute right to remain silent in the face of police questioning.

Brooks wrote: 'But I told Powell that he could very definitely have an attorney. That only a fool would . . . '

Brooks broke his pencil scribbling this remark and threw the pencil down, continuing the reading:

> . . . applying the Chapman test to this record, as we must, we are compelled to conclude there is at least a reasonable 'possibility' that the evidence complained of might have contributed to the conviction.

The detective read the court's disapproval of the many statements he took from Gregory Powell and Jimmy Smith and of his arraignment of the defendants on Wednesday, March 13, 1963, instead of March 12, within forty-eight hours. By now, Brooks was making his margin notes with a fractured pencil he had sharpened with a kitchen knife: 'I called the D.A. on Tuesday afternoon and was ready but D.A. advised wait until Wednesday morning. Said Defendant's 48-hour period was not up until after court's close Tuesday afternoon. D.A. told me no problem!'

Brooks's rage knew no bounds when he got to the end of the California Supreme Court decision:

> The submissive acceptance of an inquisitorial process was the desired result of a number of 'techniques' used on defendants. For example, Officer Brooks maintained an air of confidence in defendants' guilt and his 'conversations' with them were often characterized as merely 'a few questions' about certain 'details'; yet while they began with relatively minor topics, they soon proceeded to issues central to establishing that the crime was in fact committed by these defendants . . . and each was trapped in a web of shifting and inconsistent explanations.

'That's bad? That's bad? That's my job!' the usually unflappable detective shouted aloud in the quiet room and his own voice startled him. Then he held the coffee cup in both hands and continued:

> Most importantly, throughout these four days of questioning, Officer Brooks successfully cultivated a relationship of trust and 'friendship' with these defendants, despite the fact he must have believed they had murdered one of his brother officers. Such psychological devices for extracting confessions without crossing the line of coercion were condemned by the United States Supreme Court in *Miranda v. Arizona* (1966).

'That shows . . . that shows,' Brooks whispered brokenly. 'That proves where your head is. Where it is as far as cops are concerned!'

He turned to the last pages and read:

> We do not lightly reverse these judgments. We are well aware of the heinous nature of the crime involved, and of the strong indications of defendants' guilt.

'Strong indications!' said Pierce Brooks, and threw the sheaf of papers halfway across the kitchen.

*

'I felt that Irving Kanarek *had* to be appointed,' Judge Mark Brandler said many times to justify his decision. 'Lord, I presided during the motions for a new trial. I sat through that incident involving Mrs Bobbick. I sat through weeks of it. But it's not up to me or any other judge or the district attorney or anyone else to question the defendant's choice of attorneys, not when the attorney was thoroughly familiar with the case and had won the reversal on appeal. That was finally the thing that decided for me, not just that Smith wanted him. I couldn't let my unpleasant experiences with the man enter into it, I couldn't let the Supreme Court have another opportunity for criticism.'

A detective named Norm Moore contacted Karl. Moore had been an officer in the Police Protective League, the rather weak and ineffectual collective bargaining arm of the Los Angeles force which had only a fraction of the power of the police unions in cities like New York and Chicago.

Moore had lived near the Hettingers when Karl was a boy in Glendale, had known him most of his life, had worked Highland Park Detectives with Karl just before he resigned. Moore was nearing the end of his own police career, about to take his service pension. He was a plain-talking man and not afraid to speak his mind. Moore came to talk to Karl who by now was earning a little money helping his friend Officer Stew James, who moonlighted in landscaping and maintenance gardening.

'Listen, Karl, don't be dumb. You deserve a pension,' Moore would say to him.

'I don't know, Norm. I stole. You have no idea how much I stole.'

'Goddamnit, kid, I don't care what you stole! If you stole, it wasn't *you* stealing. Don't you see that?'

'It wasn't just the cigars, Norm. That's all the department knows about. I stole all kinds of things. I stole big things like an electric knife and a saber saw. I even stole a portable sewing machine once. I stole ... '

'Look, Karl, there're reasons that people do things. Sometimes the guy that does them is the last to understand. Now I know damn well we can get you a pension. And I know damn well that nobody deserves it more. Now all you're gonna have to do is see a few doctors ... '

'Psychiatrists?'

'Yes, psychiatrists.'

'So, you think I'm . . . '

'No. No. Look, you just see these doctors. It won't cost you anything. I'll arrange everything. You just level with them. Tell them everything.'

'Everything?'

'Everything, Karl. Damn, don't you talk to anybody?'

'You're the first one I've told about all the things . . . the things I stole.'

'Well tell them everything, for chrissake. Get it all off your chest.' Then the older man looked at his young friend thoughtfully. 'Look, kid, are you pretty depressed these days? Is it all pretty bad?'

'Yeah, pretty bad.'

'I think you're gonna feel better when you see these doctors. Maybe you can pick one out and see him once in a while. Don't you have anybody to tell your problems to?'

'No.'

'Don't you tell Helen?'

'No.'

'Why in the hell not?'

'I don't know, Norm. I just . . . I don't know. I wasn't raised that way. We didn't talk about . . . personal things when I was a kid. You just didn't burden other people.'

'There must be somebody. How about your church? Do you go to church?'

'No. Helen goes. I don't go.'

'You have to learn to talk to people.'

'Norm, now that I've told you about the stealing, I just gotta give the things back. I've got some stolen things in my garage. I've used some, some I haven't. The things seem to accuse me every day when I look at them.'

'Forget it.'

'I can't. Maybe I'd feel better if I gave them back.'

'How many places did you steal from? How many times?'

'Fifty. I don't know. Maybe a hundred. I just don't know.'

'That's what I thought. Just forget it. There's no point in humiliating yourself any more. Do you trust me?'

'Sure.'

'Okay. I say it's all right, keep the stuff. It's a bunch of petty junk anyway. For God's sake, stop worrying about it. It doesn't mean anything.'

'I can't. I feel so guilty.'

'What do you feel guilty about?'

'About the stealing. Of course. About the stealing, what else?'

He was to see doctors, seven of them. He couldn't understand it. The city just kept sending him to one doctor after another. Some of his friends suggested that the department must be hoping to hit upon a psychiatrist who would say his emotional problems were *not* service-connected. That his stealing was *not* a direct result of the murder, so the city wouldn't have to pay a pension.

The psychiatric interviews were similar. The patient would list his many physical ailments and tell of his terrible crimes and how guilty he felt about them.

'I worry a lot that I'll meet policemen I know, sir. And what they'll think of me. About my stealing. That I was a thief.'

And then he was invariably asked if he felt guilty about anything else.

'No sir, what else is there? I stole so many times!'

'Did you ever feel guilty about something *before* you began stealing?'

'No sir. There was nothing to feel guilty about. If I could just get over these feelings about the stealing. If I could just understand what made me do it.'

The diagnosis read:

It appears that Mr Hettinger is an intelligent, honest man who has a history of not having good, close, stable relationships to his parents or other people. The trauma of the Bakersfield incident threw him into a psychological regression. This resulted in the formation of incapacitating psychological symptoms. (Loss of self-esteem, obsessive thinking, compulsive stealing, diminished sexual response, withdrawing from friends.) He denies any guilt about the shooting, but evinces unconscious guilt which leads him to be self-defeating and self-punishing. He shows a remarkable lack of insight into his problems. I do not believe his compulsive stealing is a basic characteristic of him, but rather reflects his need to manipulate the environment to agree with his obsession that he is an unworthy person, to punish himself and relieve the anxiety of unconscious guilt, and to unconsciously avoid his police colleagues whom he felt looked critically at him. Had he received intensive psychiatric care earlier, it is likely much of the psychological regression could have been avoided.

Karl told another doctor: 'And I stole bigger things, sir. I stole a sewing machine! I stole . . . '

'Yes, Mr Hettinger, and is there anything that made you get these same bad feelings *before* you began stealing all these things?'

'No sir, I had bad feelings. I had dreams. I had pains. But I didn't feel the same way I feel about the stealing.'

'And how do you differentiate your feelings about the stealing?'

'I feel *guilty* about the stealing, sir. I never felt guilty before. Just bad.'

The doctor wrote:

Mr Hettinger is a well-developed, neatly dressed, pleasant and cooperative man who looks somewhat older than his stated age of thirty-two. He verbalized quite freely, however, he frequently became tearful and occasionally cried when describing the incident of the murder of his partner four years ago. He appears to be unable to escape obsessive ideas which were mostly concerned with the death of his partner. The general picture suggests an individual who is undergoing a severe depressive reaction associated with compulsive behavior which takes the form of ruminative thinking about the murder of his partner masked by the compulsive need to steal. Because of the excessive self-criticism and self-devaluation, he seems to behave in a way which is crying out for help, as well as showing his desire to be punished. The possibility of suicidal behavior should be seriously considered.

He is experiencing a considerable amount of unconscious guilt, which is not expressed overtly, but which he associates with his lack of having done all that was possible the night of the murder. As such, therefore, his present condition should be considered as service-connected. Intensive psychiatric care is recommended.

The reports were all quite similar in their findings. All recommended vigorous psychiatric treatment. One report represented the findings of several doctors of the Julius Griffin Clinic. It was a psychiatric evaluation in great depth. The doctors took an intense interest in this unusual patient, making an extensive check into his physical history as well.

The doctors discovered that the patient evinced certain text-book symptoms. His shrinkage, for example, was not a perception but real, only partially explained by his tendency to hunch his shoulders. He was unable now to stand erect. It was a classic Lilliputian reaction reflecting his view of himself.

His patient poured out his heart to Dr Griffin. He told of as many thefts as he could remember, described in detail the horror of his crimes. Tried to remember every peccadillo in his entire life. To reveal all to this confessor who broke down the barriers. He told of the erasers he and another little boy had stolen. Of the fishing plugs he and a young friend stole from a Sears store so many years ago. Of one benny he had swallowed while in the marines. He dragged it out, all of it. Every sin of his life, everything which *might* be a sin. Every inadequacy. His failure to be able to support his family, his recent impotence, which itself was a crime against the girl who had been his first and only love, and which shamed him almost as much as his crimes. Of his striking a helpless baby. All of it. All the terrible unspeakable things he had done to everyone who trusted him. And he spoke of the rumors of a coming retrial for the killers.

> The patient says that he would almost refuse to appear at another trial. He is considering running out of the state and will resist extradition back. As he spoke of this he cried. Mr Hettinger admits that he cries frequently and has episodes of severe depression. He also has great guilt feelings as to whether he deserves a pension. He realizes that from a financial standpoint it can make a great deal of difference in his future security. He has always wanted to work a farm, and he hopes that he can go into farming work. He is extremely depressed because he does not know where his life will lead him.

The Griffin report presented an interesting chapter entitled 'Patterns of Vocational Interest' which was not found in the other evaluations.

*

His interest clearly lies with the care of growing things. *Within this group, the destruction of anything is most traumatic, and of any living thing especially so.* Very often, people with this pattern of interest care excessively about the welfare of people and all living things. Often they are shy in interpersonal relationships, and show the degree of their care through service or making useful objects.

Under 'Aspects of Personality' he was described:

Mr Hettinger is a man of high-average intelligence. His shyness and determination to achieve would lead others to believe he was an extremely self-sufficient individual. He feels a need to conceal his softer feeling. His rather peculiar combination of strength and shyness would render him a dependable and basically kind individual who would work rather hard for the good of a group. He would be a very dependable husband and father.

Mr Hettinger is showing a marked and severe depression that appears to be reactive, that is, resultant from a situation or occurrence, rather than as a chronic condition. It results from his deep belief that he has failed to meet his own standards of excellence. There is the suggestion that he is near his limit for emotional stress. The patient became aware in February of this year that the two convicted criminals had appealed on the basis of recent Supreme Court rulings. The patient is extremely depressed, apprehensive, frightened, confused, and bewildered. He feels that he simply cannot go through the horror of the trial again.

A reasonably compulsive thoroughness in work habits, moral standards, and general approach to life tends to make people of this sort drive themselves exceedingly hard. And because of this, they are generally considered very reliable persons. Because of a high general level of anxiety, which is normal for them, when their tolerance for stress has been exceeded, they are subject to excessive, obsessive and compulsive thinking and behavior which exceeds the normal range. This reaction makes for a vicious cycle. The neurotically controlled behavior is beyond their understanding and leads them to behave in such a way that they further contradict their normally excessively high standards in that their behavior is beyond intellectual control.

Further into the report the doctors wrote:

We see extreme anxiety and depression with the antisocial behavior serving as a frantic signal for help. The fact that he could reconstitute in a protective environment such as an aide to the chief of police signals a competent strength. The fact that he is threatened by an impending retrial is more than he can bear. This man needs *urgent* psychological assistance and probably will need it for at least several years.

The final paragraph read:

In an unsolicited opinion by this examiner to the police department, may I humbly and respectfully suggest that a careful assay be made of procedures that are presently being used to assist officers who suffer severe physical or emotional trauma. It is unfortunate that this man was not given an opportunity for psychologically working through his fear, shame, guilt, desperation, and panic, occasioned by the event. In my opinion, this man's emotional equilibrium could have been much more stable if immediate attention and opportunity had been given to him for psychological assistance. Perhaps all he needed at that time was some cathartic ventilation and perhaps also some restoration of his self-confidence. To assume that a man can just resume a normal way of life after such an overwhelming episode is asking too much of most of us. I do not know of the arrangement which the police department has for psychological assistance, but I urge that very careful consideration be given for the *prevention* of mental and emotional disturbances arising from traumata in the line of duty.

Karl Hettinger listened to the things the doctors told him. Some of it made sense, some of it didn't. He had confidence in Dr Griffin and decided to return. Then he began worrying about the cost to his family of psychiatric visits. He had already punished his family enough and could not inflict a financial burden on them. So he stopped seeing the doctor after only a few visits. He reasoned that he was feeling a little better about things after seeing the doctors. That was all he could probably expect anyway, all he had a right to expect.

Sergeant Norm Moore waited outside at the pension hearing. The detective argued and raged at everyone who would listen.

'He should get a seventy-percent pension, goddamnit! Who can quibble? How can you measure somebody's hurt? Did you read the psychiatric report? Well, baby, those shrinks didn't even know about the goddamn memorandum. They don't have any idea that this department overtly criticizes this boy, lays the blame square on his shoulders. Did *anybody* in the hierarchy have the guts or brains to officially tell him, "You *should* have given them your gun under the circumstances. You did the *right* thing"? No, they wrote an order. You can't write that kind of order. I blew my top and called the chief's office back then as soon as I found out he was being asked to tell it to the rollcalls and letting them pick him apart. And then when he got in that shoplifting thing. It was pathetic! But the department just wrote him off like a common thief so we could maintain our integrity. Jesus Christ!'

Norm Moore knew he had won when he heard a voice in the hearing room say, 'I don't care. We're going to give this boy seventy percent.'

It is unknown whether Dr Griffin's unsolicited recommendations to the police department were ever heard in the chief's office. Internal Affairs Division still worked much the same as always, considering it their job to get antisocial policemen to resign as quickly and painlessly as possible to avoid criticism of the department. The bad-apple theory would endure.

The training which purported to prove how an ex-policeman named Karl Hettinger had allowed his partner to be killed was still regularly given. The command *never* to surrender a weapon under any circumstances was made part of the department manual. It was most official.

One man connected with the granting of the pension ironically observed, after reading all of the psychiatric reports, that the police hierarchy was in a sense much like the two condemned men who started the misery. He said the archetypal police mentality and the psychopathic mind were both utterly unable to identify with their victim in this case.

One of the pension doctors he saw, not the last, just one of many, wrote a very different kind of report on Karl Hettinger.

The doctor was George N. Thompson, well known in the Los Angeles area to lawyers who needed a psychiatric expert to testify. His report on Karl Hettinger, based on one visit, described the event of the shooting inaccurately.

He and his partner were fighting with two suspects and one of the suspects shot his partner.

The findings of the doctor were also markedly different:

> He seems to have a minimum feeling of guilt with regard to what happened on the police department and his episodes of petty theft go back into even his childhood. It may be said, of course, that his childhood petty theft episodes are not unusual, that they are in fact rather common with children. On the other hand, I can see no direct relationship between his episodes of shoplifting and any factors at his employment. It is the opinion of the examiner that his present disability is a result of the internal causes within his own personality and not due to the factors at his employment.

The paragraph under 'Treatment' was perhaps the most startling in light of what six other doctors had said.

> There is no indication for any specific treatment. Re-examination is indicated in a period of one year. Thank you for the kindness in referring this case.

After his signature, Dr Thompson did decide to add a paragraph.

> P.S. I might add that although some of the history would seem to indicate that the type of character disorder from which he suffered is kleptomania, on the other hand there is not specific indication in the examination of him that this compulsive stealing is sufficient to be diagnosed as kleptomania.

Dr Thompson's minority report would be culled from the rest and this particular doctor would be called into court to discredit Karl Hettinger.

All the faces of all the doctors had merged in his mind. He couldn't remember any of their names or any of the faces. The pensioned ex-policeman during the remainder of that year became in every sense a gardener. He did such a good job of putting his former life out of his thoughts that sometimes when he wanted to remember police work he was unable to. He would, of course, remember the events of the night in the onion field, would still relive it in his dreams. And he would still remember his crimes. They were impossible to forget. He doubted that he'd ever forget his crimes. But everything else concerning his former life he came to forget except for his service revolver. He began thinking of it more and more.

He even forgot the doctor who had won his confidence. Even Dr Griffin's face had faded from his mind. He had hated sitting in psychiatric waiting rooms with sick neurotic people, but he occasionally thought that perhaps he would go back one day if the thoughts of the crimes didn't go away. At least he didn't *dream* about his crimes. He only dreamed about the killing of Ian Campbell. He never believed the doctors who said he felt unconscious guilt about Ian. They didn't understand. He only felt guilt about the stealing.

It had confused him when one of them asked: 'Do you ever dream about the stealing?'

'No sir.'

'You only dream about the murder?'

'Yes sir.'

'The thing which you feel no guilt about?'

'That's right, sir.'

They were trying to make him say he felt guilty about Ian. It wasn't true. He'd never say that. If it were true it might mean there was a *reason* to feel guilty. That they were right. All of them. All those who implied he *killed* Ian the moment he *surrendered*.

But this was crazy. He mustn't think like this. After the retrial had come and passed, then he'd feel better. Then at least he'd stop thinking about the gun. The gun he had surrendered to Jimmy Smith. The gun he now so often longed to press to his head.

SIXTEEN

The case was given to Deputy District Attorney Phil Halpin even before the California Supreme Court reversed it. After the *Dorado* decision there could be no doubt it *would* be reversed.

Halpin was glad to get a head start. The transcript was 7,734 pages long, making it a lengthy murder case. The young prosecutor could not have dreamed at this time that the transcript would eventually contain nearly 45,000 pages, the longest in California history.

Halpin was thrilled when Joe Busch, the assistant chief of trials, picked him to try the case. Busch was the number one trial lawyer on the staff and this case would go a long way toward establishing the young prosecutor's reputation. Halpin had all the tools. He had youthful good looks, a convincing baritone voice, was articulate and bright, a most promising trial lawyer.

Phil Halpin was twenty-nine years old when the important case was given him. He was thirty-one when he quit the District Attorney's Office because of it. But in 1967 he was enthusiastic. He was ready.

The prosecutor had been divorced in December, just months before being handed the notorious case which was to him more important now from a legal standpoint, after having been reversed by the Supreme Court. He was ashamed of divorce, had always felt there was some stigma to it. And no one at the District Attorney's Office even knew when it happened. Halpin lived alone in an apartment in the hills over Silverlake. He lived quietly, did not own a television, and against the orders of his employers, did not have a telephone. The Powell-Smith case afforded him a blessed opportunity to become totally immersed in something.

His had been a teenage marriage to a girl he had known since the second grade, and now he was alone. He had two daughters and a normal share of divorcé's guilt. He wanted to lose himself in

345

this assignment. The Powell-Smith case *would* do that for him, to an extent he had not dreamed possible.

Judge Alfred Peracca was, like Judge Mark Brandler, a patient man, but more gentle and kindly. He had a thin sallow face and a long upper lip, often pursed. Judge Peracca looked like one's image of a Victorian public school headmaster. He gestured often with his large hands.

Judge Peracca's courtroom was in the Hall of Records, a decrepit building that smelled of sewer, especially on the elevator, and it was even shabbier and more dismal than the Hall of Justice.

During the pretrial motions Gregory Powell sat at the far end of the counsel table dressed neatly in a suit, and stared at the prosecutor. Next to him was Deputy Public Defender Charles Maple, then Jimmy Smith, Irving Kanarek, and Phil Halpin. For a time the young prosecutor was assisted by Deputy District Attorney Pat McCormack, eleven years his senior.

Judge Alfred Peracca had a serious heart condition. He started the case looking fresh and rested each morning during those first few weeks, but by noon he would be bone weary and find it necessary to take his medicine. He was a good and charitable judge. But even in the best of health Deputy District Attorney Halpin doubted that he could have handled what he was faced with in those interminable months.

It was estimated by the district attorney that never in Los Angeles history had so many motions been raised in one trial. There were motions to change venue, motions to sever the trial, motions to quash the whole jury list, challenges to the jury system, motions to get Karl Hettinger's personnel file, and even Ian Campbell's. News commentators who had commented on the case were subpoenaed. Smith made a motion concerning the sanity of Powell. Charles Maple, attorney for Powell, made what the prosecution would deem a ghoulish motion to exhume the body of Ian Campbell now buried five and a half years. There was a motion to appoint a psychiatrist for Hettinger. Irving Kanarek made a motion to quash the entire master panel of jurors due to neo-racism. There were motions to continue, motions to discharge attorneys, motions to reinstate attorneys, bail motions, withdrawal of bail motions, motions of mistrial, motions to film a reenactment of the crime, motions made almost daily concerning inadequate treatment by jailers.

There were hearings which would in themselves go on for days, having nothing whatever to do with the case.

'Your Honor,' Kanarek said one day, 'Mister Halpin, I am told, just made the statement that I am going to get knocked on my ass.'

'Oh come *on*, Mr Kanarek!' said Judge Peracca, arching his brows.

'Well, your Honor . . . ' said Kanarek.

'*Please*,' said the judge, a word he would utter a hundred times daily, pleading with Kanarek, pleading, it turns out, for his life. 'I will hold you in contempt for constantly reiterating. And Mr Halpin as well. Now *please* get on with the case.'

'I have not said a word to Mr Kanarek,' said Halpin. 'Not in five days and I'm not going to.'

'Your Honor,' said Jimmy Smith.

'Oh, I'm not going to hear from *you*, Mr Smith,' said the judge.

'All right,' said Halpin. 'I want the record to reflect that the court just finished admonishing me for something that's supposed to've occurred while I was at a drinking fountain and that I had no part in. I want the record to reflect that this man has stood up and made constant statements about things that I'm supposed to have said, that I'm getting admonished now by this court. I would just ask the court to quell this sort of thing.'

'All right, sit down, Mr Kanarek. You too, Mr Halpin.'

'I'm afraid for Mr Kanarek, your Honor,' said Jimmy Smith. 'Suppose he hits him?'

'Well, Mr Smith, Mr Kanarek can take care of himself,' said the judge wearily.

'Suppose he hits him?' said Jimmy Smith, seizing an issue. 'This man is defendin me for my life. You mean I don't have a right to inform the court?'

'Your Honor,' said Halpin. 'Mr Smith was also told to be quiet.'

'I should sit here idle?' said Jimmy. 'Like a piece of furniture?'

'*Please*, Mr Smith. *Please*,' said the judge.

A day later, Kanarek said suddenly: 'Your Honor, I am informed that Mr Halpin has carried a gun in his belt in the courtroom.'

Deputy District Attorney McCormack held the young prosecutor's arm and kept him in his seat as Halpin gaped in disbelief at Kanarek, who said, 'Mr Smith wants to take the stand under oath and so testify that Mr Halpin has had a gun in his belt, and he has indicated a lack of control of his temper in these proceedings. He has come to court . . . I offer sworn testimony that he had come to this courtroom with a gun!'

'Put Smith up and I'll take the stand just to the contrary!' said McCormack, who was now on his feet, no longer trying to keep his associate in his chair.

'Get him up! Get him up there!' roared Halpin.

'If we're going to have a man lose his temper . . . ' offered Kanarek.

'I've never lost my temper in this courtroom!' said Halpin.

'The record reflects Mr Halpin's temper,' said Kanarek.

'Oh, Mr Kanarek,' said the judge. '*Please. Please. Please. Please.*'

The rest of the day and the next day involved a lengthy hearing wherein Jimmy Smith took the stand to testify seeing Halpin with a gun. Gregory Powell was called by Kanarek but declined to become involved, invoking the fifth amendment. The matter was finally resolved by calling a deputy sheriff who was also involved in the alleged incident.

'About a week or so ago, Deputy, did you have a conversation concerning a pistol with Mr Halpin?' asked McCormack of the witness.

'Yes, I did.'

'Did you bring such a gun to court?'

'Yes.'

'Did he hold it and examine it for a while?'

'Yes.'

'Did he slip it into his waist?'

'Yes. Then he returned it to me.'

'And did you take it from the courtroom?'

'Yes.'

Charles Maple cross-examined this witness and determined that Halpin was considering buying a weapon and was being shown the gun by the deputy, who was also a gunsmith. Everyone was finally satisfied that Halpin was not going to shoot anyone, but Jimmy Smith and Gregory Powell were disappointed that the whole affair came to so little. But then, you never knew what could be cause for review and reversal. You just never knew.

After many months Judge Peracca would, by noon recess, be sitting at the bench with his head in his arms. His life was literally in danger and he was removed from the case, eventually undergoing heart surgery. The case was wearing out many judges.

Typewriters brought into the jail were routinely checked for hacksaw blades and vials of hydrofluoric acid used to melt steel, but on June 4, 1968, a narcotics officer received a tip from an informant that three guns would be coming into the Los Angeles County Jail inside an Underwood typewriter. As always, Greg and Jimmy were the victims of informants. They would always blame the assassination of Senator Robert Kennedy for foiling their scheme, but actually, and the plan never had a chance. As usual, if they didn't inform on each other, someone else would do the job.

Police and prosecutors would credit Gregory Powell for masterminding the aborted escape attempt, but for once he was following another's lead, at least at first.

'Come on, Greg,' said Jimmy in the jail one evening when they were making the last of their allotted calls, but Greg winked and grinned and continued the low whispering stream of erotic promises he was making to the woman on the other end of the phone.

By now Maxine had taken up with a dishwasher called Stan the Man and she had dropped from sight with the child of Gregory Powell. Neither Greg's family nor her own were to hear from her.

'Jesus,' Jimmy said in disgust. 'What's she doin, playin with herself?'

Greg stifled a chuckle and nodded and continued the low crooning string of lewd promises to the middle-aged black woman.

She was a lonely and plain-looking woman and Greg called her his Chocolate Drop. She had begun corresponding with him through another Death Row inmate. She was in love with him and wanted to marry him whether or not he was convicted again, whether or not he was sentenced to death. To Greg and Jimmy she was their hope of escape.

It was Jimmy Smith not Gregory Powell who made the initial plans. He had a friend deliver two .38 revolvers to Greg's

Chocolate Drop to keep pending further instructions. They had the woman buy metal cement and an eight-inch piece of metal the exact size of the plate on the back of a typewriter. The rest of the plan involved a black man and a blond man. They were friends of Greg and Jimmy, introduced to them by another inmate. They had their own motives, not the least of which was the thrill of being part of something so dangerous yet with no personal risk. The typewriter was bought with money Jimmy begged from his Nana and a cousin. One of the guns traded for the .38's was a .25 caliber automatic in poor condition. The blond man had to install a clip spring in it. A derringer was obtained and a revolver, which was the hardest to conceal inside the typewriter. Jimmy was listing his Nana's phone number each night to the jailers, but was actually calling Greg's girlfriend to make the arrangements.

For a dry run, Jimmy asked Greg's girlfriend to bring a rented typewriter to them and told her how to secrete hacksaw blades and Allen wrenches in the roller. She refused to deliver it, so Jimmy Smith turned to an old friend who did not know what the machine contained. The typewriter was searched casually and passed through without a hitch.

They hid the blades and wrenches in their cells inside the corrugated sides of cardboard boxes which they slit and then reglued at the opening. It had been so easy they were astonished. The guns were next. Once again, though, Greg's Chocolate Drop would only assist to a point. She would not deliver, and even thirty minutes of lewd crooning promises of what was to come when Greg escaped would not convince her. There was only one thing to do – find another unsuspecting 'friend.'

Jimmy was now so excited by the escape schemes he had the judge order the jail doctor to supply him with tranquilizers. The two partners were now scheming, planning, phoning, cajoling, threatening everyone who could possibly help them. And Greg was now involved in sexual contact with his partner almost every day. Jimmy Smith was to tell himself that at least it kept Powell's sex drives under control so he could pay attention to business – escape business.

It was only left to have a dupe bring another typewriter full of guns into the jail. Greg selected an unsuspecting inmate friend named Bayer, also defending himself, who had a girlfriend acting as his legal runner. On the evening of June 7, 1968, a Miss Grant brought a typewriter for Gregory Powell. She was caught on the informant's tip, the machine searched, and she was arrested. So was Sylvia, Greg's Chocolate Drop.

When Jimmy went to his cell the night of June 10 he was disappointed and depressed. Still, it had not been completely in vain. After all, he reasoned, Powell was in all kinds of trouble with the bulls as well as Bayer, who kicked his ass. Who knows? Somebody might take a blade and touch him off over this one. And the best part was that nobody, but nobody, even so much as guessed at Jimmy Smith's considerable part in it. That proves, he thought, that I'm smarter than that turkey-necked punk. Then Jimmy actually had to smile as he thought of the fistfight that day.

'You son of a bitch,' Bayer had said to Powell. 'You're gonna tell them my girlfriend didn't know nothing about those fucking guns. You hear me?'

'Get away from me,' said Greg, and Bayer knocked him down.

Yeah, thought Jimmy, yeah, right on his ass! And what a weak sissy punch it was too. And what did he yell when he hit the deck? What did that punk yell? thought Jimmy Smith. He opened his big mouth and screamed, 'Jimmy!' That's what he did. The punk!

Jimmy Smith and Gregory Powell, not knowing about the informer, would always believe the tightened security on June 7 was what foiled them. For on June 7 the jail was alive with deputies to guard a new arrival named Sirhan.

'If only the fuckin Ay-rab hadn't of dusted Kennedy,' Jimmy would complain to sympathetic listeners. 'It's just my luck. Everything happens to me.'

The most tragic aspect of the whole incident for Jimmy Smith was later in court when he saw a public defender playing with the typewriter exhibit, remarking upon how cleverly the revolver cylinder had been removed from the gun and made to look like part of the works of the machine. Jimmy's eyes popped when he saw the lawyer poke beneath a roller with a pencil and dislodge a quantity of white powder. At first the lawyer didn't notice, and Jimmy's heart stopped. Jesus! he thought. The black man had said he'd stick some in there and he'd done it. There it was, jammed far beneath the roll, now spilling out. Maybe a gram, maybe two! The deliverer of the guns had made good his promise of a bonus in the typewriter to calm the nerves of Jimmy Smith during the escape.

Jimmy's tongue felt fat and red raw when the lawyer reached inside and tore the little bindle and powder spilled down the side of the machine.

'What's this?' asked the lawyer, blowing the precious crystalline substance away. 'Fingerprint powder?'

'Jumpin fuckin Jesus,' Jimmy Smith moaned aloud. 'My whole life's just one big junkyard of misery and bad luck!'

Once during recess, Phil Halpin told his associate, Pat McCormack, about another case he tried with Kanarek.

'Judge Walker got so mad at Kanarek he flew off the handle and declared a mistrial,' said Halpin. 'The case lasted five days. It should've been a simple half-day case.'

'What eventually happened?' asked McCormack.

'We settled for a misdemeanor plea.'

'Then his tactics paid off for his client, didn't they?' said Maple, who had been listening.

'But damn it, the system can't accommodate such tactics,' Halpin argued.

Phil Halpin sometimes thought he would punch the next person who told him he shouldn't let the case bother him. Of course it bothered him. He was involved precisely because no one else cared. The case had been around too long. The defendants and counsel knew of course that this would happen, and this was part of the strategy. It could theoretically be done in almost every capital case except that most lawyers have to make a living. Charles Maple was a salaried public defender, and Irving Kanarek didn't seem to care about making a living. And Halpin thought of Kanarek the first day in court, with a fresh haircut, an expensive suit, and a clean shirt and tie. He would wear the suit until he was ready to discard it. Then he would start again.

'It's an era of violent change in the law,' Halpin complained to his superior. 'All judges are afraid of being reversed. And they all say they can handle guys like Kanarek.'

Halpin's boss, Joe Busch, smiled placatingly. 'Where's your sense of humor, Phil?'

'I've lived with him too long to laugh,' said the young prosecutor.

*

Judge Arthur Alarcon was perhaps not as gentle as his predecessors in this case, but his reputation as a jurist was impeccable. This judge was younger, healthier, more exuberant. He was descended from lawyers and professional men, and Spanish gentry of the Southwest. The books on criminal law, evidence, and procedure by Fricke and Alarcon netted him a tidy royalty and could be found in any law office in the state.

He was a no-nonsense judge known to be fair to both defense and prosecution. The Powell–Smith murder case was becoming the butt of derisive courtroom humor. If anyone could get it moving again it was Arthur Alarcon.

On January 23, 1968, in Department 78, the usually smiling green eyes of Judge Alarcon were not smiling. Almost at once Irving Kanarek, as was his way, interrupted the judge who admonished him for it. The following day, after further argument, the judge warned Kanarek three times that his bailiff would enforce courtroom decorum. After being told to sit down by the judge, and some argument on that score, Judge Alarcon asked: 'Have you ever tried a death penalty matter as a trial lawyer?'

'No, your Honor,' was the reply.

When court reconvened, Judge Alarcon said: 'I conferred with the presiding judge of the court and I also studied a previous case, *People versus Flanagin*, in which Mr Kanarek was the attorney of record, in which there was a motion for a new trial. Among the reasons indicated by Judge Hauk for a new trial was the inadequacy of the representation of Mr Flanagin who was charged with possession of marijuana, one cigarette. There was a time estimate by the people of one day. The case took some two weeks to try.

'Mr Kanarek has indicated that he has never appeared as the attorney in a death penalty case. I have considered Mr Kanarek's behavior before this court, his attempts to urge a motion for a severance for defendant Smith. In view of the very serious nature of the charge in this case, in view of the fact that a previous jury had found this defendant guilty of murder in the first degree and recommended the death penalty, this court has a special duty to see that Mr Smith gets the finest representation possible.

'In view of Mr Kanarek's lack of experience in any death penalty case as a trial lawyer, in view of the finding of Judge Hauk that Mr Kanarek was not competent to represent a defendant in a

one-day marijuana possession, the court will vacate the order of
Judge Brandler appointing Mr Kanarek. Mr Kanarek is now
relieved as the attorney of record. Mr William A. Drake is
appointed.'

During the three months Drake was Jimmy Smith's attorney, he
hardly spoke a dozen words at a time with the defendant, who
refused his services and was sullen, hostile and threatening.

Drake was to say, 'Heaven forbid any of my motions being
granted! Smith doesn't want them granted. Just as Powell doesn't
want his granted. Like all condemned men, they want motions
denied to have issues raised on appeal. But is that so surprising?
Does anyone expect a person fighting for his life to cooperate with
the forces implicitly attempting to destroy him?'

Jimmy Smith filed a written declaration:

Judge Alarcon does not wish me to have a fair trial, but wishes me
to be found guilty and executed, which is the real reason he
purportedly replaced my attorney, Mr Kanarek. Further, it is my
belief that Judge Alarcon wishes to use my cadaver as a stepping
stone to become governor of California.

In May the California Supreme Court heard the new appeal of
Irving Kanarek and would say of Judge Alarcon's ruling:

The outright removal of counsel on the ground of an alleged
'incompetency' is more of a threat to the independence of the bar
than is arbitrary misuse of contempt power. As Mr Kanarek
persuasively argues, 'If the advocate must labor under the threat
that, at any moment, if his argument or advocacy should incur the
displeasure or lack of immediate comprehension by the trial judge,
he may be summarily relieved as counsel on a subjective charge of
incompetency by the very trial judge he is attempting to convince,
his advocacy must of necessity be most guarded and lose much of
its force and effect.'

The Supreme Court of the state thus decreed that membership in
the bar, not a trial judge's opinion, would determine an attorney's
fitness to try a case. Irving Kanarek was reinstated, Judge Alarcon

was replaced, and once again, the Powell–Smith murder case was making legal history.

Irving Kanarek's co-counsel was perhaps the most sympathetic with the relentless controversial lawyer.

'This isn't the first trial where the factual issues are lost,' Maple would argue, in defense of both Kanarek and himself. 'The court system itself has become the antagonist. After the Alarcon incident, where he was declared incompetent, Irving Kanarek began to feel the *system* was against him personally. He has to protect himself by the only device he knows – multiple objections. On all grounds.'

It was appellate issues which interested Deputy Public Defender Charles Maple, who was a graduate of Harvard Law School and proud of it. 'That's where the real changes in the law occur,' he said to his client Gregory Powell. 'Few cases are won on appeal, but the effects are so far-reaching that one is attracted to appellate issues. An advocate tries to find and set up in the trial record legal issues that are of constitutional proportions.'

But if Phil Halpin wondered if Charles Maple was a dupe of his client, he need not have. Charles Maple had been a trial lawyer for many years and knew exactly what his client was about.

'Powell's buying time,' he said. 'His only hope is time, change in the law, appellate decisions, abolition of capital punishment – and escape. Down here he has a chance to escape, up there he has none.

'I think he's as intelligent as Chessman was. These cons are good lawyers. And they have real lawyers advising them and helping them between trials. *Of course* they seize upon any device to perpetuate their lives or win their freedom. Who wouldn't?

'*Using* me? Because he makes a motion to represent himself? They *all* do it. If it's denied they have another issue. If it's granted they say "Okay I need Charlie Maple as advisory counsel. I'm not in a position to do legal research. I'm not a trained lawyer." In effect, "I want Charlie Maple as my lackey and runner." Powell knows no lawyer worth his salt would do it. It's a ploy. He's set up an issue that's impossible to satisfy.

'On Death Row they have exchanges of briefs, motion exchanges, habeas corpus exchanges. If they're not good lawyers when they go up there, they're pretty fair when they come down.

'One case involving the murder of a police officer was reversed three times. It was never reversed on guilt but there were developments of law during intervening trials.'

Yet even Charles Maple began to complain of Irving Kanarek, who would call the public defender at three or four in the morning to discuss a motion he wished to raise the next day. Kanarek would often work all night for his client, Jimmy Smith. Maple eventually had his number changed because of it.

Charles Maple, when arguing with the prosecutor, in trial or out, had what he thought to be an impregnable position. He said to Halpin: 'File one additional count of kidnapping with injury. That carries life without parole. We'll plead guilty to murder *and* the second charge. We'll gladly accept the sentence of life without possibility of parole. I've told you people a dozen times, we'll cop out if he can just live out the rest of his life in prison. *If you just don't kill him.* Now I ask you, is that an unreasonable proposition?'

The district attorney had sent the star witness a huge carton of court transcripts to read, to refresh his recollection about his testimony at the 1963 trial. The witness never touched it. His wife rewrapped it and sent it back. It was so large it cost five dollars for fourth-class postage.

One night in the summer of 1968, when Karl was fishing with Jim Cannell, Helen Hettinger heard voices in the yard. It was close to midnight. She saw two men in the dark and heard one say: 'He must be home, the outside light is on.'

'Who is it?' she asked, very much afraid, now that the trial was about to begin, remembering the threats and crank letters of the first trial.

'I'm an attorney here to see your husband, Mrs Hettinger,' said a voice outside the door.

'I don't know who you are,' said Helen, peeking through a window at a man. 'You'll have to call. You'll have to come back. Please go away.' She tried to describe him later. He was a short man and heavy. He talked politely but she was frightened.

When her husband heard, he envisioned assassins sent by the killers. He didn't know whether or not to call the police.

A few days later the men returned at the dinner hour. One of the men apologized to Karl for upsetting his wife and asked to have a talk. The man reminded the witness that he was an officer of the court and that the witness should speak to him.

'Leave me alone,' said Karl. 'You talk to me at the District Attorney's Office or in court. Don't you *ever* come to my house again at night and scare my family.'

The man left. The witness trembled. He would soon meet the man in the courtroom. It was Irving Kanarek.

Karl Hettinger was subpoenaed to pretrial motions, for by now Charles Maple and Iving Kanarek had learned certain things about him.

It had been five years since the two defendants had seen him. He was thin and nervous now while they had each filled out and were relaxed, bored in fact, by the months in court. He walked and sat hunched over. His eyes were usually downcast and he wet his lips often. He looked very different to his former hunters.

'Have you been in treatment or therapy with any doctors?' asked Maple of the witness.

'Objection,' said McCormack.

'Sustained,' said the new judge, Thomas Le Sage.

'Are there any doctors you have discussed the events in Kern County with?'

'Objections,' said McCormack.

'Sustained.'

'Your Honor, if I may,' said Kanarek, 'I'd like you to have his mental capacity examined.'

'Sustained!' said the judge, not waiting for the objection.

'Well, may I . . . ?' said Kanarek.

'Sustained! I have already made the ruling as to what you are going to be able to inquire into at this time.'

'Your Honor, may I approach the bench?'

'Not on the same subject. No.'

'Your Honor, the witness's competency is the issue.'

'It is *not* the issue at this time.'

'But it *will* be,' warned Kanarek.

Judge Thomas Le Sage had known very little about the marathon case when he took the bench the first day. He knew that the case had already caused a Supreme Court reversal for Judge Brandler and another for Judge Alarcon. He knew that Judge Peracca's heart had almost given out because of it. He knew that nobody wanted the case and that's why he, the newly appointed junior judge on the Superior Court bench, was being given it. There was no publicity connected with the case now, only drudgery and frustration. Other judges joked about it.

Judge Le Sage was very dignified, perhaps excessively so. His first day on the bench, ignorant of the months of exhausting motions, he said: 'Well, gentlemen, are there any pretrial motions?'

Irving Kanarek beamed. Charles Maple grinned. Phil Halpin groaned and snapped his pencil in two.

Prosecutor Halpin had been frustrated by Charles Maple's deliberate ways, by his completeness compulsion, and his willingness to often join Kanarek in futile motions, but he respected Maple's ability as a trial lawyer and so had only felt mild anger with Powell's defender. And strangely, after these many months he was even losing his sustaining rage at Kanarek. His spirit was becoming deadened.

'Nobody cares about the case anymore,' Halpin complained to his superiors. 'Each new judge we get is mildly amused at my frustration. There's no concept of what's gone before. No one cares now that a cop was killed so brutally. Pat McCormack's talking about leaving to become a court commissioner. I'll be left alone then. For the duration!'

'Don't let Kanarek bother you so much,' was the stock reply.

'It's not Kanarek, damnit. It's the system that indulges him! Hundreds of these killers'll win their freedom one day after all the witness and judges and prosecutors either die, or disappear, or give up!'

'Listen, Phil, after it's over . . . '

'Sometimes I get a crazy feeling that it'll *never* be over. There's no finality in the law itself. This is a war of attrition in no man's land, and the obstructionists are winning. Judge Peracca almost died. Do you know how many judges have sat on this case? How many people have been involved? How many years? Are we going to be able to *drag* these civilian witnesses back if and when we ever get to trial?'

'Phil, you're overwrought. You'll . . . '

'The American system of justice in the laughing stock of the English-speaking world and totally incomprehensible to the rest of the world!'

'Take a real break this weekend, Phil. Take the wife out to dinner. Go to a show.'

Oh God, thought Halpin, realizing that they didn't know about his divorce and how alone he was.

The finish of Phil Halpin came as a result of several incidents. Before he was to break there were minor skirmishes.

Once McCormack removed his thick glasses and invited Kanarek outside to fight. Kanarek merely backed up, blinked patiently, shook his head and set his heavy lips in disgust that McCormack could be so vulgar. His face said he would not be provoked. He would endure.

There were lighter moments. Once one of those courtroom regulars, called 'railbirds' by the lawyers, began passing notes to all counsel in the case with his recommendations as to how the case should be handled. He began to torment and follow Irving Kanarek. Kanarek tried to make a citizen's arrest on the railbird for jaywalking in front of the courthouse but the police refused to accept the prisoner. The court ordered the man to stop hounding the lawyer.

Phil Halpin began worrying about his emotional health when he couldn't even join the others in laughing at the incident. 'I've lived with him too long to laugh,' Halpin said.

It became intolerable when Halpin discovered that McCormack was quitting the District Attorney's Office. Now Halpin was truly all alone.

Then on March 6, 1969, there was a routine squabble while a juror was being challenged for bias by the defense. The court had indulgently permitted each juror to be challenged for bias separately. This day there was only the one prospective juror in court. She was on the stand. Irving Kanarek suddenly looked toward Phil Halpin and said he objected to a certain remark.

'I'm addressing my remarks to the court,' said the young prosecutor.

'He was directing his face at Mr Maple,' said Kanarek to the judge. 'And he just said, "Fuck you."'

'I didn't say any such thing!' Halpin sputtered, going white, leaping to his feet. 'I did not say any such thing! I will not permit this man to say this about me!'

Judge Le Sage, a God-fearing man, looked shocked. The woman juror looked as though she might faint.

Phil Halpin's fists were clenched, his eyes were popping, and he was staring at Irving Kanarek, who only blinked and retreated a step and looked toward the judge with the same patient bulldog expression Halpin had seen and hated every day for a year and a half. Then Halpin leaped forward, grabbing Kanarek by the front of the shirt.

'I will not permit this man to say this about me!' Halpin shrieked.

'Just a minute!' yelled Jimmy, getting up.

'Sit down!' the bailiff shouted, running forward.

'Hey!' yelled Jimmy.

'I didn't say any such thing, and I never have said such a thing, and I won't tolerate it!' shouted Halpin, shaking Kanarek, who would not resist.

'He said it just before he grabbed me!' said Irving Kanarek.

'He struck him!' screamed Jimmy Smith.

'Just a moment,' pleaded Judge Le Sage.

'The bailiff! The judge! Mr Maple! Everyone in court witnessed it!' screamed Jimmy Smith.

'I want this man instructed to keep quiet!' shouted Phil Halpin, releasing Kanarek.

'All counsel sit down!' said the judge.

'If you put your hands on *me* like that, we're goin to war!' said Jimmy Smith.

'Sit down!' said the judge.

'I'm a man!' said Jimmy. 'M-A-N!'

'All counsel be seated!' said the judge.

'That's right!' said Jimmy Smith to Halpin who was watching Irving Kanarek stand calmly, hands outstretched toward the judge, in a gesture of supplication for wrongs done him.

'Just a moment!' said the judge.

'Your Honor . . . ' said Kanarek.

'Just a moment, Mr Kanarek!' said the judge. 'Be seated!'

'Your Honor . . . ' said Kanarek.

'Just a moment! Be seated!'

'We gotta do somethin about this, your Honor. Period! He struck him!' said the outraged Jimmy Smith.

'Just a moment,' pleaded the judge.

'What *is* this?' yelled Jimmy. 'He actually attacked my counsel!'

'Just a moment, Mr Smith!'

'I would like to be heard!' said Halpin.

'Just a moment!' said the judge.

'I can't believe it!' yelled the incredulous Jimmy Smith.

'The court is going to call a recess in this case at this time,' said the judge, 'and counsel are instructed, and the parties are instructed, and the bailiff is instructed to advise the court of any . . . '

'Your Honor . . . ' said Kanarek.

'Just a minute, Mr Kanarek! You are to be seated and say nothing! Not to in any way communicate with one another in any way by words or by show of emotions!'

'Your Honor . . . ' said Kanarek.

'Just a moment, Mr Kanarek! You are instructed not to say anything!'

'Yes, your Honor,' said Kanarek finally.

'The court is calling a recess,' said the judge, 'until counsel can reflect and recover their sense of poise.'

When court proceedings resumed in fifteen minutes the judge said, 'In connection with the last matter the court instructs all counsel involved not to address the court concerning this in any way while the court is making the following orders: Mr Halpin and Mr Kanarek are to file affidavits of the occurrence before Thursday at four o'clock. The court will consider whether one or both counsel are to be held in contempt. The bailiffs, sheriffs, clerk, and reporter are instructed not to discuss the matter.'

'Your Honor,' said Kanarek, 'if I may, I would ask that Mr Halpin be arrested immediately for a misdemeanor committed in the presence of the court in violation of the penal code of the state of California. In the presence of peace officers he committed an assault.'

Phil Halpin was defeated. He not only walked off the case, but he resigned from the District Attorney's Office and vowed never to enter a courtroom again. It was little consolation that before his resignation was final, Irving Kanarek would be removed from the case.

It was Jimmy Smith who eventually proved the undoing of Irving Kanarek. He filed a declaration that same month asking the court to curb his attorney's examination of jurors on racial grounds:

> This defendant does not subscribe to or identify with any black militant organizations. Attorney Kanarek does not know anything whatsoever about defendant's racial background as a member of the black race.
>
> Respectfully submitted,

Another declaration that same month by Jimmy said:

> It is respectfully suggested that said attorney has lost all conception of the legal conduct and performance required of an attorney in a capital case, that is, if such knowledge ever existed for said attorney.
>
> Respectfully submitted,

During a noon recess in March, six years after the murder of Ian Campbell, Prosecutor Joe Busch walked from the court with Kanarek and said to him, 'Irving, you ought to analyze your approach through a psychiatrist. Your personality alienates people.'

Irving Kanarek apparently ignored the advice, but later when the judge took the bench, Kanarek wanted it made part of the official record that the district attorney said he should see a psychiatrist.

Toward the end of the month, Jimmy Smith was growing more hostile toward the lawyer and finally he assaulted him.

Prosecutor Joe Busch said, 'When Jimmy Smith threw his chair and shoved Kanarek away, it probably was the final straw which influenced the judge.' Judge Le Sage once again listened to a motion from Jimmy Smith to relieve his counsel. A Pasadena attorney named Charles Hollopeter was brought into the courtroom to talk to Jimmy Smith. Kanarek maintained that his client was distraught and could not make a clear choice to have him relieved.

'There is no foundation for the court to proceed,' Kanarek argued vehemently.

'You are again instructed to be seated, Mr Kanarek, while the court inquires of Mr Smith,' said the judge.

'Well, your Honor, I most certainly do object under *Cooper versus the Superior Court.* . . . '

The judge listened to the argument and said, 'You are again instructed to be seated.'

Then the attorney argued that the court's apparent decision to remove him would violate his client's rights.

'Now if your Honor *deems* that I am being muzzled, and that I am being gagged by the sheriff, if your Honor shall deem that this record is to exist, that this record can reflect that I am being muzzled as if the bailiff here restrained me, and put me down, and strapped me to the chair, and put a gag on me, if your Honor deems that this record shall so reveal that, then, of course, I think your Honor gets the point of what I am driving at. May that be deemed, your Honor?'

'The record will not show that at all, Mr Kanarek!' said the judge, a bit shocked. 'The court is at all times willing to hear you in full! And I again request you as a member of the bar to be seated while the court proceeds.'

The attorney argued still further that the judge had no right to inquire of Jimmy Smith as to replacing his lawyer. Finally Kanarek said, 'May it be deemed that I have been gagged and that your Honor has used the full force, that your Honor has used all of his power inherent, statutory, constitutional, explicit, in keeping me from using the English language, keeping me from speaking. May *that* be deemed, your Honor?'

'The record will not so show, Mr Kanarek!' said the judge. 'The court is just making the request that you be seated and the court is advising you again that you will be extended the fullest opportunity to argue the matter.'

'That isn't the point, your Honor.'

'And you are again kindly instructed to be seated.'

'Well, your Honor, I don't personally wish in the sense of the word to fly in the face of the court's order, but I deem, based upon my study of the law, that your Honor is proceeding illegally . . . . So all I am saying is, let it be deemed, so that the issue will be framed. The issue will then be framed for minds wiser than ours. That's the theory of our appellate courts. They are supposed to rectify mistakes and tell us when we are wrong . . . . May *that* be deemed, your Honor?'

'No! The court can't make that order because that is not representative of all of the factual situation here, Mr Kanarek.'

'Your Honor certainly has the power to have the bailiff take me into custody. Your Honor has that power. I am not interested in flying in the face of the court's order. I am not interested in antagonizing the court.'

'Mr Kanarek,' said the judge patiently, 'the court is not in any way antagonized by your arguments here. You are again instructed to be seated while the court proceeds further with this hearing.'

'Well, your Honor, may it be so deemed? Because if I keep on doing this, will there be some point when your Honor is going to have the bailiff take me into custody?'

'You may be seated, Mr Kanarek. The court does not intend to make that order. You are instructed to be seated.'

'Well then, your Honor, may it be deemed as I have requested, your Honor? I don't . . . I mean I wish to . . . I wish to obey all the court's orders that are valid. But most respectfully, your Honor, it is my most respectful contention that it is not a legal order.'

'Mr Kanarek, again the court reminds you that you are on your feet and addressing the court contrary to the express order of the court!'

'Well, your Honor, since your Honor is determined that your Honor will not arrest me, then I have no alternative, your Honor. I want the record to reflect, however, that it is my position, that I am *deemed* arrested, *deemed* that an Order to Show Cause In Re Contempt has been issued, *deemed* that the court has used all of its powers of gagging me and so forth, because the reason this is so vital, your Honor, is because what Mr Smith says is potentially usable by the district attorney against him, plus the fact that he has the right that I have psychiatrists examine him.'

'You have made the motion a number of times, have you not, Mr Smith?' the judge finally said to the defendant.

'Yes. I have made the motion to relieve Mr Kanarek.'

'And you still wish the court to grant you that relief?'

'That is my wish, your Honor. Please.'

'The people wish to speak on the motion?'

'Your Honor, I object to that!' said Kanarek.

Judge Le Sage finally said: 'It has become evident that there is a growing lack of confidence and communication between the defendant Smith and his counsel. This lack of confidence has led to physical encounters between the defendant Smith and his counsel in the open courtroom . . . . The court is of the opinion, and finds, that the constitutional right . . . '

'Your Honor!' said Kanarek.

'Just a moment . . . to counsel will be denied to the defendant Smith if he is required further to be represented by Mr Kanarek. Mr Kanarek is hereby relieved.'

The order was entered on April Fool's Day, 1969. The next day the judge ordered the joint trial split in two.

When Phil Halpin read of Judge Le Sage pleading with the attorney to sit and be silent, he reeled in his chair. The prosecutor was dizzy, hot, disoriented.

Deemed to have sat down. *Deemed* to have been silenced. It suddenly struck the prosecutor. They had all gone over the brink into madness. The physical fact was meaningless, only the *words* mattered. A man need not sit if it was *deemed* that he sat. He need not speak if it was *deemed* that he spoke. Lawyers could stipulate to anything. They could deem that lie was truth, that fantasy was real. Who could refute it? Only the words mattered.

He lay in his bed alone in the darkness one night and reviewed his dead dreams and the wreckage of his life much like a young man named Jimmy Smith had done exactly six years before when Jimmy was exactly his age.

He was a failure in his personal life and his career. He wondered if his marriage had been sacrificed to the whorish vocation he had pursued so determinedly. He had worked fifteen hours a day, seven days a week, to get through law school. He was sick, racked with a flu virus and a relentless lung-tearing cough. He thought about getting blind drunk until he passed out, but was too weak to get up from the bed.

I tried to tell them, Halpin thought feverishly. Every judge was the same. They just didn't understand that no defendant is going to cooperate with his destroyers. You couldn't make the defendants submit by giving them whatever they wanted. He couldn't make the judges understand that. And they wouldn't believe that the antics of the defense amounted to a clear-cut attempt to stop the system itself. How many judges had told him, 'I can *handle* it, Mr Halpin?'

He thought of the good offer he had had to go into private practice. He had passed it up for this case. He had been obsessed by this case. And it wasn't just the murder. True, he had grown up in the same Glendale neighborhood as Karl Hettinger and knew many of the same people. True, he had two girls like Ian Campbell had. True, he hated what Kanarek and Maple were obviously trying to do to Hettinger. But in the end he was totally incapable of knowing who were the knaves and who were the fools. What he really hated and feared was what this case had done to his conception of his profession, and himself.

They made movies and wrote books about courtroom drama. It was all a hoax. There was no drama. Not in a real courtroom. It was all a cruel joke and an incredibly silly thing to devote one's life to.

'We're not remotely concerned with a search for truth, he thought. The advantage to the defense is the passage of time and the courts permit it and pervert justice. Who respects the system? Not the defendants surely. Not the public. Not *me*. God knows!

When lawyers were permitted to obstruct and even stop this pathetic process, then it was not even madness. There was some dignity in madness.

Eyes watering, trembling with chill, unshaven, he stared at himself in the mirror, mouth open, and imagined instead the baggy-eyed bulldog countenance of Kanarek, mouth agape, saying: 'Not so, your Honor! Not so!'

In the dark, in a pool of sweat, he broke into a coughing spasm. It was so bad he felt his lungs would rip apart. He struggled up and staggered into the bathroom. Even the judge had the flu. It was going around. Then he realized who had it first. Where it had come from. Irving Kanarek!

Halpin bummed around for one year and lived off the eight thousand dollars which represented his life's savings and total material worth in the world. He returned to the District Attorney's Office the next year when he needed a job. He decided it would be better than pumping gas, and he knew nothing else.

Phil Halpin didn't talk much about the case when he returned. When pressed, he would only say, 'At the end, I would've made *any* deal with Powell and Smith if I'd had the power. I would've let them go. Dropped all charges. Released them. If only I could've put their two lawyers in the gas chamber.'

Irving Kanarek filed a fifty-four-page declaration relating the nature of the case: 'In March of 1963 Officer Ian Campbell passed away in Kern County . . . ' He stated his years of involvement as Jimmy Smith's counsel. He asked that the court pay him $200,000 exclusive of expenses incurred.

Judge Dell granted him a fee of $4,800 and wrote:

> It's the intent of the court to discharge in full any and all claims made by you in connection with your services pursuant to Section 987a, penal code, in representing Mr Smith. It's the court's position that your negotiation or retention of such warrant will constitute an acquiescence, and of course, termination of responsibility, notwithstanding the fact that you claim substantially larger amounts.
>
> Very truly yours,

There was considerable jubilation when Irving Kanarek was gone. Public Defender Charles Maple immediately altered his own tactics and it looked as though the jury would be selected and the trial would finally begin.

But Charles Maple was *not* jubilant. He thought often of his indefatigable colleague, and actually missed his eccentric ways. The two old file cabinets containing stacks of Kanarek's documents were removed from the court, there was no bulging battered briefcase full of law books and vitamin pills and even an occasional sandwich. Maple recalled the time he went to lunch in the unkempt bachelor's old red convertible so crammed with books and papers there was hardly room to sit and listen to Kanarek's discussion of vitamin therapy.

It was courtroom legend that Irving Kanarek had been an engineer and the inventor of an exotic rocket fuel from which he drew royalties, enough to sustain him. Perhaps it was true, thought Maple, but it was clear that the dogged little lawyer was a poor man. And his payment by the court angered Maple most of all.

'I've never argued that he was a good trial man,' Maple said. 'A trial lawyer has to think on his feet and conduct himself with proper decorum, and can't dwell too long on a point without losing his effectiveness. But I truly believe he's a good appellate lawyer. He sees issues, and given time, writes about them pretty well. And by God, he gives all he's got to a case he believes in.

'I, as a public defender, got my paychecks, but Irving's in private practice. There isn't a lawyer in the country who would've devoted every day to a case for eighteen months with little if any thought to financial remuneration. Bill Drake was actually in court less than two weeks altogether when Judge Alarcon fired Kanarek. Drake was paid forty-five hundred, Irving Kanarek was paid forty-eight hundred. I would guess about one dollar an hour for what he put into this case.

'If he's insensitive and obstreperous, a judge with backbone can punish him through contempt proceedings. If he persists, jail him. Don't punish him like this. Not in this cowardly underhanded way. Forty-eight hundred dollars? For his years of work?'

It was not laughable to Charles Maple when his colleague left the courtroom on that last day. Maple helped him cram his documents into the slumping briefcase. Kanarek made several trips to the old red convertible before he had it all. He was sweating when he made the last trip, mouth open, jowls sagging, as he puffed out of the courtroom. He was carrying books in both arms. He looked back, blinking like a long-suffering pit bull. The books were dreadfully heavy. He walked slowly under the great weight of the law.

SEVENTEEN

The star witness was by now convinced he had found the solution to his problems. He would escape. He would run away to Oregon and go into farming there. An old friend from his Pierce College days had done it. The friend was lamed by polio and yet he had the courage to strike out anew and make good in farming. It was the thing he had always wanted. If his crippled friend could do it, why couldn't he, a healthy man, also do it?

And what was the use of remaining here? Waiting. The retrial hadn't started in 1967 as they had promised. It had not started in 1968 as they also had promised. The pretrial motions might drag on into 1970. Or forever. They couldn't bring him back. If they tried to extradite him he'd fight them.

It was impossible to stay here. He had applied for a job driving a school bus. He thought he'd like that. He liked children and everyone always said how patient he was with them.

But he didn't hear from the city of Glendale about his bus-driving application and assumed they must have checked on him and learned he had been a thief. That must be it. He never considered any other possibility. He had been a fool to put in the application. When you were once a thief no one could trust you. He didn't blame them. He deserved it. He wouldn't bother filling out any more applications. He'd just sell his house and move to Oregon, to a new life for all of them.

So he put his house on the market and people came.

'Karl, you've got the price too high,' his wife said in agreement with the realtor. 'You've got to be a little flexible. Some of the offers are good if you'll just come down a little.'

'No, I think it's a fair price and we should stick to it.'

'Karl, I'll follow you wherever you say. If you want to try Oregon I'll go. But you'll never sell the house if you won't come down on the price.'

'No, I think we should stick to our price.'

'Are you sure you really *want* to go to Oregon?'

'Of course I'm sure. Do you think I wanna mow lawns all my life? Do you think that's all I'm good for?'

'I think you could do anything you wanted. If you'd only try. If you could only develop a little of your old confidence.'

'I *have* confidence, Helen.'

Once they actually drove to Oregon. It was to be a holiday as well as a business trip to see the country. On the trip he drove out of his way to the town where Adah, now remarried, lived with her husband and children. He decided he wanted to see her and to see Ian's children. He drove through the town but at the last minute changed his mind. Despite Helen's protests he drove on. It would be better not to. It was a stupid idea in the first place. What made him think of it? Why would they want to see him? Of anyone in the world, why would they want to see him?

When they returned, Helen said, 'Let's take the last offer. It was a good price. Let's get out of here. Let's go where nobody can ever drag you into another courtroom. Let them turn Powell and Smith loose. Who cares? I don't care. Let's get out.'

'I want to, Helen.'

'Then lower the price so we can do it.'

'I'm sorry, Helen. But I . . . I think it's a fair price.'

He occasionally saw policemen from the old days. Once he saw his old vice squad partner, John Calderwood. Calderwood waved to him and looked as though he might want to talk. But Calderwood was laughing. It looked like he was laughing. He knew what it must be. The pension. That was it. They were all laughing about that. They all thought he'd cheated the city out of the money. They probably said things. What right did a thief have to a pension? What right did a . . . coward have. He believed some of them called him that. He remembered that once Calderwood told him he had invited Ian and his wife over to dinner just before it happened. There had to be a reason for Calderwood telling him that. He wouldn't just say it. He was telling Karl he thought Karl had killed his friend. That's what he was saying.

There were so many things to worry about. To fear. There were the girls at the bank. He knew they wondered. It was clear in the way they looked at him. Why would a healthy man in his early thirties be cashing a pension check? They probably called him a cheat or worse.

The pastor of the Lutheran Church came to his home. He tried to talk Karl into coming to services. The word of his depressive state, more severe than it had ever been, had gotten to the minister. Karl refused. He had little faith in religion. He went only once when his daughter Laurie sang in the choir. He would go anywhere for his children.

The desert racer frightened his friends and family. They became afraid for him.

'Take it easy, Karl,' his childhood friends Terry McManus and Ray Henka told him. 'This racing is for kids. You're ten years older than anybody out there.'

'I'm careful,' he answered. 'Look at my football helmet. Look at my shin guards and shoulder pads. Did you ever see a more careful rider? Or a sillier-looking one?'

They smiled grimly and watched him race his old battered 100 cc Hodaka which could not possibly compete with the newer more powerful bikes of the young desert racers. He was almost old enough to be the father of some of the youngsters who competed in the trail bike class. He rode thousands of miles on the Mojave Desert. He seldom missed a Sunday. He rode from Twenty-nine Palms to Parker Dam. He rode scrambles. He rode any and every kind of desert race.

Henka and McManus followed him into racing. Now that he was not a policeman, he was seeing more of the old friends from his college days, like Bill Wittick, who gave him a racing helmet with the words 'One Lap Karl' painted on it. But none of his friends had the daring to stay with him as a rider. They were much too old and wise to take the risks the youngsters and Karl Hettinger would take.

He rode his little trail bike to a novice trophy at Red Rock Canyon and he gained a little confidence. He raced from Barstow to Las Vegas. He raced anywhere he could. Helen was his pit crew. On Sundays he was a different man astride his sliding, bucking machine, the sand in his face, the smell of hot gasoline in his nostrils. He could leave the other self far behind.

Bill Wittick one day said to him, 'Karl, you're too good a poker player not to know that you got nothing left after you play your hole card.'

'What's that got to do with motorcycles?'

'What I mean is, you shouldn't try to go through whatever you're going through by yourself. Why don't you tell Ray or Terry or me what's bothering you? What we can do for you?'

'Nothing, Bill. Nobody can do anything for me.'

'Is it the trial coming up? Is it real bad, Karl? The anticipation?'

'I told you, Bill, nothing's bothering me,' he said and actually trembled at the mere mention of the trial.

'You're too strong and stubborn for your own good,' Wittick said. 'If you were weaker you'd let us help you. The strongest scrambler needs a pit crew, sweetheart.'

But his close civilian friends had no more success than his close policeman friends. He wouldn't talk to any of them. He couldn't. He could only roar across the desert on a little motorcycle. For a few hours each week he could outrun them all. All the ghosts. All the devils. All his pursuers.

There were the infrequent camping trips when he would join his other friends for a day during the week. Officers Jim Cannell and Stew James got weekdays off and they would have Lake Isabella practically to themselves.

Occasionally there would be a spark, just a tiny one.

'Stew, bring the milk for the baby,' James's wife said.

'I can't. I'm doing two other things.'

'What's that?'

Karl answered for his friend, saying, 'Sitting and chewing gum.'

It was a small joke, but it was a bit of the old Karl Hettinger returned.

Once he even got blind drunk with Dick Howard when Stew James, who was a Boy Scout leader, got lost in the mountains, and they had hunted him frantically all night. When he showed up the next morning they got drunk in relief, the first time Karl had been drunk since college.

Then there were evenings in camp when other policemen were there and Cannell told his autobiographical stories without pause:

'And so my mother was a Hollywood girl and had to get married about nine or ten times and I ended up in the McKinley Home for Boys along with our good friend and colleague Officer

Stew James sitting to your left or to your right depending on where you sit around this lovely fire. . . .' He would pause only to drink beer.

'And you talk about precocious. I ran away and hitchhiked to El Paso and bought a phony I.D. and ended up in the big towns all over the South. Nobody gives a shit about you down there. And I was a cab driver in St Louis at fourteen, a pimp at fifteen. I been shot. I been everywhere, done everything. I went to the South Pole in Operation Deep Freeze just to say I been there. I've won the game. I believe when you're dead, eight people stand around you and you may as well've never been there. So you gotta live, my friends. Some of these young coppers today never been anywhere, never did anything. We got two in Hollywood that ain't ever been laid!'

His wife, the Jehovah's Witness, at this point left her husband and went to the tent.

'These new kids we're recruiting nowadays are scary. Scary! Either super-cop-type goons or just the opposite.'

And then Cannell realized he was talking about police work and that was a subject they never discussed on camping trips so he changed the subject without taking a breath. 'I got some records. You should hear these records, you vulgar bastards,' he said loudly for the benefit of the Jehovah's Witness, who felt policemen were vulgar.

'Who's talking about records?' one of the policemen asked.

'I am. It's a good thing to talk about and I don't care how you feel about classical music. These symphonies would turn you on. I mean turn you *on*.'

And on he went. Keeping things going, performing always. They all enjoyed Cannell, especially Karl. But Karl was reminded of something, of another broad-shouldered policeman, a quiet one, who enjoyed classical music.

Karl would usually leave them at dusk to make the long drive home. He didn't like to camp overnight unless his family was along.

'Which way you going home, Karl?' Cannell would invariably ask.

'Think I'll go home the river way, the Bakersfield way,' Karl would invariably answer, and Cannell would himself become depressed to think that his friend always chose to travel by way of

Bakersfield, going close to where it happened – to a place which looked in the winter, at night, as desolate as the moon, especially near the foot of Wheeler Ridge, where onions grew.

When Karl was gone, Cannell said, 'Helen told me they once went to Bakersfield to find the tractor driver.'

'Tractor driver?'

'The black guy. The guy who ran with him that night. He went to visit him but couldn't find him. I wish he wouldn't always go home that way.'

Stew James, the one they called the worrier, looked into the campfire with his anxious, worried eyes and posed a question to all of them.

'All right. Six years ago Powell and Smith killed Ian Campbell. They didn't kill Karl Hettinger. *Who* is killing Karl Hettinger? That's what I can't figure out. That's what I want to know.'

And all the faces stared at each other in the firelight, but no one could answer him.

It had been a hard day and he was glad his family had gone to bed early. He went to the kitchen for another beer. He had to sleep tonight. He had to be fresh tomorrow. It was the most important day of his life. Then he felt the diarrhea coming on just thinking about it. In the kitchen he saw a part of the newspaper which he hadn't read tonight. That was unusual. He usually read the newspaper through from beginning to end. He began reading absently as he opened the beer. It was an article of the year in review done by a very hopeful reporter who believed the new decade was bound to be better than the last and see the end of riots, cult murders, assassinations. He decided the reporter was a fool.

Then he noticed an article on Apollo 12 which was scheduled to go next week. The picture showed the photos of the moon's surface taken by Apollo 11. There was something about these photos which made the gardener look at them carefully. The surface was bleak and gray, flat and desolate. It was lonely, so lonely. It looked like a place where the dust would choke you and the wind would howl in your ears in the darkness. He sat down at the kitchen table. He saw an article about the naval hearings of last January, about the surrender of the USS Pueblo. He looked at the picture of Commander Lloyd Bucher. It was a kind and weary face. He felt very strange while he read. It was about a high-ranking naval officer condemning Bucher for surrendering the Pueblo, for not fighting to the death. Suddenly he was sweating and cold and trembling. He had to stop reading. Reading was making him dizzy. He threw the paper in the trash and got a blanket to put around himself.

Tomorrow would be November 6, 1969. The people would make the final argument to the jury in the Jimmy Smith trial. It would then be over. It would really be over. Then they could decide whether to give him death or life or something else. He didn't care what they did. He didn't care if they released him. He only wanted it to be right. That everything would be right. He only wanted it to end. That's all the gardener wanted. That's all he dared hope for.

EIGHTEEN

'It was a different ball game this time around,' the prosecutor complained. 'The statements weren't admissible anymore. I couldn't get in any of those copouts to Pierce Brooks, those terrible damaging statements that showed what kind of people they were.'

Pierce Brooks, retired L.A.P.D. captain, now chief of police in Springfield, Oregon, just sat in court each day, having been subpoenaed from so far away, and shook his head sadly at the reversal which brought him back to Los Angeles.

'We could've just put Karl Hettinger on back in 1963,' he said, 'Karl and the coroner. A two-witness trial. One week. If only we'd known what the Supreme Court was going to do.

'I was just trying to be thorough. To get at *all* the truth. Why don't they tip you off when they're going to make monumental changes? And when they do, why does it apply *ex post facto* to cases already tried? Just because they're still on appeal? That doesn't make sense. *Ex post facto* isn't legal in any other human endeavor. Why?

'They criticized *me* for getting too much of the truth. For too many confessions,' said Brooks. 'We're only here now to protect the court record, to say the right things so a higher court can't reverse the case. Nobody cares about the truth.'

In the early summer of 1969 the chief witness against Gregory Powell was a different man from the one who had testified in the emotion-charged 1963 trial. His husky voice was stretched to a rasp. He stared lifelessly at the wall in the rear of the courtroom. His testimony was flat and riddled with: 'I don't recall.' He had managed to retain enough of the salient points to be what former detective Pierce Brooks would still say was a fine witness. But the detail had all vanished, along with names, faces, other events from his past. It was hard for him to believe he had ever been something other than what he was.

'Would you say,' asked Deputy Public Defender Maple, 'that the trousers they wore that night were darker than the jackets?'

'I don't recall.'

'Were the jackets of a black leather type?'

'I recall them being a leather type. I don't recall if they were black or brown.'

'Did they have collars? Did they come up close to the neck line?'

'I don't recall.'

'Do you recall whether they were buttoned or zippered?'

'I don't recall.'

'Were they long-sleeved?'

'Yes, as I recall.'

'Were they open at the neck?'

'I don't recall.'

'Would you be able to tell me, sir, if Mr Powell had a shirt on beneath it?'

'I don't recall.'

And still later:

'Did you tell Officer Brooks that Jimmy Smith was facing you with a .38 in his hand but you weren't sure which .38 it was?'

'I don't recall saying that.'

'Did you tell Brooks that Powell stood to Smith's right, facing Campbell?'

'I don't recall.'

'When you were being interviewed by Brooks, were you aware that Mr Powell had been arrested?'

'I don't recall.'

The witness sat at the stand, his sunburned face like a baked apple, and listened in resignation to the district attorney explaining the well-worn chart to these jurors, as it had been explained to other jurors.

'The green arrow points to a farm area called the Coberly West. The blue arrow is the Mettler ranch. The white arrow is another Mettler ranch. The light blue arrow is a place called the Opal Fry ranch. The *red arrow* depicts the place where a police officer died.'

Red arrow, thought the witness. There was something about a red arrow once. What was it? The witness scratched his memory and planted the seed but it was not like scratching friendly soil. Most of the time his memory would not bear. The soil of his mind was inhospitable and barren like the choking dusty soil of the southern San Joaquin Valley when no water is brought to it.

Then he remembered! A red arrow. It had made him cry once. He had seen a big one on a billboard. He thought it made him cry but maybe something else made him cry. He never was sure what made him cry. He had to force himself to think of something else. Quickly, what? The yard. Yes, the one with the stately pine he loved. No, tomatoes. That was it. Fat. Red. Meaty. Dew-wet and succulent. Acres and acres of tomatoes. His own land and he kneeling in the midst of tomatoes. In the sunshine, and it never turned to night. He loved the tomatoes. The tomatoes were so quiet.

Charles Maple believed that Gregory Powell may have been a victim of a brain injury. And he never thought of brain injury without thinking of Donald, the son of an old friend who suffered an injury to the dominant side of his brain at age seven. Charles Maple handled legal matters for the boy and helped him be placed through the years in one institution after another, pitying the explosive, suffering boy, even taking him into his own home at some personal risk.

He believed in his client and was now glad that the Public Defender's Office had given him entirely to the defense of Gregory Powell. The case had become an important part of his life after so many months of back-breaking work.

Charles Maple was accustomed to caring for people with special needs. Both his wife and pretty daughter were partially crippled by hip disease, though, ironically, his other child, Charles Junior, was an accomplished ballet dancer. Maple was an elder in the Presbyterian church and he proved in his life that he was totally committed to the tenets of faith. But he was aware that there could exist a paradox of injustice when an advocate had *too* much faith.

A video-taping was done in 1969 through Maple's requests, at the UCLA Neuropsychiatric Clinic. It was the first time that a video-taped interview involving sodium amytal was done in Los Angeles and later introduced in court.

His opponent, Deputy District Attorney Joe Busch, said, 'Oh,

378

the sodium amytal is a lovely defense tactic all right. It enables our boy to tell only the part of the story he wants, without being subjected to cross-examination.'

A jury eventually saw the interview. Gregory Powell was on a bed in a room with draped walls. The bottle containing the drug was gradually dripping into his arm to achieve the desired effect. There were two nurses and three doctors in the room. During the interview the defendant spoke in drunken, halting speech to a Dr Suarez:

'What happened in the car?' the doctor asked the patient.

'There were several colloquies in the car,' said Greg thickly.

'What?'

'Colloquies,' said Greg, annoyed at the doctor's unfamiliarity with legal jargon.

'What did you discuss with Jimmy?'

'Nothing. Jimmy was my subordinate,' said the patient testily. Greg then smiled drunkenly and said, 'I told Hettinger to get his ass in the car. If he didn't he was gonna be in reeeel trouble.'

Then Greg closed his eyes when talking about the shooting. He wore a tiny gold cross around his throat, on a very short chain so that the cross was at the base of his long neck. The gold cross was clearly apparent to anyone viewing the taping.

'Just explain what happened,' said the interviewer, as the patient's speech got more halting.

'Hettinger got out at my demand and moved past . . . Hettinger . . . past Campbell . . . so the . . . outside Campbell, and I raised my gun and shot Campbell. . . . '

'What?'

'I walked around the back and I raised my gun and I shot Campbell.'

'How did that happen?'

'I don't know. The gun had gone off by accident before . . . It had a real funny hammer and trigger . . . pull it halfway back and let it snap forward and it would fire . . . I don't know . . . '

'Well, how did you feel about Campbell getting shot?'

'Felt pretty sick.'

'What do you mean?'

'He was a nice guy. He'd been joking with me about the marines and everything and I just . . . uh . . . he was one hell of a nice person.'

Later, when asked if he meant to shoot Karl Hettinger, Greg would deny that he had, saying, 'Oh, hell, I could shoot a jackrabbit in the head thirty yards away when he'd jump out of the brush, no sweat.'

During the final address to the jury, Deputy District Attorney Raymond Byrne, the co-prosecutor, commented on the tape which had been introduced by Maple over the objections of his client Gregory Powell, who was unsatisfied with it.

'At the end of the video tape, may I inquire,' said the prosecutor, 'would a person call for his lawyer? I would perhaps suggest that Mr Powell had reasons, and he can certainly out-psyche the psychiatrist. And I would wonder why Dr Suarez never asked *anything* about the Little Lindbergh Law. The video tape, I would suggest, was a work of art. Hopefully it may even be put in for an Emmy Award.

'I would submit at this time that Mr Maple, for his client, will inquire of you whether or not it was not Mr Powell who did the shooting, but perhaps his mother – that familial concept. Or perhaps the shot was really fired by defendant Powell's sisters. Or perhaps by his brother. Or by the parole officer. Or by the institution. Or the state of California. Or by the world. Just so you lay the blame of the dead man, the killing of that person, on somebody *other* than defendant Gregory Ulas Powell, because he is, of course, the only one here who is charged with murder.

'I submit that you tell this defendant, Gregory Powell, by your verdict that his act that night was nothing less than a calculated expression of accumulated rage that resulted in the cold-blooded, willful, premeditated killing of Ian Campbell, a police officer and human being. I thank you very much.'

In June 1969, Helen Hettinger worried as she watched her husband stare at the television news without hearing. Gregory Powell had once more been found guilty. The penalty trial too was finished and the jury was expected to bring in a swift verdict of death.

'Come on, Karl. Let me make you a sandwich.'

'No, I'm not hungry.'

'You didn't eat any dinner at all.'

'No thanks, Moms.'

'Well, the Powell trial's over. You won't have to go again on that one.'

'No. Once more with Smith. That's what the D.A. promised me the other day and maybe once more in Smith's penalty trial. Maybe two times more. Maybe.'

'Maybe they'll reverse it again,' said Helen angrily and was instantly sorry she said it.

'I don't think I'll ever do it again, Helen,' he said, making his wife afraid.

'Let me fix you a sandwich or something,' she said.

It was later that evening when the phone call came.

'I'm sorry, Karl,' said the voice on the telephone. 'The jury hung. There was a holdout. It's a mistrial. We'll have to try him again on the penalty part of it. Damn! We'll have to select a new jury and of course they'll know nothing of the guilt trial so we'll have to go into the whole thing in detail again. Damn, I'm disappointed.'

The new penalty trial would be as thorough as a guilt trial. All the witnesses were once more subpoenaed. It lasted four and one half months.

Prosecutor Raymond Byrne was a tall man with a long face and a crooked nose that bent to the left. He was forever pushing his drooping glasses up. He began staying during recesses or returning early to talk with Gregory Powell, who seemed to enjoy the conversation and the cigarettes and the sunflower seeds the prosecutor gave him.

On August 2, 1969, Byrne learned that it was Gregory Powell's thirty-sixth birthday. When Byrne came back from lunch that day he carried with him a cupcake.

'Happy birthday,' said Byrne, giving it to the defendant.

From then on, the defendant talked freely to the prosecutor about life on Death Row, the thousands of things they planned, how they made home brew and whiskey. And his recent conclusion that homosexuality was at the root of his behavioral problems. And of his ambivalence toward his father and particularly toward his mother, who ruled the family in his absence.

One day Byrne asked the defendant, 'How do you *really* feel about those kids, Greg?'

'What kids?' asked Greg.

'Campbell's kids,' said Byrne. 'And about Campbell?'

'To tell you the honest truth, Mr Byrne, I feel nothing.'

'Nothing?'

'Nothing about them. Any of them. But I *will* send them money if I ever make any. Because I said I would.'

There it was, Byrne thought. The sociopathic personality. Greg looked at the tall prosecutor apologetically. 'It was as though he *wished* he could say he felt something about the murder,' Byrne explained. 'Wishing he could feel something, because he knew other people did. It was the first time I ever truly understood sociopathy. It just wasn't there. He just had no remorse in his makeup.'

Deputy District Attorney Sheldon Brown was chosen to help prosecute Gregory Powell in the new penalty trial. He was dark and slim and walked as though his feet were always sore.

Brown found Greg ingratiating. 'Good morning, Mr Brown. Did you have a nice evening?'

Public Defender Charlie Maple, always resourceful and thorough, admitted to a completeness compulsion. When he had first learned that Karl Hettinger had been pensioned off the police force, he subpoenaed the personnel and pension files of the former officer. He had been astonished to learn that the former policeman had resigned under fire on a shoplifting charge.

The public defender read all of the psychiatric reports. He was by his own admission more 'personally involved' in this case than any he had ever tried. He called Dr George N. Thompson, whose report was so unlike the others. The report that said Karl Hettinger mentioned having been involved in a 'gun battle' with two robbery suspects before he started shoplifting.

'You made a diagnosis,' said the judge to Dr Thompson, 'that there is some type of a character disorder involved in this case, which some people might call kleptomania? And you have some mixed feelings as to whether it amounts to kleptomania or not. Is that right?'

'Well, I wouldn't refer to it as mixed feelings, but that I was attempting to, for the Pension Board, clarify the diagnosis to some extent. And so I discussed it further. That was an afterthought apparently, that one might consider a diagnosis of kleptomania. However, I thought it was probably not true kleptomania.'

'Do you have any recollection,' asked Prosecutor Sheldon Brown, 'in reading those files, of anything being inconsistent in the file with relation to what Mr Hettinger told you in the interview?'

'No.'

'If the court please, I have here a department pension report, city of Los Angeles, and I ask it be marked for identification. Does that report appear to be one that you examined during your preparation of the report on Mr Hettinger?'

'Well, I can't recall,' said the doctor, 'if this was in the file or not. If it was, I can only say that I know this doctor was one of the examiners before the board.'

'If it *was* in the file then I take it that you would have reviewed it?'

'Yes.'

'Now you will notice in the clinical history of this report that it's apparently over a year prior to your seeing Mr Hettinger. It sets forth a statement of facts as to how Mr Hettinger's partner actually met his death.'

'Yes.'

'Was there a statement in this other report that he and his partner were kidnapped and taken in the vicinity of Bakersfield where his partner was killed and he managed to escape?'

'Yes.'

Charles Maple was anxious the day he examined Karl Hettinger on the inconsistency he had found in the Thompson report.

'In discussing the shooting out there in Kern County when Officer Campbell was killed, did you ever tell anyone that you were involved in a *gun battle* with two robbery suspects? That you and your partner were fighting the two suspects when one of the suspects shot your partner and your partner died?'

'I don't recall saying those words,' said the witness. 'Words were said to a medical secretary as to why I was there. She stated words to the effect, "Was it a gun battle or a robbery?" or something to that effect. I didn't wish to go into it with her, knowing I would be seeing a doctor shortly.'

'What doctor is that you are speaking of?' asked Maple.

'I don't recall his name.'

'Do you recall what time of year it was?'

'No.'

'And did you relate to the doctor your version of what happened up there in Kern County?'

'I don't recall.'

Out of the presence of the jury Maple argued to the court: 'It turns out that he has a prior history of theft that's concealed from the police department. Never disclosed when he became a policeman! These kinds of matters it seems to me do have a bearing.'

Deputy District Attorney Brown told his superiors:

'Maple believes Hettinger made an ego-enhancing statement to Dr Thompson about a gun battle to impress Thompson, so it wouldn't jeopardize his pension chances. Even though Thompson had prior reports on Hettinger and knew all the facts. It's absolutely absurd, but the judge let him call Thompson as a witness.'

'Now, sir,' Maple said to the doctor, 'in the course of that particular interview, did Mr Hettinger in substance and effect state to you that on March 10th, 1963, he began to develop an emotional problem and did he state that he was in a *gun battle* with two robbery suspects, and that he and his partner were fighting with the suspects when one of the suspects shot his partner and his partner died?'

'Yes, so far as I know, he said *something* to that effect,' said Dr Thompson. 'Perhaps not in those exact words.'

'Would you state in *substance* those were Mr Hettinger's words, in *substance* rather than exact words?'

'Yes. In substance. There may have been some alterations by the typist, but in general I would say that is what he said.'

'All right,' said Sheldon Brown impatiently, when it was his turn. 'When a girl is in with you typing an interview as it goes along, a good deal of discretion is left to her, clarifying what is told to her, is that correct?'

'Yes.'

'A typist just types in substance, in effect paraphrasing what has been said. Is that correct?'

'That's correct.'

'Is there any doubt in your mind,' asked Maple on redirect examination, 'that Mr Hettinger, in answering the questions in your interview, stated what we have referred to as the version recorded in your notes?'

'Not in a *general* way,' said the doctor, 'but I *do* think some of the wording may be an interpretation of the typist.'

'If you felt the typist didn't accurately reflect what was being said, you would have asked her to correct it, is that not true?' asked Maple, now seeming impatient with the equivocation of his own witness.

'I think not in this instance. I was really not concerned with the details of the shooting at all. Only the fact of it, you see?'

'Dr Thompson was one of the best known psychiatric witnesses in the country,' Sheldon Brown said later. 'We've faced him dozens of times. It was obvious he was going to hurt Maple's cause more than help it. Using Thompson like he did engendered great sympathy for Hettinger. Yet Maple seized on anything that might save Gregory Powell's life.

'It was strange,' Brown went on. 'Maple's an ethical attorney, but he so completely identified with Gregory Powell that he was ignoring physical facts. The Dr Thompson affair was a case in point. The whole ludicrous gun battle thing. It was obvious the doctor didn't know what the hell had been said on this issue and cared less. Maple said Thompson "dragged" the story of that ordinary childhood pilfering out of Hettinger, and Maple implied that Hettinger was some kind of lifelong thief.

'Christ, Hettinger had confessed to every one of the shrinks about this pitiable kid stuff. And in much greater detail. Maple just seemed to've *rejected* the other psychiatric reports as though to him they didn't exist. Thompson's report, he felt sure, proved Hettinger to be a lifelong thief and liar. I don't think even Gregory Powell bought his thesis. Nobody on the jury did, I'm sure.

'Maple had been on the case too long. He was giving money to Powell, getting him clothes, acting as liaison, confessor, psychiatrist to Powell and his whole family. He was even trying unsuccessfully to get Powell and his girlfriend married, that poor black woman Powell had conned into helping him escape and who got herself in trouble for it. He just set logic aside as far as his client was concerned. The devastating Lindbergh statement, all of it. He said Hettinger and Smith were both lying about that. He *believed* in Gregory Powell implicitly. I don't think any defendant ever got a more faithful defense.'

Charles Maple had very pale blue eyes and almost no eyebrows. The church elder had a black dot on the iris of his right eye and a powerful voice which was as much at home before a congregation as a jury. He gestured with his hands when he talked, and when he frowned, his hairless brows made the pale blue eyes penetrate. Maple had the oratorical style of an evangelist, but the Harvard lawyer did not damn, he defended. He offered forgiveness and could extend his compassion to all men, all but the one whose testimony had to be a lie.

The defender was positive beyond any doubt that Gregory Powell was innocent of a calculated murder, that he stood to die in the gas chamber because of one man. Charles Maple had no charitable thoughts for that man, and his voice trembled prophetically when out of the jury's presence he discussed the star witness.

'It's ironic,' Maple said, 'that Hettinger, like Powell, has been a sociopath all his life. He's been a thief since an early age. He reverted under stress to his preadolescent behavior. And he'll lie and manipulate to make himself look better, just like any guiltless sociopath. He said there was a girl secretary in Dr Thompson's office and that she took the gun battle story from him. It's a boldfaced lie. Why did he lie? I don't *know* why he lied, I only know that he did.

'Why did he steal? Because he might as well make it while he could. He'd been Chief Parker's chauffeur. His career was ruined because of innuendo. He stole because he felt it was owed him. He feels no guilt, not for stealing, not for being inadequate as a policeman during his confrontation with Powell and Smith. He's a true sociopath. The kleptomania thing was all an act. He doesn't know the meaning of guilt!'

Gregory Powell had a much more fatalistic view of the witness:

'In 1963 he was a shadow figure to me. Not quite real in many senses. In later years it was very odd, this man was about the most important figure in my life, and I knew hardly anything about him.'

Joshua Hill was one of the six men brought down from San Quentin Death Row by Gregory Powell to testify for him as a character witness. Hill was a young man who had shown a startling vocabulary when he was a baby, and despite his limited education, he had an adult I.Q. of 140. But Joshua Hill had not been toilet-trained until he was seven years old. He was known on the row as 'Batman' because of a homemade cape he wore. When the Batman mood would strike, he often challenged one or several guards and had to be tear-gassed and subdued. It was said by Hill's mother that the Batman's father had beaten him up countless times during the boy's formative years to make him tough. If true, the father had done his job well. Joshua Hill was very tough.

On two occasions Hill seriously assaulted other condemned men, fracturing the skull of the row's only Eskimo resident, and slashing the throat of another, though neither attack was fatal. Hill was darkly handsome, liked to pass for an Italian mafioso, and called himself Vito Giuseppe Cellini during his days on the street.

It was a terrible disappointment to some of the other inmates in the county jail high power tank when jailers found a handcuff key in Joshua Hill's mouth. His Green Beret training and natural fierceness had induced some of them to bet that even without a gun he could escape a courtroom if unmanacled. Perhaps the gamblers were even more disappointed than Gregory Powell and Joshua Hill.

In the spring, jail deputies had found Gregory Powell in possession of an Allen wrench which was used to try to remove steel screens from jail windows. Now, in the summer of 1969, after all escape plans had been thwarted one by one, it was a short-tempered Gregory Powell who was as usual being led from the courtroom to the jail elevator at day's end.

He saw an old man, one of the railbirds who spend their days as spectators in criminal cases, prowling the corridors and court-rooms of the Hall of Justice, Hall of Records, and County Court House.

This particular railbird was interested in Gregory Powell and came almost every day to see him. The railbird never spoke, he only grinned, and after a few weeks of this, Greg's head would spin on its swivel a dozen times each morning, glaring at the old man in the back of the courtroom. The railbird grinned like Death.

Greg began to complain about the spectator but was told that the railbird was harmless and had a right to be here. Finally the old man began following Greg when the guards took him in chains from the courtroom to the jail elevators at day's end, and

Greg started to listen for the shuffling footsteps behind him, and his head would swivel and he would peek between the deputies. The old man would be there. Grinning at him. Like Time. Like Death.

One very hot summer day, deputies made a written report that prisoner Gregory U. Powell went berserk at the elevator landing and tried to attack a seventy-year-old male Caucasian for no apparent reason. The prisoner was restrained, but continued screaming and spitting at the man. The prisoner was led away, beside himself with rage the rest of the night. The next day the old man was in the courtroom. Grinning.

'All of the confessions were inadmissible,' Prosecutor Sheldon Brown said. 'It was truly forgotten by now, even by counsel, what these defendants had once said in their confessions. Hettinger was a broken man, anyone could see that. Campbell was so forgotten he may as well never have lived.'

Sheldon Brown had been mildly amused at how Greg had played to the three blacks on the jury, telling in part how he had taught a black man to write. When the jury for Gregory Powell returned with a verdict of death in only six hours, one of the blacks told Brown that they only stayed out that long to make it look respectable.

Still another escape attempt was planned. It was to take place in the day room of the new county jail. Several court prisoners were thought to be involved, including another police killer, and, ironically, an ex-Los Angeles policeman turned saloon owner, on trial for the insurance murder of his wife.

Jimmy Smith was inexplicably transferred to the old county jail before the plotters were caught and the tools confiscated. They had been informed on. Jimmy was furious because his transfer made it look like *he* was the snitch. Jimmy decided it was an elderly inmate who went to the attorney room that night.

It seemed that somebody *always* snitched him off. But it couldn't have been the old man. The bulls knew too much. No, it had to have been somebody close to him. Jimmy was put in the old Sirhan cell and forever separated from his partner. At least I won't be with Powell anymore, he thought. At least he'd been spared that. Something good had come of it after all.

*

When Jimmy Smith had acquired a typewriter, Gregory Powell got a better one. In that Powell had testified under sodium amytal, Jimmy Smith also had to do it.

The interview was not taped, however, and was unspectacular, at times incoherent. There was one interesting moment. Jimmy's testimony was disjointed and self-contradictory, but during the crucial moment of describing the shooting, he said:

'The body was jerkin and his right hand was jerkin. . . . I'll never forget . . . I don't see the picture of Powell doin this. I don't see the picture of anybody doin this.'

Jimmy Smith during his interminable wait for a second trial kept a diary to allay his frayed nerves and ulcerated stomach. At times he wrote outraged declarations:

Hundreds of days in court with more to come! Ten different judges have tried to stick their hooks in me, but I have managed to last. And for what, I don't know. But I will never cower before any master! Nor bend to any threat!

At other times the defendant waxed poetic:

Poor fool that I am. Never again will friendly eyes caress my eyes . . . nor love sweet lips kiss me. No hand grip mine in a display of mutual love. I, the grave, have claimed you, Jimmy, as my love. Farewell, Jimmy! I'm your everything!

In the high power tank Jimmy Smith always got the gossip from the row. It was the same of course: the appeals, how the truly skillful jailhouse lawyers were doing with their own retrials, the sexual tidbits – who was doing it to whom – the new dudes who had arrived, the bad dudes to watch out for.

One new one to watch out for was a skinny youth named Robles who had slashed a sleeping motorist and a hitchhiker in separate assaults to get himself into prison in the first place. Neither died, but while in the Santa Clara County Jail, Robles became involved in a bizarre escape scheme. He convinced his cellmate to pretend to hang himself with a bedsheet. Robles would hold his friend up and when the guard came in to cut him down, they would both overpower the bull. Unfortunately for the cellmate, Robles hanged his friend during the rehearsal. When guards came in to cut the dead man down they found Robles giggling and singing, 'Hang down your head, Tom Dooley.'

Finally, in yet another prison murder, Robles fatally cut the throat of an inmate with a toothbrush/razorblade weapon, at last guaranteeing himself a trip to the row.

'Jumpin fuckin Jesus,' said Jimmy Smith when he heard the story. 'We got a couple here about as bad almost. One sucker jist don't care about nothin but jackin off and fightin the bulls. But that one you was talkin about reminds me of the guy that's down here now as a witness. Jesus,' said Jimmy, as he thought of the swarthy inmate who had carved lines in his own face beside the eyes and above the lips and rubbed in carbon black to make permanent cat's whiskers. He would occasionally come to jailers and say he was 'getting the feeling' again. He warned that he wanted to kill and should be returned to the row. They were all in module 2500, the high power tank.

'Looky here,' said Jimmy Smith. 'It ain't right, them lettin some crazy fuckin killers run loose like that, man.'

'Hell no,' the other con agreed. 'Ain't nobody gonna be safe. Somebody better get his shit together cause everybody's gonna have to be carrying blades all the time.'

'Jesus,' said Jimmy Smith, thinking of going back there. No one cared on the row what your crime had been as long as you left the other cons alone and minded your own business. But dudes like this one murdered cons without reason!

They both agreed. It just wasn't fair to the other condemned men. No one on Death Row would be safe as long as they had this killer among them.

Attorney Charles Hollopeter at first characterized his client as paranoid. He seemed to Hollopeter more concerned with jail conditions than with saving his life. He would only talk to his lawyer on the subject of certain motions: for hotter food, for having certain jail lights extinguished, for having other lights left on.

Jimmy Smith had twice attacked Irving Kanarek, but now for some reason Jimmy seemed equally unhappy with Hollopeter. That made it unanimous. Jimmy thought every lawyer assigned him was incompetent or worse.

Prosecutor Dino Fulgoni was a short, muscular, craggy man. A physical man, very different from the older, dapper Pasadena lawyer with his curiously homespun delivery, who had a way of putting himself on trial and coming off sympathetically to juries. Fulgoni believed his opponent, Charles Hollopeter, was as good a trial man as there was in the Los Angeles area.

But Fulgoni was not terribly pleased with his own star witness, who was willing to say only what he saw and remembered and no more. 'Hettinger was not,' Fulgoni complained, 'willing to make those small inferences we all make and must make without seeming equivocal. His testimony was cold, unemotional, too objective.'

It was true. The witness showed absolutely no emotion during the 1969 retrials. He was a different man to all of them: the defendants, Pierce Brooks, the dozens of civilians and police witnesses who had heard his demonstrative testimony in 1963. But if the witness seemed emotionless, his wife was not. Helen Hettinger broke out in a serious case of hives for the first time in her life. Her body was covered with itching maddening welts until the trials were over.

One of the days he stood in a courtroom corridor waiting to be called, the witness was approached by a black man who looked very familiar.

'What say, Karl?' said the bony farm worker with the scarred mahogany face. 'Remember me? Emmanuel McFadden?'

'Emmanuel? Is it you?'

'It's me.' The farm worker grinned. 'You look different, Karl. A little skinny.'

'Getting old, Emmanuel.'

'How they been treatin you, Karl?'

'Okay, Emmanuel. I stopped in Bakersfield once on my way to the lake. I looked for you.'

'Yeah, I heard there was a white man lookin for me. I jist figgered it was you. I hear tell you ain't a officer no more.'

'No.'

'You okay, Karl?'

'Sure, Emmanuel. Sure.'

Jimmy Smith's attorney believed Hettinger to be honest in what he said, but found him rather contemptible. 'A beaten dog,' was the term he used out of court.

As to the witness's stealing, Hollopeter had a less complex and psychological interpretation than Charles Maple had, but came to much the same conclusion. Hollopeter would allude to the old psychiatrist joke: 'Your son, madam, takes things because . . . ' 'Yes, Doctor?' 'Because he is a thief, madam.'

Hollopeter knew Hettinger was vulnerable to psychological attack but avoided the tactic, not because of any compunction, but because he feared it would engender sympathy for the witness. Hollopeter would say of the witness: 'A real tough guy never would've let this happen in the first place. He could've crushed Powell's skull with a jack or a hubcap. If he were a tough guy.'

The witness had no friends at the defense table, not at either trial.

Early in the trial, Hollopeter made the following comment at the bench:

'I have one criticism, your Honor, and I welcome the chance to state it. I do object to counsel referring to the witness as "officer." The man, I am informed, is not an officer.'

'Actually that's true,' said Fulgoni. 'I'll call him *Mister* Hettinger.'

In October, 1969, the witness was asked by Hollopeter:

'When you got back to the shooting scene did you see Mr Campbell?'

'Yes.'

'Where was he?'

'He was lying in a ditch.'

'Was he alive at that time?'

'I don't think he was,' said the witness, his voice mechanical, his eyes vacant.

'When you rode in the ambulance with Mr Campbell, did you notice any wounds on his body?'

'I just saw a lot of blood. Just a lot of blood.'

'Mr Hettinger, what is your present occupation?'

'I am in landscaping and maintenance.'

'You are no longer a police officer?'

'No.'

'When did you terminate your employment as a police officer?'

'In May of 1966.'

'How many times have you testified about what occurred on the night of March 9th and the morning of March 10th, 1963?'

'Six or eight times,' said the witness. 'I can't quite remember.'

'When you testify, does it bring back the events to your mind?'

'Sometimes they sort of fade away between trials. And then I get my memory refreshed,' said the witness haltingly, staring at the wall in the back of the courtroom.

'You told us you looked over your shoulder and you could see flashes that seemed to be gun flashes and then appeared to be teardrop-shaped flashes. Mr Hettinger, were you very greatly shocked by this event?'

'I would say that my state of mind . . . I was in disbelief . . . I couldn't believe that it had happened.'

'How old were you at that time?'

'About twenty-eight or so.'

'Were you in good physical condition?'

'Yes, I think so.'

'When you ran, you intended to escape if you could, didn't you?'

'I wanted to,' said the witness bleakly.

'Did you think you'd reach a place of safety?'

'I don't know what I thought.'

'Why did you slow down to look around?'

'I think . . . I was still in a state of . . . disbelief. I couldn't believe that this had happened. And call it a . . . curiosity, if you must. I looked back. I couldn't believe it.'

When the star witness was finished he approached the district attorney during recess. 'This is the last trial, isn't it?' he asked, but Fulgoni was talking with Hollopeter about something that had happened in chambers. 'Sirs,' said the witness, 'is this . . . ? Do I have to come . . . ? Is it really going to . . . end now?'

But they didn't hear him. They were arguing about Gregory Powell testifying for Smith, and Fulgoni was angry.

The witness was suddenly too tired to stand there. He left the two lawyers and went to his truck. He drove home, but his wife was at work. He wanted to tell her it was probably over. He didn't think he'd have to testify anymore. Not unless it was reversed again.

He put on his jeans and old shirt. It was too late to go to work, so he just went out to his own yard and knelt in the flowers. There were two brown snails feeding on his newly planted yellow gazaneas. He carried the snails across the yard to the street and released them.

Then he went back to his gazaneas. He found one more fiery than the rest. It was folded and tucked, elliptical. He touched its petal. He wanted to help it open. Then he looked at the little gazanea. It was like a fiery teardrop. He thought he heard a quiet whimpering behind him and spun around, but there was nothing. Then his eyes were clouded and burning and he realized he was crying. It had just happened. There was no warning anymore. That scared him more than anything ever had – that a man could just start crying without warning.

The time had come for Gregory Powell to make good on a long-standing promise to Jimmy Smith. Fulgoni knew what Powell was going to do. It was at this point obvious to the prosecutor. It could never work at a court trial but it could in front of juries of ordinary people, unsophisticated in the ways of law and legal stratagems.

Fulgoni believed it could work, in part because all the testimony had pointed to Smith being the ineffectual follower of an arrogant and dangerous leader. Then there was the mere physical appearance of both men. Powell was always strange-looking, could make his face cruel and deadly with little effort. Fulgoni argued at the bench for all he was worth.

'He's going to take the Fifth, your Honor. You know it. I know it. Maple knows it. Mr Hollopeter knows it!'

'I don't know it,' said Hollopeter. 'Your Honor, I won't dignify that statement that I'm acting in bad faith. I deny it.'

'If we could have a hearing outside the presence of the jury,' said Fulgoni, 'then if he took the Fifth . . .'

But Judge Shepherd finally decided that Section 913 of the California evidence code forbade the trier of fact to draw an inference from a privilege exercised in a prior hearing, that Jimmy Smith was entitled to call any witness he wished without a prior adjudication as to whether or not the witness would testify. Fulgoni was stopped. The code section was relatively new and there was no case law Fulgoni could rely upon to force a hearing to stop the witness from doing what the prosecutor knew he would do.

So Gregory Powell made good on his promise to his partner, became somewhat of a folk hero to the other Death Row inmates who knew what he was going to do. And six and one half years of court proceedings came down to just three questions with the responses known far in advance by almost everyone but the uninitiated jurors.

The first question by Hollopeter was: 'Mr Powell, how many times did you shoot Officer Campbell?'

'I object to that question as being asked in bad faith at this point!' said the raging Fulgoni.

'Overruled,' said the judge.

The defense witness conferred with his lawyer, Charles Maple, before answering.

The second question was: 'Oh the evening of March 9th, 1963, did you decide to kidnap two officers in the vicinity of Carlos and Gower?'

The witness conferred once again with his lawyer before answering.

The third question was: 'Did you retain Officer Campbell's gun in your possession until *after* the shooting?'

The answer was exactly the same in all three cases.

Gregory Powell stared at each juror, his blue eyes going flat and icy. His lips tightened, and Jimmy Smith said later it was just like Greg used to be, just like when he had a gun in his hand and cocked it. This time it didn't scare Jimmy, it thrilled him. Jimmy could have kissed him. At that moment, for the only time in his life, Jimmy Smith *loved* Gregory Powell, whose answer was:

'On the advice of my counsel and based on the Fifth and Fourteenth amendments of the Constitution of the United States I refuse to answer that question on the ground that the answer may tend to incriminate me.'

'I have no further questions,' said Hollopeter.

The defense could at last rest, and did. Fulgoni looked at the jurors, at some who recoiled when Gregory Powell left the stand and stared them down as he swaggered away.

'Damn!' said Fulgoni. 'Damn!' and he pounded his big-knuckled fist into his palm.

*

In truth, though, Dino Fulgoni was beaten by more than a trick. Ian Campbell had been slain almost seven years before November 6, 1969, when the jury returned the verdict of life imprisonment for Jimmy Smith. It was difficult indeed to bring in a verdict of death for a crime so far in the forgotten past. Time, which had always been Jimmy Smith's relentless foe, at last became his ally.

There was one last moment in the end of the penalty trial which some observers thought as important to the defendant as Gregory Powell, and as time itself. It was Jimmy Smith's final statement to the jury.

'If you are permitted to continue life, what do you plan to do?' Hollopeter had asked him.

'I don't know. Somethin constructive if I can.'

'I don't understand you,' said Hollopeter to his client.

'Well, I don't wanna make no promises I can't keep. Or say that I'm gonna do somethin or don't do somethin. It's very easy to say it . . . . I don't know whether I wanna live.'

'All right, you may cross-examine,' said Hollopeter to Fulgoni.

'There's one more thing I wanna say that I forgot,' Jimmy interrupted.

'All right, go ahead and say it,' said Hollopeter.

'I don't know as to whether I wanna live or not. I wasn't even brave enough to stop that guy from killin that man. I coulda stopped him. I didn't have the guts. I'm sorry he's dead. I can't bring him back. What can I do? Take my life. I ain't never lived. I ain't nothin . . . ' And then the defendant began weeping and bowed his head and thought of his partner and all that each had done for, and to, the other. And of the self they saw in each other. Jimmy Smith knew he would never be rid of Gregory Powell. 'I'm better off dead,' he sobbed. 'What's the use of livin? I ain't even a man.'

'Your Honor, could I suggest a short recess?' asked Hollopeter.

'I request it too, your Honor,' said Fulgoni softly.

In November 1969 Defendant Gregory Powell wrote:

Your Honor, it is with singular trepidation I begin this task of writing. Consistent with my express philosophy, i.e. life-oriented, I have no alternative but to attempt to persuade this court that I believe my case is one in which there would be justification for applying a great legal tradition, justice tempered with mercy.

During the course of the trials, numerous psychiatrists, all learned and sincere men I am sure, testified that in their opinions, I was a sociopath. This dehumanizing label sounded very scientific and precise, yes, and evoked fear when I first heard it applied to myself on the witness stand. But given time to think, reflect, and most of all, study, in an attempt to define sociopathy as applied to myself, I found myself questioning the validity of this label. I do not claim I am a sociopath, but many of the symptoms attributed to sociopaths do in fact indicate this may be a proper label, at least in general.

This is not a plea for release into a society where everyday living demands that even the average person be able to display at least adequate judgment. This is a supplication, if you will, a plea to be allowed to live out my natural life span in prison. To live in an environ where a person is not faced with making judgmental decision, but a completely controlled environment.

During the trial I read the FBI rap sheet which listed all of the offenses for which I have been convicted. This reading was necessarily an incomplete composite of my life from 1949 to the present, for I was unable to bring out in my testimony the effect the numerous convictions and imprisonment had upon me. The almost continual incarceration from the time I was fifteen to the present caused what is commonly known as a very institutionalized person. Due to this, the terror a normal person would feel at the prospect of spending the rest of his life in prison is absent when I think of imprisonment. In fact, conversely, the terror would be felt by me only if I were to have to face the prospect of life outside the prison.

'The institutional man,' a reader of the document would say of the defendant's description of the most misunderstood phenomenon of penology. 'How many institutional men are there in and out of prison who can *never* function except in a controlled environment? And who dares admit it? And what institution is open to them?'

'They kept kicking me out,' Gregory Powell had said when he was nineteen years old. 'I kept stealing cars and going back, but they just kept kicking me out of prison.'

*

1970 was only four weeks away when the twelfth Superior Court judge to sit on this case formally sentenced the defendant, Gregory Powell, to death. The sentence was entered in one of the last of the 159 volumes of a 45,000-page court transcript, the longest in California history. There would, of course, be another automatic appeal, but most observers now saw little chance for a reversal.

In 1970, once again on Death Row, Gregory Powell requested and received a book entitled *What Happens When You Die?*. There was, however, another condemned man named Robert P. Anderson who would have something to do with whether Gregory Powell would die in the gas chamber.

NINETEEN

The new decade was not unkind to Gregory Powell, who had adopted his middle name and insisted that everyone, even his family, refer to him as Ulas Powell.

'His alter ego speaking,' his attorney Charles Maple would say. Maple didn't believe there was much chance of the case being reversed again, but kept in touch.

The condemned man would receive a little money from friends or relatives, a few dollars here and there. Once he received a considerable sum from an anti-capital-punishment group in Italy. He had by now thoroughly convinced himself that repressed homosexuality had been at the root of his problems his whole life long, and that he had been born to be a homosexual. His added weight, darker hair, and glasses softened his appearance considerably. He granted interviews and even drew cartoons for gay newspapers and was quoted as saying he had a distinct aversion to gas and would never be executed.

He joined a convict religious cult and edited their political writings, and he waited for a decision in *People versus Anderson*. In March 1972 the California Supreme Court declared that capital punishment was cruel and unusual. Gregory Powell escaped the row forever and went to the yard.

Now he was free to live his life in the kind of non-threatening, controlled environment he said should always have been his. He was not unhappy. He kept busy with his many letters. After all, the family needed him more than ever, and sometimes *he* even lent *them* money. Douglas had matured, given up hard drugs, and stayed out of trouble, profiting from his brother's mistakes. But Lei Lani had fared badly and there was still friction between Sharon and his parents. His letters went to each of them: sisters, brother, mother and father. He directed them in business, household, and personal matters. He advised, scolded, commanded. If he had one wish it would be that they lived close

399

enough to visit him more often. It was too hard to properly administer a family by letter. But he sighed and decided he'd just have to do his best. They couldn't possibly manage without him. He was, and always would be, the rightful head of the family.

TWENTY

When Jimmy Smith was finally sentenced to life imprisonment he thought he'd won. The day they drove him up to Folsom Prison he wasn't sure. Not when he saw it, more horrible than he ever remembered it. Not when those hateful iron gates yawned, and the gray, slimy, cold granite walls swallowed him.

He had one bitter but torch-bright consolation: Gregory Powell still faced death and might one day be smoked. Then, after *People versus Anderson*, he realized Powell had beaten him again. Powell was in the yard at San Quentin living reasonably like a human being while Jimmy languished in the hell of Folsom. He could never win from Gregory Powell.

Without the slightest hope of another successful appeal, Jimmy permitted the man who had won his first reversal to try again. He once more accepted the services of Irving Kanarek. In early 1973, the relentless lawyer filed an opening brief in the case of appellant Jimmy Lee Smith. One of the arguments read as follows:

> During pre-second trial proceedings, strong indications of Hettinger's mental instability were uncovered. In view of this and the circumstances of the killing (appellant's extreme terror rendering his memory of the events at the time of the killing very confused; the fact that it was night and that visibility was poor; the fact of guns 'floating around') it is probable that Hettinger, himself, was responsible for the firing of four shots at Campbell although he may have mistaken Officer Campbell for one of the defendants in so doing.

One day one of the old guards in the yard said to Jimmy, 'You know, Jimmy, when you left here back in '63 I figured you'd come back to us. Most of em do. But not for killin a cop. I didn't think you'd come back to us for good.'

Jimmy listened to the bull and smiled grimly and looked across the yard at the Flea, who was older now, more demented, more loathsome than ever – pawing at motes and gleaming dust specks in the silver dusk – snuffling around the yard, foul-smelling, looking for a man, any man, for anything the man cared to do to him, just for human contact.

'Sir,' said Jimmy Smith to the guard, 'I was *born* for Folsom Prison.'

TWENTY-ONE

The new decade at first did not seem much different from the last. At least not to Karl Hettinger. In fact, 1970 was one of his worst years. He waited every day for the phone to ring telling him the appeal had been heard and the case reversed. For a time he thought it was inevitable. During this year, the city of Los Angeles reduced his pension to sixty percent. He was strangely glad about the reduction though he never told anyone.

During the retrial, the defense or the prosecution, he was never sure which, had taken his service revolver from him to introduce into evidence, even though his gun had not been one of the murder weapons and had not been introduced into evidence during the 1963 trials. He tried to get the gun back but the District Attorney's Office told him he would have to wait until the conviction was affirmed by a higher court – perhaps two or three more years he was told. So they even took that away from him. They even robbed him of that, he thought.

The first year passed, and was not a very hopeful year. Not to the world and not to him. But as the world settled into the 1970's, Karl Hettinger noticed that his mind did not falter quite so much. He remembered a little better. There were not quite so many blank pages. One day he saw an ad in the paper, and he did something extraordinary. He took some vocational testing at UCLA and began talking about going back to college. Then, most extraordinary of all, in the fall of 1971 he did it.

He enrolled in three courses: plant identification, landscaping, and landscaping construction principles.

But these were not the same agricultural students of two decades ago. Many of these were future ecologists, apparently more sophisticated.

'I don't know if I can compete with these younger students, Helen,' he said, immediately regretting the big decision.

'Of course you can, Karl. Don't quit now. Please don't. It's the first . . . Karl, it's been a lot of years since you've gone out in the world like this.'

'Well, maybe I'll stick it out. Maybe I'll just see if I can do it. Maybe I won't flunk the courses at least.'

'You won't, Karl. I know you won't.'

He was enrolled in night classes. And after gardening all day he would sit silently hunched at his desk in the college classroom with all the dirt carefully scrubbed from his fingernails, perhaps the neatest of the students, certainly the one with the shortest haircut and the most sunburned face. At thirty-seven, he was one of the old men on campus.

'They cheated on the finals,' he told his wife in despair one day.

'Who cheated?'

'So many of them. The students. They cheated!'

'Well, don't let it bother you.'

'It's just so hard to take. It just got me so upset. I went to the instructor later to complain, to tell her about all the cheating. She said if I needed a good grade because of the G.I. bill or something, don't worry about it, she'd *take care* of it!'

'Now, don't let that bother you, Karl,' his wife said. 'Don't let it hurt your confidence.'

'I can't help it, Helen. It's like *stealing* for a grade.'

He anxiously awaited the mailman after that school term. He was sure that he would get a less than average grade. He was bitter, felt once again betrayed. He had been honest, and they had stolen for their grades. He finally convinced himself that even a failing mark was more than a mere possibility. When the mail arrived, he saw that he had earned two A's and a B.

It was the first success he had tasted in nine years. It was glorious. A triumph.

Other changes were taking place as the months passed and time further separated him from the horror of the last trial. The dreams came less frequently. More potency returned. He still walked round-shouldered, but he didn't dig his nails into his palms, and he didn't sit hunkered forward with his hands between his knees.

Karl Hettinger began talking about truly striking out into a more challenging life. It would have to be something to do with the land, of course. He loved the sun and the earth and living things too much to ever leave them. Finally, in 1972, he heard from an old college friend who had lived with him in the dormitory of Pierce College so many years ago. It was a job offer. A real job. A hard, sixty hour a week job in the fields supervising other workers, managing a huge plant nursery. Supervising other men!

Karl almost turned his friend down. He had frightened skirmishes with himself. He tossed in the night. But it wasn't the same kind of debilitating fear, it was just a fearful excitement now, a bearable kind of anxiety. There was so much to think about. He began talking to Helen.

'Well, if you want to do it, Karl, then do it. Do something. Take a first step. We'll be with you. We'll be with you whatever you decide.'

One evening during dinner he offered a wishbone to his son. 'I have a riddle for you,' he said as they each tugged at an end of the wishbone and Karl ended up with the winning piece.

'What's that, Daddy?'

'Tell me how your dad can wish for something that makes it impossible for *either* wisher to lose.'

All three of his tow-headed children giggled and offered some guesses, and even Helen joined in as they all pondered the father's riddle.

'Give up?' he said finally.

'We give up, Dad.'

'I wished,' he said slowly, 'that each of you,' and he looked at his wife, 'would get *your* wish. Now, since I've won, that means you *will* get your wish. We all win.'

The remote little lake would have been almost too clear to fish, even had the bluegills been big enough, but they weren't. They were stunted because there were so many fish and not enough food in that clear water.

The lake was near Donner Pass in the High Sierras, near the place where George Donner and his party met their horrible destinies a century before. The Donners hadn't been able to plunge forward through the blizzard to the lake, but it was early summer now and there was no snow on the ground. The pine needles were very long and rubbery, the moss was fragrant from the dew.

Each day Karl felt more confident, and now, here where your breath snapped in your lungs, and the moss was dappled with sunlight, he felt that almost anything was possible. He shucked his light jacket and Helen sat down and plunked a few pebbles into the pool.

'What in the world are you up to?' she asked as he peeled off the T-shirt and rubbed his arms and chest, smiling up at the piny evergreens. Next he opened his belt and in a second was out of his pants, shoes, and socks. He whooped into the lake, lunging forward, sending ripples far out across the quiet water. He dove beneath the water, surfaced happily, perhaps even joyfully, and turned just as Helen dove naked from the bank, coming up screaming with excitement and shock from the cold.

Karl struggled forward in the cool mud and splashed her with icy water, rubbing her arms and back as they knelt in the mud. 'It feels like the fish're kissing us,' she said.

'What?'

'The bluegills. Look at them.'

They were surrounded by stunted bluegills, whose tiny mouths, much too small to bite a hook, were pressed against their hips, backs, arms, legs.

'They're nibbling us,' said Karl. 'They're hungry. They nibble anything that enters the water.'

'I say they're kissing us, damn it,' said Helen, brushing her hand over his cheek, touching the pearly drops of mountain water, seeing no fear in his eyes.

'Okay, Helen, they're *kissing* us,' said Karl.

They were pressed together in the clear water. The sun had touched the High Sierras and there wasn't a flake of snow left on the ground.

Karl took the job with his friend. It meant great decisions in 1973, like selling his house. He did it. It meant uprooting the family and moving away. They did it. Most of all it meant going to live near a place he had tried to escape for ten years.

Karl Hettinger was destined to face his devils. He could never escape irony. The friend's acreage was near Bakersfield, just a few miles from a place where onions grow so thick you can smell them from the Maricopa Highway. Just a few miles from that place where a policeman ran through the fields with a farmer one cold and bitter night under a late and lonely moon near the foot of the Tehachapi Mountains, near a place called Wheeler Ridge, near a place they marked with a blood-red arrow.

TWENTY-TWO

During the same year, another man, a slender man with deep, close-set eyes, took a wristwatch to a jeweler for repair and cleaning. It was a stainless steel, fifteen-dollar watch, not worth repairing, once seen on the wrist of a dead man. The slender man thought of good things when he looked at the watch: of a young piper atop the bowsprit of a bug-eyed ketch called *Jolly Roger*, of playing in Hancock Park, or of bicycling all the way to Griffith Park and hiking up to the observatory when they were children together.

'This is just a cheap watch,' said the jeweler. 'Why waste money on such a watch?'

'I've been wearing that watch ten years, since 1963. I like it.'

'It's your money,' said the jeweler, writing up a ticket. 'But what a waste. I could sell you a new watch twice as good and save you money to boot.'

'I want *that* watch. Now please give me a ticket so I can go.'

During that year there were several young lads growing up in the San Fernando Valley of California, ordinary boys, unknown to each other, who had all derived their Scottish name from the same source.

There was an Armenian obstetrician practicing in that valley, a former piper, who seldom played his pipes anymore, and who, when asked by expectant parents to suggest a suitable name for a child, would invariably answer: 'Yes, if it's a boy I have a suggestion for you. It's really a fine name. My favorite, I guess. And very short. Three letters. It's from the Gaelic. It means John. . . . '

And there were two girls on a holiday from northern California visiting their grandmother in Hollywood.

It was exciting to visit their grandmother. There were surprises. They might go to the Music Center to a concert. Or to a ballet.

The grandmother still believed in culture *and* discipline for children but of course could not teach discipline to grandchildren.

The grandmother was over seventy now, and lived alone, but was an active woman, handsome, with a throaty chuckle, far younger-looking than her years. She was still self-employed as an auditor and bookkeeper, and did not own a car in the world's most mobile city. She was content to ride the bus which was relaxing and afforded her time to read. Everyone secretly admired her self-sufficiency and great strength.

The adolescent girls were very tall and long-legged like their parents. The younger, Lori, had a resolute chin and jaw, and held her mouth in a way that was very familiar to the grandmother. And strangely enough, she had familiar mannerisms, a way of answering, a shrug of the shoulders. Her sister, Valerie, the darker of the two, did not have that mouth and chin and jaw, but she had blue-gray eyes and a brooding look when thoughtful or troubled. Yet the grandmother could always say something which she knew would make it vanish and then the face would brighten magically. That too was very familiar to the grandmother.

On a quiet day while in her grandmother's bedroom, Valerie glanced at the framed portrait taken in olden days of the man she knew was her grandfather. He was in a graduation gown with a gold sash of the Manitoba Medical College.

And on the other wall was a larger picture of a smiling young man, not the most representative likeness perhaps, but one of the few he had taken in his later years.

She looked at the picture and it was hard to remember him. There were just glimmers. Very faint. Some memory perhaps of an event, a moment. It reminded her of something she wanted to tell her grandmother so she ran into the kitchen.

'Guess what, Grandma Chrissie? I forgot to tell you. I'm going to take clarinet lessons.'

'That's marvelous, Valerie,' said the grandmother, placing a pan on the stove and turning to face the girl. 'But whatever made you choose the clarinet? I thought you might play piano.'

'I decided on the clarinet because I think it'll help me later when I learn to play the pipes.'

'The what?'

'Bagpipes. Don't you think it would be nice if I could play the bagpipes?'

'Well, Valerie,' Chrissie said, and she faltered. There was a sudden tightness in her chest. Instantly. Without warning. And her breath shortened. Chrissie looked at the eyes, now dove gray, and for a moment lost the thread of the child's remark. Then she thought she heard them: piercing, wailing – now plaintive, then majestic – the skirl of pipes! She could almost smell the grass in Hancock Park and the tar from the great pits. She wanted to rush to the window to catch a glimpse of a tall boy marching . . . .

'What's wrong, Grandma Chrissie?'

'Well, Valerie . . . I . . . '

Chrissie Campbell turned her back to the girl and put her palms to her face under the glasses, which was her way, and massaged her eyes and sighed heavily until it was gone. No one had ever seen her weep. Not ever.

'Don't you think it would be nice to play the bagpipes, Grandma Chrissie? Don't you like the idea?'

'I like the idea, Valerie,' said Chrissie Campbell finally, taking careful measured breaths until it all stopped. She turned, calm now, only a little pale, and held the child's face in her hands and smiled. And like another child's face of years past, the brooding look vanished and the face became lighted and the eyes went more to the blue.

'Yes, darling,' said Chrissie. 'I think that's a lovely idea.'

AFTERWORD

In 1978 the Community Release Board of the state of California decreed that Gregory Ulas Powell and Jimmy Lee Smith shall be released from prison no later than March 1983.